Modeling Spatio-Temporal Data

Several important topics in spatial and spatio-temporal statistics developed in the last 15 years have not received enough attention in textbooks. *Modeling Spatio-Temporal Data: Markov Random Fields, Objective Bayes, and Multiscale Models* aims to fill this gap by providing an overview of a variety of recently proposed approaches for the analysis of spatial and spatio-temporal datasets, including proper Gaussian Markov random fields, dynamic multiscale spatio-temporal models, and objective priors for spatial and spatio-temporal models. The goal is to make these approaches more accessible to practitioners, and to stimulate additional research in these important areas of spatial and spatio-temporal statistics.

Key topics discussed in this book include:

- Proper Gaussian Markov random fields and their uses as building blocks for spatio-temporal models and multiscale models.
- Hierarchical models with intrinsic conditional autoregressive priors for spatial random effects, including reference priors, results on fast computations, and objective Bayes model selection.
- Objective priors for state-space models and a new approximate reference prior for a spatio-temporal model with dynamic spatio-temporal random effects.
- Spatio-temporal models based on proper Gaussian Markov random fields for Poisson observations.
- Dynamic multiscale spatio-temporal thresholding for spatial clustering and data compression.
- Multiscale spatio-temporal assimilation of computer model output and monitoring station data.
- Dynamic multiscale heteroscedastic multivariate spatio-temporal models.
- The M-open multiple optima paradox and some of its practical implications for multiscale modeling.
- Ensembles of dynamic multiscale spatio-temporal models for smooth spatio-temporal processes.

The audience for this book are practitioners, researchers, and graduate students in statistics, data science, machine learning, and related fields. Prerequisites for this book are master's-level courses on statistical inference, linear models, and Bayesian statistics. This book can be used as a textbook for a special topics course on spatial and spatio-temporal statistics, as well as supplementary material for graduate courses on spatial and spatio-temporal modeling.

Modeling Spatio-Temporal Data

Markov Random Fields, Objective Bayes, and Multiscale Models

Edited by
Marco A. R. Ferreira

CRC Press
Taylor & Francis Group
Boca Raton London New York

CRC Press is an imprint of the
Taylor & Francis Group, an **informa** business
A CHAPMAN & HALL BOOK

First edition published 2025
by CRC Press
2385 Executive Center Drive, Suite 320, Boca Raton, FL 33431, U.S.A.

and by CRC Press
4 Park Square, Milton Park, Abingdon, Oxon, OX14 4RN

CRC Press is an imprint of Taylor & Francis Group, LLC

ISBN: 978-1-032-62209-5 (hbk)
ISBN: 978-1-032-62342-9 (pbk)
ISBN: 978-1-032-62344-3 (ebk)

DOI: 10.1201/9781032623443

Typeset in Latin Modern font
by KnowledgeWorks Global Ltd.

Publisher's note: This book has been prepared from camera-ready copy provided by the authors.

To Allison, Lucas, and Rafael

Contents

Preface

Several important topics in spatial and spatio-temporal statistics developed in the last 15 years have not received enough attention in textbooks. *Modeling Spatio-Temporal Data: Markov Random Fields, Objective Bayes, and Multiscale Models* aims to fill some of this gap by presenting important subjects such as proper Gaussian Markov random fields, dynamic multiscale spatio-temporal models, and objective priors for spatial and spatio-temporal models. The audience for this book are practitioners, researchers, and graduate students in statistics, data science, machine learning, and related fields. This book can be used as a textbook for a special topics course on spatial and spatio-temporal statistics. This text can also be used as supplementary material for graduate courses on spatial and spatio-temporal modeling. Prerequisites for this book are master's-level courses on statistical inference, linear models, and Bayesian statistics. We have strived to make the chapters in this book self-contained. Thus, in addition to the Table of Contents and Bibliography, each of the chapters contains its own Table of Contents and Bibliography. In summary, this book provides an overview of a variety of recently proposed approaches for the analysis of spatial and spatio-temporal datasets. The goal is to make these approaches more accessible to practitioners, and to stimulate additional research in these important areas of spatial and spatio-temporal statistics.

Contents

Proper Markov random fields may be used as building blocks for highly structured spatial and spatio-temporal models. In Chapter 1, Proper Gaussian Markov Random Fields, we present two different specifications of proper Markov random fields, as well as a proper bivariate Markov random field known as a two-fold conditional autoregressive model. In addition, this chapter shows how to use proper Markov random fields to build spatio-temporal models and multiscale random field models.

Hierarchical models with intrinsic conditional autoregressive (ICAR) priors for spatial random effects are widely used for the analysis of spatial areal data. Spatial areal data arise when the region under study is subdivided into subregions and the data are usually observed as totals or averages for each subregion. In Chapter 2, Gaussian Spatial Hierarchical Models with ICAR Priors, we present several recent developments for this class of models. In particular, we review objective priors that provide a reference analysis where the data speak for themselves. This chapter also reviews recent results on fast computations, equivalence results, and objective Bayes model selection. Finally, the chapter presents the R package ref.ICAR, that allows practitioners to perform objective Bayesian estimation and model selection for Gaussian spatial hierarchical models with ICAR random effects.

One important aspect in the practical implementation of spatio-temporal models is the choice of priors for the hyperparameters. Because many spatio-temporal models are built as highly structured multivariate state-space models, results for state-space models may inform us about appropriate priors for large classes of spatio-temporal models. Unfortunately, the literature on objective priors for state-space models is sparse. In Chapter 3, Objective Priors for Spatio-Temporal Models, we briefly review this literature. We then explore the relationship between state-space models and hierarchical models with ICAR random effects to motivate the specification of approximate reference priors for spatio-temporal models. To demonstrate these ideas, we present an objective approximate reference prior for the hyperparameters of a spatio-temporal model that includes a time-varying overall mean level and dynamic spatio-temporal random effects. We illustrate the application of the resulting methods with an analysis of a dataset on unemployment.

The recent COVID pandemic has made undeniably clear the need for statistical models and methods for the analysis of epidemic processes. In Chapter 4, Spatio-Temporal Models for Poisson Areal Data, we use proper Gaussian Markov random fields as building blocks to create spatio-temporal models for Poisson areal data. These models are useful for the analysis of epidemic processes. We illustrate the application of these models with an analysis of the AIDS epidemic in the state of Rio de Janeiro, Brazil.

Chapter 5, Dynamic Multiscale Spatio-Temporal Thresholding, proposes thresholding methods in the context of dynamic multiscale spatio-temporal models for Gaussian univariate data. These models decompose large spatio-temporal datasets into many much smaller multiscale coefficients that can be analyzed in parallel. Thus, these models lead to computations that are scalable, parallelizable, and fast. In this chapter, we develop thresholding methods that allow the latent multiscale coefficients to either change over time, be fixed over time different than zero, or be equal to zero vectors. This then informs decisions on whether neighboring subregions may be clustered and treated as a single subregion. As the ability to cluster subregions may lead to dimension reduction, these methods may be useful for data compression. We illustrate the use of multiscale thresholding methods with an application to agriculture production.

Across scientific disciplines, there is great interest in statistical methods for the assimilation of computer model predictions with observed data. In Chapter 6, Multiscale Spatio-Temporal Data Assimilation, we present a Bayesian hierarchical model for such data assimilation in both time and space. In this hierarchical model, the joint distribution of the two data types is modeled through a common spatio-temporal latent process. The model accommodates the change of support problem and handles spatially and temporally distributed additive bias associated with computer model predictions. We illustrate the application of these models with an assimilation of fine particulate matter data.

With the ability to collect increasingly more spatio-temporal data, comes the need for models and methods for the analysis of multivariate spatio-temporal data. In Chapter 7, Multiscale Heteroscedastic Multivariate Spatio-Temporal Models, we extend dynamic multiscale spatio-temporal models from univariate to multivariate Gaussian data. In addition, these novel models accommodate heteroscedasticity, allowing the covariance matrix to vary spatially. We illustrate the application of these models with an analysis of tropospheric temperature.

Multiscale models tend to produce latent mean process estimates that are spatially blocky. This may be ameliorated with an ensemble approach, where several models based on shifted multiscale structures are fitted and a model averaging approach is used to produce latent mean process estimates. In this case, the usual intuition would be to apply the full Bayesian machinery to compute model posterior probabilities and use them as weights to implement Bayesian model averaging. Unfortunately, the use of model posterior probabilities is actually not a good idea when the true process is smooth and the multiscale process is used as an approximation. The reason for that is explained in Chapter 8, A Model Selection Paradox with Implications to Multiscale Modeling. This chapter considers the M-open case when the true model is not among the models being entertained in the analysis. In that case, for regular parametric models, as the sample size increases the posterior probability of the model closest to the true model in terms of Kullback-Leibler divergence will converge to one. But what happens if there are multiple models that minimize the Kullback-Leibler divergence from the true model? In that case, while the posterior probability of the set of models that minimize the Kullback-Leibler divergence from the true model converges to one as the sample size increases, there is no limiting distribution for the posterior probabilities of models within this set. We call this the M-open multiple optima paradox. We illustrate this paradoxical behavior with four examples: a toy example to introduce the main concepts, two examples related to inference on phylogenetic trees, and one multiscale analysis example on wavelet nonparametric regression.

Following up on the previous paragraph, while dynamic multiscale spatio-temporal models lead to fast computations, the spatial portion of the resulting covariance functions are usually blocky step functions. In Chapter 9, Ensembles of Dynamic Multiscale Spatio-Temporal Models, we present a method that leads the spatial part of the covariance functions to be smoother. Specifically, our method is based on an ensemble of dynamic multiscale spatio-temporal models defined on a collection of multiscale structures. Because of the model selection paradox presented in Chapter 8, the weights used in the ensemble method have to be carefully specified. In Chapter 9, we discuss how to specify these weights in a way that avoids the paradox presented in

Chapter 8. We illustrate the use of ensembles of dynamic multiscale spatio-temporal models with an application to temperature data.

Acknowledgments

I am deeply grateful to the people who helped me make this book a reality. First of all, I would like to thank my co-authors who agreed to contribute to the research and writing of the chapters in this book. I have learned a lot from my co-authors. In addition, I much appreciate the support of those at Chapman & Hall/CRC, in particular, Sherry Thomas for keeping me on track, Shashi Kumar for help with LaTeX, and Rob Calver for his insights and general production support. Finally, I would like to thank my wife Allison and our sons Lucas and Rafael for their joy, patience, and love.

Blacksburg, VA, USA — Marco Ferreira
April 2024

Editor

Marco A. R. Ferreira is a professor in the Department of Statistics at Virginia Tech. Dr. Ferreira has served the statistics profession by being on the editorial boards of multiple scientific journals including the journal *Bayesian Analysis*, several committees of ISBA and ASA, as well as scientific committees of numerous domestic and international conferences. His current research areas include dynamic models for time series and spatiotemporal data, multiscale models, objective Bayesian methods, stochastic search algorithms, and statistical computation. Major areas of application include bioinformatics, economics, epidemiology, and environmental science. Dr. Ferreira's research has been funded by grants from the industry, the National Science Foundation, and the National Institutes of Health. He has published important scientific papers in top journals such as the *Journal of the American Statistical Association*, *Journal of the Royal Statistical Society*, *Biometrika*, and *Bayesian Analysis*. At the time of this writing, he has advised over 15 PhD students and postdocs who work in academic, industrial, and governmental positions.

Contributors

Mohamed Elkhouly
Cleveland State University
Cleveland, Ohio

Marco A. R. Ferreira
Virginia Tech
Blacksburg, Virginia

Thais C. O. Fonseca
Universidade Federal do Rio de Janeiro
Rio de Janeiro, Brazil

Christopher T. Franck
Virginia Tech
Blacksburg, Virginia

Elizabeth B. Hypólito
Instituto Brasileiro de Geografía e Estatística
Rio de Janeiro, Brazil

M. Alejandra Jaramillo
Universidad Militar Nueva Granada
Bogota, Colombia

Erica M. Porter
Clemson University
Clemson, South Carolina

Ana G. Rappold
U.S. Environmental Protection Agency
Chapel Hill, North Carolina

Juan C. Vivar
U.S. Food and Drug Administration
Silver Spring, Maryland

1

Proper Gaussian Markov Random Fields

Marco A. R. Ferreira

Virginia Tech

We discuss the use of Markov random fields as building blocks for highly structured models for Gaussian areal data, that is, data that are available in a partition of the geographical domain of interest. Specifically, we discuss the role that *proper* Gaussian Markov random fields (GMRFs) may play in the construction of several classes of spatio-temporal and multiscale models. Proper GMRFs are much more natural than improper GMRFs, also known as intrinsic conditional autoregressive (ICAR) models, when used as building blocks for highly structured models due to three facts. First, the use of proper GMRFs defines a proper joint probability model for the observations and the latent process of interest. Second, such a proper joint probability model permits decision-theoretic-based model selection to be implemented. Finally, the existence of a proper joint probability model often permits special structures to be explored for developing efficient Bayesian computation.

DOI: 10.1201/9781032623443-1

1.1 Introduction

As pointed out by Banerjee et al. (2014), when used as priors for spatial random effects, proper and improper GMRFs do not differ much. However, proper GMRFs are much more natural than improper GMRFs when used as building blocks for highly structured models.

This advantage of proper GMRFs over improper GMRFs is due to three reasons. First, the use of proper GMRFs defines a proper joint probability model for the observations and the latent process of interest. Conversely, the use of improper GMRFs has typically required the computational implementation of sum-zero constraints that may obscure the joint probability model for the observations and the latent process. Second, because there is a proper joint probability model for the observations and the latent process, formal Bayesian model selection based on decision theory can be implemented. In particular, when using proper GMRFs building blocks, Bayes factors, conditional Bayes factors, and posterior probabilities can be computed and used for model selection and model averaging. Finally, the existence of a proper joint probability model often allows the development of efficient Bayesian computation based on the exploration of special probabilistic structures.

Recently, Keefe et al. (2018) have developed formal sum-zero constrained GMRFs – also called sum-zero constrained ICAR models – which are proper and, thus, inherit the benefits mentioned in the previous paragraph. In addition, in a series of papers, Ferreira (2019), Ferreira et al. (2021), and Porter et al. (2023) have established equivalence results between the use of the sum-zero constrained GMRFs of Keefe et al. (2018) and the use of the widely utilized Besag's improper GMRFs (Besag, 1974; Besag et al., 1991) with computationally imposed sum-zero constraints. In the development of sum-zero constrained GMRFs, Keefe et al. (2018) used the Type II proper GMRFs that we present in Section 1.2.3. Objective Bayes priors (Keefe et al., 2019) and objective Bayes model selection (Porter et al., 2023) have been developed for Gaussian hierarchical models with sum-zero constrained GMRFs as priors for spatial random effects. Details about the sum-zero constrained GMRFs of Keefe et al. (2018) and related developments can be found in Chapter 2 (Porter et al., 2024).

The remainder of this chapter is organized as follows. Section 1.2 discusses some different specifications of proper and improper Markov random fields, including a twofold conditional autoregressive model useful for bivariate processes on a lattice. Section 1.3 discusses spatio-temporal models for areal data. Section 1.4 reviews multiscale random field models. Section 1.5 discusses inference for highly structured models based on proper GMRFs. Section 1.6 concludes the chapter with a discussion.

1.2 Gaussian Markov Random Fields

The use of Markov random fields for the analysis of areal data was introduced in a landmark article by Besag (1974) and has become widespread since the work of Besag et al. (1991), Besag and Kooperberg (1995), and Besag et al. (1995). Many different types of Markov random field models have appeared in the literature. We start this section discussing an improper MRF, and after that we focus on two proper GMRF models that have been previously used for the construction of highly structured spatio-temporal and multiscale models.

1.2.1 Improper GMRFs

Improper GMRFs, also known as Besag's intrinsic conditional autoregressions, have a long history of applications in statistics (e.g., see Waller et al., 1997; Rue and Held, 2005; Banerjee et al., 2014, and references therein). Consider a region of interest that is partitioned in n subregions. Within a subregion i, interest lies on a variable of interest Z_i. Besag et al. (1991) proposed an improper GMRF that can be defined in terms of full conditional distributions for $Z_1, \ldots, Z_n$, where the full conditional distribution of Z_i is the distribution of Z_i given $Z_1, \ldots, Z_{i-1}, Z_i, \ldots, Z_n$.

Let N_i be the set of neighbors of subregion i and d_i be the number of neighbors of subregion i. Besag et al. (1991) considered a model such that the full conditional density of Z_i is

$$f(z_i|z_{\neq i}) \propto \exp\left\{-\frac{\tau}{2}(z_i - \bar{z}_i)^2\right\}, \tag{1.1}$$

where $\bar{z}_i = d_i^{-1}\sum_{k\in N_i} z_k$. Note that if we add a constant a to each z_i, the constant a cancels out from the full conditional density given in Equation (1.1). Thus, $Z_1, \ldots, Z_n$ don't have a well-defined mean and, as a consequence, the joint distribution of $\boldsymbol{Z}$ is improper. However, it is possible to represent this joint distribution by what Sun et al. (1999) have called a partially informative normal distribution with density given by

$$f(\boldsymbol{z}) \propto \exp\left(-\frac{\tau}{2}\boldsymbol{z}'\boldsymbol{B}\boldsymbol{z}\right), \tag{1.2}$$

where $\tau\boldsymbol{B}$ is the precision matrix and $\boldsymbol{B}$ is an $n \times n$ matrix with ith diagonal element equal to d_i, and ijth off-diagonal element equal to -1 if $i \in N_j$ and 0 otherwise.

There are at least three ways to modify the specification of Besag's ICAR model to make it proper. Section 1.2.2 presents the way proposed by Sun et al. (1999), and Section 1.2.3 presents the way proposed by Leroux et al. (2000) and considered by Pettitt et al. (2002) and Ferreira and De Oliveira (2007). A third way proposed by Keefe et al. (2018, 2019) uses a limiting argument to impose a sum-zero constraint and is presented in Chapter 2 (Porter et al., 2024).

1.2.2 Type I Proper GMRF

Sun et al. (1999) considered proper Gaussian MRFs of the form

$$f(\boldsymbol{z}) \quad \propto \quad \exp\left(-\frac{1}{2\delta}\boldsymbol{z}'\boldsymbol{B}\boldsymbol{z}\right), \tag{1.3}$$

where $\delta^{-1}\boldsymbol{B}$ is a precision matrix, and $\boldsymbol{B}$ has ith diagonal element equal to a_i and ijth off-diagonal element equal to $-a_i\beta_{ij}$. They considered two specifications for the elements of the precision matrix.

In the first specification for the precision matrix, Sun et al. (1999) assumed $a_i = d_i$ and

$$\beta_{ij} = \begin{cases} \rho/d_i, & \text{if } j \in N_i, \\ 0, & \text{otherwise,} \end{cases}$$

with $|\rho| < 1$. As a result, $\boldsymbol{B} = \boldsymbol{D} - \rho\boldsymbol{C}$ where $\boldsymbol{D} = \text{diag}(d_1, \ldots, d_n)$ and $\boldsymbol{C}$ is the adjacency matrix. That is, the ijth element of $\boldsymbol{C}$ is equal to 1 if $i \in N_j$ and is equal to 0 otherwise. When $|\rho| < 1$, the matrix $\boldsymbol{B}$ is diagonally dominant and therefore positive definite. When $\rho = 1$, the model under this first specification becomes the improper GMRF described in Section 1.2.1 and proposed by Besag et al. (1991). Min et al. (2010) used this proper GMRF specification as a prior for spatial random effects in a spatial logistic linear mixed model.

In the second specification, Sun et al. (1999) assumed $a_i = 1$ and

$$\beta_{ij} = \begin{cases} \rho, & \text{if } j \in N_i, \\ 0, & \text{otherwise.} \end{cases}$$

In this case, $\boldsymbol{B} = \boldsymbol{I}_n - \rho\boldsymbol{C}$. Let $\lambda_1 \leq \cdots \leq \lambda_n$ be the eigenvalues of $\boldsymbol{C}$. Sun et al. (1999) showed that $\lambda_1 < 0$ and $\lambda_n > 0$. Moreover, Sun et al. (1999) showed that if $\lambda_1^{-1} < \rho < \lambda_n^{-1}$, then $\boldsymbol{B}$ is positive definite. In that case, the density in Equation (1.3) is proper and can be written as

$$f(\boldsymbol{z}) \quad = \quad (2\pi)^{-n/2}|\boldsymbol{B}|^{1/2}\delta^{-n/2}\exp\left(-\frac{1}{2\delta}\boldsymbol{z}'\boldsymbol{B}\boldsymbol{z}\right), \tag{1.4}$$

Finally, if $\rho = 0$, then $\boldsymbol{B} = \boldsymbol{D}$ and $Z_1, \ldots, Z_n$ are independent.

1.2.3 Type II Proper GMRF

The Type II proper GMRF specification was first proposed by Leroux et al. (2000) for the estimation of disease rates in small areas. The following parameterization of the Type II proper GMRF was independently proposed by Pettitt et al. (2002) and Ferreira (2002):

$$\boldsymbol{Z} \sim PGMRF(\boldsymbol{\mu}_z, \boldsymbol{P}),$$

where $\boldsymbol{Z}$ follows a proper GMRF process with mean vector $\boldsymbol{\mu}_z$ and precision matrix $\boldsymbol{P}$. Thus, the density of $\boldsymbol{Z}$ is proportional to

$$\exp\left\{-\frac{1}{2}(\boldsymbol{z}-\boldsymbol{\mu}_z)'\boldsymbol{P}(\boldsymbol{z}-\boldsymbol{\mu}_z)\right\},$$

where $\boldsymbol{P} = \tau(\boldsymbol{I}_S + \phi\boldsymbol{M})$, τ is a scale parameter, $\boldsymbol{I}_S$ is the $S \times S$ identity matrix, and $\phi \geq 0$ controls the degree of spatial correlation. In addition, the matrix $\boldsymbol{M}$ is such that

$$(\boldsymbol{M})_{k,l} = \begin{cases} m_k & \text{if } k = l, \\ -g_{kl} & \text{if } k \in N_l, \\ 0 & \text{otherwise,} \end{cases}$$

where $g_{kl} > 0$ is a measure of similarity between regions k and l, $m_k = \sum_{l \in N_k} g_{kl}$. Pettitt et al. (2002) proposed this model for point-referenced data and assumed that g_{kl} is a function of the distance between k and l. Ferreira (2002) proposed the above specification for areal data with regions being neighbors if they share a border and with $g_{kl} = 1$ if $k \in N_l$. Ferreira (2002) used Type II proper GMRFs as building blocks for multiscale random fields; we review these multiscale random fields in Section 1.4. In this parameterization, the spatial dependence increases with ϕ; when $\phi = 0$ the regions are independent, and when $\phi \to \infty$ the proper GMRF approaches the intrinsic autoregressive model (Besag et al., 1991; Besag and Kooperberg, 1995).

Leroux et al. (2000) proposed the Type II proper GMRF with a precision matrix equal to $\boldsymbol{P} = \sigma^{-2}((1-\lambda)\boldsymbol{I}_S + \lambda\boldsymbol{M})$, with $\sigma^2 > 0$ and $0 \leq \lambda < 1$. Leroux et al. (2000) assumed that the elements of the matrix $\boldsymbol{M}$ are defined such that regions are neighbors if they share a border and, thus, $g_{kl} = 1$ if $k \in N_l$. Hence, the parameterization proposed by Leroux et al. (2000) is equivalent to the parameterization considered by Pettitt et al. (2002) and Ferreira (2002) with $\tau = (1-\lambda)/\sigma^2$ and $\phi = \lambda/(1-\lambda)$. The package CARBayes (Lee, 2013), available for the statistical programming language R (R Core Team, 2023), implements hierarchical models with spatial random effects that follow the Type II proper GMRFs with the parameterization proposed by Leroux et al. (2000).

There have been key developments using Type II proper GMRFs with the parameterization proposed by Pettitt et al. (2002) and Ferreira (2002). Ferreira and De Oliveira (2007) developed reference priors for the case when a Type II proper GMRF is used directly as a model for observations. Keefe et al. (2018) proposed a sum-zero constrained GMRF based on Type II proper GMRFs. Keefe et al. (2019) developed reference priors for the parameters of Gaussian hierarchical models with spatial random effects that follow the sum-zero constrained GMRFs of Keefe et al. (2018). Results from Ferreira (2019), Ferreira et al. (2021), and Porter et al. (2023) show that for Gaussian hierarchical models, the use of the sum-zero constrained GMRFs of Keefe et al. (2018) is equivalent to the use of the improper Besag GMRFs discussed in Section 1.2.1 with either centering on the fly or with conditioning on the sum-zero constraint. Objective Bayes estimation based on the reference priors proposed by Keefe et al. (2019) and objective Bayes model selection (Porter et al., 2023) are implemented in the R package ref.ICAR (Porter et al., 2018). A review of objective Bayes analysis for Gaussian hierarchical models with spatial random effects following the sum-zero constrained GMRFs of Keefe et al. (2018) appears in Chapter 2 (Porter et al., 2024).

1.2.4 Twofold Conditional Autoregressive Model

Kim et al. (2001) introduced a twofold conditional autoregressive model for bivariate processes on regular and irregular grids. Consider an $n \times 2$ matrix $\boldsymbol{Z} = (\boldsymbol{Z}_1', \boldsymbol{Z}_2')$, with each column corresponding to one of the dimensions of the process on the grid, with $\boldsymbol{Z}_1 = (Z_{11}, \ldots, Z_{1n})'$ and $\boldsymbol{Z}_2 = (Z_{21}, \ldots, Z_{2n})'$. In addition, let $\boldsymbol{C}$ be an adjacency matrix with ijth element equal to 1 if $i \in N_j$ and equal to 0 otherwise. Moreover, let $\boldsymbol{D} = \text{diag}(2d_1 + 1, \ldots, 2d_n + 1)$. Kim et al. (2001) defined the model through the full conditional distributions of each individual random quantity on the grid. Each full conditional distribution is assumed to be Gaussian with an expected value given by

$$
\begin{aligned}
E\left(Z_{li}|\boldsymbol{Z}_{-(li)}\right) &= \frac{\rho_l}{2d_i+1}\sum_{j=1}^{I} C_{ij}Z_{lj} + \frac{\rho_3}{2d_i+1}\sqrt{\frac{\delta_l}{\delta_{3-l}}}\sum_{j=1}^{I} C_{ij}Z_{3-l,j} \\
&\quad + \frac{\rho_0}{2d_i+1}\sqrt{\frac{\delta_l}{\delta_{3-l}}}Z_{3-l,i} \\
&= \frac{1}{2d_i+1}\left(\rho_l\sum_{j\sim i} Z_{lj} + \rho_3\sqrt{\frac{\delta_l}{\delta_{3-l}}}\sum_{j\sim i} Z_{3-l,j}\right. \\
&\quad \left. + \rho_0\sqrt{\frac{\delta_l}{\delta_{3-l}}}Z_{3-l,i}\right),
\end{aligned}
\tag{1.5}
$$

and variance

$$
Var\left(Z_{li}|\boldsymbol{Z}_{-(li)}\right) = \frac{\delta_l}{2d_i+1}, \tag{1.6}
$$

where δ_1 and δ_2 are dispersion parameters and ρ_0, ρ_1, ρ_2, and ρ_3 measure the spatial dependence between the several random quantities on the grid.

Equations (1.5) and (1.6) imply that $\boldsymbol{Z}$ follows a multivariate Gaussian distribution with mean vector $\boldsymbol{0}$ and precision matrix given by

$$
\begin{pmatrix}
\delta_1^{-1}(\boldsymbol{D} - \rho_1\boldsymbol{C}) & -(\delta_1\delta_2)^{-1/2}(\rho_0\boldsymbol{I} + \rho_3\boldsymbol{C}) \\
-(\delta_1\delta_2)^{-1/2}(\rho_0\boldsymbol{I} + \rho_3\boldsymbol{C}) & \delta_2^{-1}(\boldsymbol{D} - \rho_2\boldsymbol{C})
\end{pmatrix}
\tag{1.7}
$$

Kim et al. (2001) showed that when compared with a usual conditional autoregressive model, the twofold conditional autoregressive model provides better results in terms of the mean squared error of the estimates. In addition, Kim et al. (2001) applied the twofold conditional autoregressive model to a dataset on male and female mortality in the state of Missouri due to lung cancer. This application to real data showed favorable performance of the procedure based on the twofold conditional autoregressive model in terms of posterior variance.

General specifications for multivariate processes on a grid have been proposed in the literature. For example, Mardia (1988) proposed a multivariate Gaussian Markov random field and applied it to image processing. Gelfand and Vounatsou (2003) and Carlin and Banerjee (2003) proposed the construction of multivariate Gaussian Markov random fields through the specification of the full conditional distributions for each multivariate quantity at a given position on the lattice. Assunção et al. (2002) and Gamerman et al. (2003) proposed the use of multivariate Gaussian Markov random fields as priors for spatially varying regression coefficients. Zhang et al. (2006) extended multivariate conditionally autoregressive processes to a multivariate spatio-temporal CAR process. More recently, Gao et al. (2023) proposed a framework to detect boundaries of spatial differences for multivariate areal data.

1.3 Spatio-Temporal Models for Areal Data

Consider a collection of regions indexed by the integers $1, 2, \ldots, S$, forming a regular or irregular grid within a geographical domain of interest. Assume that this grid is endowed with a neighborhood system $\{N_s; s = 1, \ldots, S\}$, where N_s denotes the collection of regions that are neighbors of region s. For each time t and region s, $t = 1, \ldots, T,\ s = 1, \ldots, S$, the variable of interest y_{ts} is observed. Denote the observed field at time t as $\boldsymbol{y}_t = (y_{t1}, \ldots, y_{tS})'$.

1.3.1 Common Mean Level Model

One of the most traditionally used spatio-temporal models (see Banerjee et al., 2014, and references therein) assumes that at each time point, the mean level is common to all regions. Specifically, the common mean level model is

$$\begin{aligned} \boldsymbol{y}_t &= \delta_t \mathbf{1} + \boldsymbol{x}_t + \boldsymbol{\epsilon}_t, \qquad \boldsymbol{\epsilon}_t \sim N(\mathbf{0}, \boldsymbol{V}_t), \\ \delta_t &= \delta_{t-1} + \omega_t, \qquad w_t \sim N(0, \nu), \end{aligned}$$

where $\boldsymbol{x}_t = (x_{t1}, \ldots, x_{tS})'$ is a vector of time-specific spatial random effects at time t, and δ_t is the common mean level at time t for all regions. Traditionally, the vector of spatial random effects $\boldsymbol{x}_t$ is modeled with an improper GMRF prior with density proportional to $\exp\{-0.5\tau \boldsymbol{x}_t' \boldsymbol{M} \boldsymbol{x}_t\}$, where the matrix $\boldsymbol{M}$ is singular. In order to guarantee posterior propriety, a sum-zero constraint

$\sum_{s=1}^{S} x_{ts} = 0$ is imposed computationally for all t within an MCMC algorithm in what is known as centering on the fly.

Epidemiological processes may be modeled by substituting the Gaussian observational equation with a Poisson observational equation and a logarithm link function such as in the model

$$\begin{aligned} y_{ts} &\sim Poisson(\lambda_{ts} E_{ts}) \\ \theta_{ts} &= log(\lambda_{ts}) \\ \boldsymbol{\theta}_t &= \delta_t \mathbf{1} + \boldsymbol{x}_t + \boldsymbol{\epsilon}_t, \qquad \boldsymbol{\epsilon}_t \sim N(\mathbf{0}, \boldsymbol{V}_t), \\ \delta_t &= \delta_{t-1} + \omega_t, \qquad w_t \sim N(0, \nu), \end{aligned}$$

where the expected number of cases E_{ts} is assumed to be known and may be computed based on the demographic characteristics of region s at time t.

Note that in the common mean level model, the spatial random effects are assumed to be specific to each time point. In particular, this assumption implies that a higher risk for one region when compared to other regions at a certain time point does not lead to higher risk for that region in subsequent time points.

1.3.2 Region-Specific Linear Trends

Sun et al. (2000) proposed a model for disease mapping that assumes a Poisson distribution for the number of deaths in age group g and region s at time t, that is

$$y_{tsg} \sim Poisson(\lambda_{tsg} E_{tsg}),$$

where E_{tsg} is the population size of age group g in region s at time t.

In addition, this model assumes a logarithm link function

$$\log(\lambda_{tsg}) = \alpha_g + Z_s + (\beta_g + W_s)(t - \bar{t}) + \epsilon_{tsg},$$

relating the logarithm of the risk with age and region effects as well as time. Here, $\bar{t}$ is equal to the average of the time points, and ϵ_{tsg} are i.i.d. $N(0, \sigma^2_\epsilon)$ and allow for extra-Poisson variability. In addition, α_g is the effect of the gth age group and Z_s is the effect of the sth region. Moreover, the logarithm of the risk is assumed to depend linearly on time. In this linear relationship, the slope is the sum of a coefficient β_g corresponding to the gth age group and a coefficient W_s corresponding to the sth region. Sun et al. (2000) treated α_g and β_g, $g = 1, \ldots, G$, as fixed effects and assumed for them independent Gaussian priors. In addition, they assumed that the region-specific intercept Z_s and slope W_s, $s = 1, \ldots, S$, are spatial random effects, where $\boldsymbol{Z} = (Z_1, \ldots, Z_S)$ and $\boldsymbol{W} = (W_1, \ldots, W_S)$ follow Type I Gaussian MRF priors as discussed in Section 1.2.2. Therefore, the temporal trend is assumed to be linear and to vary through the regions with spatially correlated intercept and slope.

1.3.3 Highly Structured Latent Spatio-Temporal Processes

Vivar and Ferreira (2009) introduced a class of spatio-temporal models with highly structured latent processes. Specifically, they assumed that the observed fields $\boldsymbol{y}_1, \ldots, \boldsymbol{y}_T$ are conditionally independent given a latent random field process $\boldsymbol{x}_1, \ldots, \boldsymbol{x}_T$. Moreover, the observed fields and the latent process are related through the observational equation

$$\boldsymbol{y}_t = \boldsymbol{F}_t'\boldsymbol{x}_t + \boldsymbol{\epsilon}_t, \qquad \boldsymbol{\epsilon}_t \sim PGMRF(\mathbf{0}_S, \boldsymbol{V}_t^{-1}), \tag{1.8}$$

where $\mathbf{0}_S$ is an S-dimensional vector of zeros and the observational error vectors $\boldsymbol{\epsilon}_1, \ldots, \boldsymbol{\epsilon}_T$ are independent. The matrix $\boldsymbol{F}_t$, usually known or unknown up to some few parameters, linearly relates the latent random field process to the field observations. The covariance matrix $\boldsymbol{V}_t$ encodes the spatial dependence of the observational errors. Vivar and Ferreira (2009) assumed the existence of Markovian relationships among the elements of $\boldsymbol{\epsilon}_t$ that lead to the sparsity of the precision matrix $\boldsymbol{V}_t^{-1}$.

Vivar and Ferreira (2009) assumed that the temporal evolution of the latent random field process is described by the system equation

$$\boldsymbol{x}_t = \boldsymbol{G}_t\boldsymbol{x}_{t-1} + \boldsymbol{\omega}_t, \qquad \boldsymbol{\omega}_t \sim PGMRF(\mathbf{0}_S, \boldsymbol{W}_t^{-1}), \tag{1.9}$$

where the system innovations $\boldsymbol{\omega}_1, \ldots, \boldsymbol{\omega}_T$ are independent. The matrix $\boldsymbol{G}_t$ describes the spatio-temporal evolution of the latent process, while the matrix $\boldsymbol{W}_t$ describes the spatial covariance structure of $\boldsymbol{\omega}_t$. A mean level field $\boldsymbol{\mu}_x$ containing the temporal stationary expected behavior of the latent process can be easily incorporated in Equation (1.9).

Many other spatio-temporal models for areal data previously proposed in the literature are included in the class of models given by Equations (1.8) and (1.9). For example, Wikle et al. (1998) and Garside and Wilkinson (2003) related observations obtained in an irregular grid to a latent process in a regular grid through the observational equation. Moreover, they assumed the latent process evolves through time according to a transport and diffusion evolution equation. Finally, they assumed spatial independence of the elements within $\boldsymbol{\epsilon}_t$ and $\boldsymbol{\omega}_t$ and, consequentially, they assume that V_t and W_t are diagonal matrices. Kaiser et al. (2002) set $\boldsymbol{x}_t$ constant through time equal to $\boldsymbol{\beta}$, let $\boldsymbol{F}_t$ be equal to time-specific regressor values for each t, and considered several different parameterizations for the precision matrix $\boldsymbol{V}_t^{-1}$. Finally, Knorr-Held (2000) proposed a model whose latent process follows a time evolution that is a particular case of what Vivar and Ferreira (2009) called a first-order temporal trend model.

Equations (1.8) and (1.9) define a highly flexible class of models for spatio-temporal areal data. Different models corresponding to distinct scientific assumptions can be obtained by appropriately specifying the matrices $\boldsymbol{F}_t$, $\boldsymbol{G}_t$, $\boldsymbol{V}_t$, and $\boldsymbol{W}_t$. Often in practice, these matrices are parametrized by a low dimensional hyperparameter vector $\boldsymbol{\theta}$. For example, a first-order polynomial trend model assumes $\boldsymbol{F}_t' = \boldsymbol{I}_S$ and $\boldsymbol{G}_t = \rho\boldsymbol{I}_S$, $\forall t$. As a result, the latent vector

$\boldsymbol{x}_t = (x_{t1}, \ldots, x_{tS})'$ is the conditional expectation of the observations at time t and can be interpreted as a mean level field. By setting $\rho = 1$ this model corresponds to a multivariate random walk for the mean level field.

As another example, in a second-order polynomial trend model, there are two latent fields at each time point: $\boldsymbol{x}_{1t} = (x_{1t1}, \ldots, x_{1tS})'$ and $\boldsymbol{x}_{2t} = (x_{2t1}, \ldots, x_{2tS})'$. In such a model, $\boldsymbol{x}_{1t}$ is the conditional expectation of the observations (or mean level field) and $\boldsymbol{x}_{2t}$ is the gradient field at time t. This specification implies a stochastic temporal trend at time t for each region s. Specifically,

$$\boldsymbol{x}_t = \begin{pmatrix} \boldsymbol{x}_{1t} \\ \boldsymbol{x}_{2t} \end{pmatrix}, \quad \boldsymbol{F}_t = \begin{pmatrix} \boldsymbol{I}_S \\ \boldsymbol{0}_S \end{pmatrix},$$

$$\boldsymbol{G}_t = \begin{pmatrix} \rho_1 \boldsymbol{I}_S & \rho_1 \boldsymbol{I}_S \\ \boldsymbol{0}_S & \rho_2 \boldsymbol{I}_S \end{pmatrix} \quad \text{and} \quad \boldsymbol{W}^{-1} = \begin{pmatrix} \boldsymbol{W}_1^{-1} & \boldsymbol{0}_S \\ \boldsymbol{0}_S & \boldsymbol{W}_2^{-1} \end{pmatrix},$$

where $\boldsymbol{W}_1^{-1} = \tau_1(\boldsymbol{I}_S + \phi_1 \boldsymbol{M})$ and $\boldsymbol{W}_2^{-1} = \tau_2(\boldsymbol{I}_S + \phi_2 \boldsymbol{M})$. The fact that the precision matrix is block diagonal allows the system equation to be partitioned in two equations: one for the level field and another for the gradient field.

Another possible spatio-temporal behavior is when an abnormal increase in the conditional expectation for a certain region at a time point causes an increase of the conditional expectation for neighbor regions in the following time points. Such behavior can be modeled by what Vivar and Ferreira (2009) call a spatio-temporal contamination model that assumes $\boldsymbol{F}_t = \boldsymbol{I}_S$ and $\boldsymbol{G}_t = \rho(1 + \beta h)^{-1} \boldsymbol{H}$, where $\boldsymbol{H}$ describes the spatio-temporal interactions among across-time neighbor regions. Specifically, $(\boldsymbol{H})_{kl} = 1$ if $k = l$, $(\boldsymbol{H})_{kl} = \beta$ if $k \in C_l$, and $(\boldsymbol{H})_{kl} = 0$ otherwise. Here, $\beta > 0$ is a contamination coefficient, ρ measures temporal persistence, C_l is the set of across-time neighbors of region l, and $h = \max_l \sum_{k=1, k \neq l}^{S} \mathbf{1}(k \in C_l)$ is the maximum number of across-time neighbors.

Spatio-temporal transport and diffusion may be modeled by assuming $\boldsymbol{F}_t = \boldsymbol{I}_S$ and $(\boldsymbol{G}_t)_{kl} = g_{kl}$ if $l \in T_k$, and $(\boldsymbol{G}_t)_{kl} = 0$ otherwise. Here T_k is the set of transport-diffusion neighbors of region k. Wikle et al. (1998) and Garside and Wilkinson (2003) have proposed particular cases of this model with first- and second-order transport-diffusion neighborhood structures, respectively, and diagonal covariance matrices $\boldsymbol{V}_t$ and $\boldsymbol{W}_t$. Note that $E(x_{tk} | \boldsymbol{x}_1, \ldots, \boldsymbol{x}_{t-1}) = \sum_{l \in T_k} g_{kl} x_{t-1,l}$. Hence, g_{kl} is the contribution of $x_{t-1,l}$ to the conditional expected value of x_{tk} given the latent process prior to time t. When the contributions g_{kl} are asymmetric, the resulting spatio-temporal covariance function is also asymmetric. This model may be useful for modeling pollutant diffusion and transport by wind.

Finally, due to the ability of the state-space framework to combine different model structures, characteristics of the models discussed above can be combined into more complex spatio-temporal models. Chapter 4 illustrates these concepts with an application of state-space models to the analysis of the AIDS epidemic in Rio de Janeiro, Brazil (Ferreira and Vivar, 2024).

1.4 Multiscale Random Field Models

Several processes in nature are better thought of as being defined in many levels of resolution, with different dynamics dominating each resolution level. For example, the distribution of ground-level ozone is controlled by dynamics at various scales: micro-scale atmospheric processes, such as boundary layer effects near buildings and differential heating; mesoscale processes, such as thunderstorms; and large-scale weather systems, such as frontal passages and persistent high-pressure systems (Boubel et al., 1994). Ferreira (2002) and Ferreira et al. (2005) introduced a class of Gaussian multiscale random fields useful for modeling such processes. Their multiscale framework can be used for the coherent combination of spatial areal data available at different levels of resolution. Please see chapter 10 of Ferreira and Lee (2007) for a detailed account of this class of Gaussian multiscale random fields.

In the multiscale random field model of Ferreira (2002) and Ferreira et al. (2005), each level of resolution is connected with the immediately finer level through a linear function plus Gaussian noise. Their model construction starts with a proper Gaussian Markov random field process for each level of resolution, and then uses Jeffrey's rule of conditioning (Jeffrey, 1988) to revise the implied distributions and ensure that the probability distributions at the different resolution levels are compatible and form a valid joint probability distribution. Analogously to Gaussian multiscale time series analysis (Ferreira et al., 2006), this multiscale construction results in a class of models for random fields that exhibits a variety of spatial correlation functions based on a parsimonious parameterization, has the ability to combine information across levels of resolution, and may emulate long-range spatial dependence.

This multiscale framework is related to deterministic and stochastic multigrid methods. Deterministic multigrid methods, first introduced by Fedorenko (1961), are used for the numerical solution of differential equations: the system of differential equations is discretized over a range of resolution levels, an iterative algorithm such as Jacobi or Gauss-Seidel is used to obtain a solution at a given resolution level, and that solution is used as starting point at the next resolution level (e.g., see Briggs et al., 2000). At coarser resolution levels, computations deal with larger scale characteristics of the problem and ignore small-scale characteristics, and as a result they converge faster but provide less accurate solutions. Multigrid relaxation uses the less accurate solution at a coarser resolution as a starting point for the algorithm at the immediate finer resolution, and computations at this scale further refine the solution. Deterministic multigrid methods may also use an aggregated version of a solution from a finer scale as starting point for the algorithm at the immediate coarser scale, as this may help escape sub-optimal solutions. As discussed by Terzopoulos (1986) in the context of image analysis, multigrid methods converge faster than their single-scale counterparts.

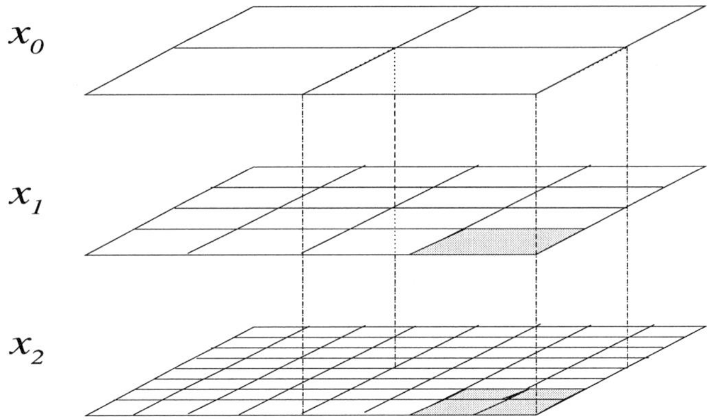

FIGURE 1.1
Multiscale random field with three levels of resolution.

Stochastic multigrid methods were introduced by Goodman and Sokal (1989) to accelerate the convergence of Monte Carlo algorithms for the solution of statistical physics problems. Liu and Sabatti (2000) adapted stochastic multigrid methods to Bayesian computation. In order to simulate from a model defined on a fine level, Goodman and Sokal (1989) and Liu and Sabatti (2000) use a Markov chain Monte Carlo (MCMC) algorithm with proposals for deterministically aggregated blocks. As discussed by Goodman and Sokal (1989) and Liu and Sabatti (2000), when compared to MCMC algorithms based on fine level single-region updates, stochastic multiscale computations lead to much faster convergence.

1.4.1 The Model

In this section, we review the multiscale random field model of Ferreira (2002) and Ferreira et al. (2005). Figure 1.1 represents an example of a multiscale random field with three levels of resolution. In this example, at the coarsest and intermediate levels of resolution, each region corresponds to four regions of the immediate finer level. In general, we consider a multiscale random field with L levels of resolution, where $\boldsymbol{x}_l$ denotes the vectorized l-th level of resolution with n_l elements, $l = 0, \ldots, L-1$. Let (l, j) denote the j-th region of the l-th resolution level and let x_{lj} be the corresponding variable of interest. We assume that region (l, j) has b_{lj} neighboring regions at the same level of resolution, one parent region at the immediate coarser level $l-1$, and $m_{l+1,j}$ descendant regions at the immediate finer level $l+1$. Denote by N_{lj} the set of regions of the l-th level that are neighbors of (l, j) and by D_{lj} the set of regions of the $(l+1)$-th level that are descendants of (l, j).

In this multiscale construction, $\boldsymbol{x}_0$ is the coarsest level and increases in l mean progressively finer levels. The progression from coarser to finer levels is interpreted as an increase in the resolution analogously to the wavelets framework. In addition, in the multiscale framework of Ferreira (2002) and Ferreira et al. (2005) each level has a meaningful practical interpretation. For example, their multiscale model can be used to combine spatial information at different geographic aggregation levels, such as the progressively coarser zip (postal) code level, county level, and state level.

A key feature of this multiscale model construction is that each equation is conditional on different information, as the equations would be incompatible if conditioned on the same information. Specifically, denote by I_l the initial knowledge about the behavior of the process at the l-th level, and by G_l the accumulated knowledge from the coarsest level up to the l-th level x_l. Specifically, $G_0 = I_0$ denotes the knowledge about the generator mechanism of the coarsest level x_0 and the accumulated knowledge up to level x_0 as well. In addition $G_1 = I_0 \bigcap I_1$. In general, recursively define $G_l = G_{l-1} \bigcap I_l$.

Ferreira (2002) and Ferreira et al. (2005) define the general Gaussian multiscale random field model in a hierarchical manner; conditional on all the information G_{L-1} used to construct the model, the true dependence direction is from coarser to finer levels. Specifically, they assume that conditional on the initial information I_l about level l, x_l follows the Type II proper GMRF process discussed in Section 1.2.3

$$\boldsymbol{x}_l | I_l \sim N(\mu \mathbf{1}_{n_l}, \boldsymbol{\Sigma}_l), \tag{1.10}$$

where $\boldsymbol{\Sigma}_l^{-1} = \tau_l(\boldsymbol{I}_{n_l} + \phi_l \boldsymbol{M}_l)$ and the role and definitions of τ_l, ϕ_l and $\boldsymbol{M}_l$ are analogous to the role and definitions of τ, ϕ and $\boldsymbol{M}$ in Section 1.2.3.

Moreover, assume that conditional on I_l there exists a known linear transformation mapping the level $\boldsymbol{x}_l$ to the immediate coarser level $\boldsymbol{x}_{l-1}$, plus Gaussian noise ($l = 1, \ldots, L-1$):

$$p(\boldsymbol{x}_{l-1} | I_l, \boldsymbol{x}_l) = N(\boldsymbol{x}_{l-1} | \boldsymbol{A}_l \boldsymbol{x}_l, \boldsymbol{U}_l), \tag{1.11}$$

where $\boldsymbol{U}_l = \delta_l \boldsymbol{I}_{n_{l-1}}$ and the sum of the elements of each row of $\boldsymbol{A}_l$ is equal to one. This representation is useful in a number of situations as for example when the relationship between coarser and finer levels occurs on irregular grids, or when the number of descendants is not constant, or when the aggregation weights are not constant.

Bayes' Theorem can then be applied to Equations (1.10) and (1.11) to obtain the conditional distribution of $\boldsymbol{x}_l$ given I_l and $\boldsymbol{x}_{l-1}$, $l = 1, \ldots, L-1$. These conditional distributions can then be combined to yield the joint multiscale model for $(\boldsymbol{x}_0, \ldots, \boldsymbol{x}_{L-1})$ given all the knowledge G_{L-1} (Ferreira, 2002; Ferreira et al., 2005)

$$\begin{aligned} p(\boldsymbol{x}_0, \ldots, \boldsymbol{x}_{L-1} | G_{L-1}) = {} & N(\boldsymbol{x}_0 | \mu \mathbf{1}_{n_0}, \boldsymbol{Q}_0) \\ & \prod_{l=1}^{L-1} N(\boldsymbol{x}_l | \mu \mathbf{1}_{n_l} + \boldsymbol{B}_l(\boldsymbol{x}_{l-1} - \mu \mathbf{1}_{n_{l-1}}), \boldsymbol{\Sigma}_l - \boldsymbol{B}_l \boldsymbol{W}_l \boldsymbol{B}_l'), \end{aligned}$$

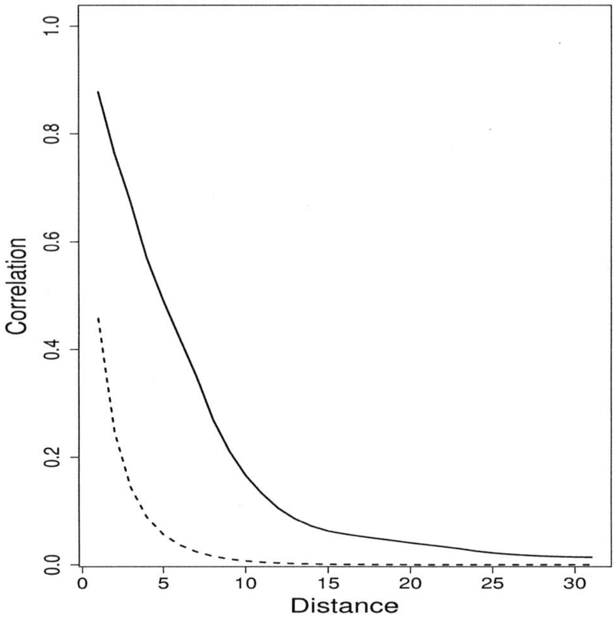

FIGURE 1.2
Marginal spatial correlation at the finest resolution level of a multiscale random field with four levels or resolution (solid line) and spatial correlation of the corresponding Markov random field at the finest level (dashed line).

where $\boldsymbol{B}_l = \boldsymbol{\Sigma}_l \boldsymbol{A}_l' \boldsymbol{W}_l^{-1}$, $\boldsymbol{W}_l = \boldsymbol{A}_l \boldsymbol{\Sigma}_l \boldsymbol{A}_l' + \delta_l \boldsymbol{I}_{n_{l-1}}$, and $\boldsymbol{Q}_l = \boldsymbol{\Sigma}_l - \boldsymbol{B}_l(\boldsymbol{W}_l - \boldsymbol{Q}_{l-1})\boldsymbol{B}_l'$.

Multiscale random field models can be used to induce long-range behavior at the finest resolution level. This characteristic of multiscale random field models is in sharp contrast to Gaussian Markov random fields, which are able to elicit only fast-decaying spatial correlation functions. For example, Figure 1.2 shows the marginal spatial correlation function at the finest resolution level of a multiscale random field with 4 levels of resolution (solid line). At each level of resolution, the *initial model* is a first-order proper GMRF as in Equation (1.10). The values of the parameters are $\tau_0 = 1$, $\phi_0 = 1$, $\tau_1 = 1$, $\phi_1 = 2$, $\delta_1 = 0.01$, $\tau_2 = 1$, $\phi_2 = 4$, $\delta_2 = 0.01$, $\tau_3 = 1$, $\phi_3 = 8$, and $\delta_3 = 0.01$. For comparison purposes, Figure 1.2 also shows the spatial correlation function of the corresponding Markov random field at the finest level with $\tau = \tau_3 = 1$ and $\phi = \phi_3 = 8$ (dashed line). While the spatial correlation of the proper GMRF model decays exponentially with distance, the spatial correlation of the Gaussian multiscale random field decays in a much slower manner. Therefore, in the multiscale random field model, the spatial correlation at the coarser levels induces long-range dependence at the finest level.

1.4.2 Multiscale Random Fields for Permeability Fields

An important application of multiscale random field model is in soil permeability estimation. Permeability is a measure of how easily fluid flows through porous media, and its estimation is crucially important in areas such as environmental cleanup and petroleum production. Lee et al. (2002) proposed the use of Markov random fields as priors for permeability estimation and developed an efficient MCMC algorithm to implement the analysis. Ferreira et al. (2003) developed a Bayesian multiscale framework for the coherent assimilation of data available at different levels of resolution and obtained from different sources for the estimation of 1-D permeability fields. Ferreira et al. (2005) extended that work to a multiscale framework for the estimation of 2-D permeability fields. A detailed account of that work can be found in Chapter 16 of Ferreira and Lee (2007).

The multiscale random fields discussed here are useful as priors for latent fields generated by multiscale processes where coarser levels dominate finer levels. For example, in the case of permeability fields, at a small scale of resolution permeability depends on the space between rocks, whereas at an intermediate scale permeability depends on the sizes of the rocks, and at a large scale permeability depends on the geometry of the field such as the existence of faults and barriers.

In the estimation of permeability fields, there are usually two types of data: static data and dynamic data. Static data arise from many sources such as core samples, well tests, and geological studies. A key aspect is that different sources of static data provide information at different resolution levels. For example, core samples obtained when drilling a well provide precise information at a fine resolution level but only at the location of the well. In addition, static data from well tests give information at an intermediate resolution level about the permeability field near the well. Finally, geological studies such as seismic surveys provide information at a coarse level of resolution, but usually covering the whole region of interest.

Dynamic data arise as measurements of variables of interest as water or oil are extracted from wells. For example, dynamic data such as the time series of pressure and amount of oil produced at the different wells may be obtained from the history of oil production from a petroleum field. As another example, in the area of aquifer cleanup dynamic data may be obtained through tracer experiments. As an example of such an experiment, injection wells and producing wells are drilled at different points of the aquifer. After that, water is injected into the injection wells at a constant rate and extracted through the producing wells. After this system reaches a steady state, a tracer is injected in the injection wells and the dynamic data are obtained as the time series of tracer concentrations at the producing wells.

While the static data may be regarded as noisy measurements of the permeability field at different scales of resolution, the permeability field and the dynamic data are related through highly nonlinear physical laws. Given the permeability field and other physical characteristics of the aquifer or oil field, a system of differential equations describing these physical laws may be solved to predict the dynamic data (Gelhar, 1993). Computer codes that numerically solve this system of differential equations are called fluid flow simulators (King and Datta-Gupta, 1998; Vasco and Datta-Gupta, 1999).

The framework of Ferreira (2002) and Ferreira et al. (2005) uses an MCMC algorithm to assimilate the static data and the dynamic data to explore the posterior distribution of the permeability field and other parameters. Each iteration of the MCMC algorithm is performed in a cascade way, from coarser to finer levels. First, all the information about the static data is sent from finer to coarser levels by analytically integrating out the finer levels. Then, the dynamic data are used to simulate the coarser levels by using the fluid flow simulator running at those levels. Finally, the finer levels are simulated conditional on the simulated coarser levels. Besides the faster running time of the fluid flow simulator at the coarser levels, the MCMC for the coarser levels converges much faster than for the finer levels because the marginal posterior distribution of the permeability field at coarser levels has fewer local maxima than the marginal posterior distribution of the permeability field at finer levels. In this approach, the results at coarser levels guide the simulation at finer levels, and as a consequence, the convergence at finer levels is faster than the convergence of algorithms based on traditional GMRF models. Moreover, the final result is a coherent analysis of the permeability field at different levels of resolution.

The framework of Ferreira (2002) and Ferreira et al. (2005) has similarities with the renormalization group approach to image analysis proposed by Gidas (1989). Gidas (1989) starts with a model for the image (in gray scale) at the finest resolution level, and his objective is to find the maximum a posteriori of the image at the finest level. With that objective in mind, Gidas (1989) defines conditional distributions at coarser levels given finer levels that may depend on different temperatures. Then, he proceeds from finer to coarser levels recursively obtaining approximate marginal densities at each level. After obtaining the marginal density at the coarsest level, he obtains the maximum a posteriori at the coarsest level. He then proceeds from coarser to finer levels, obtaining the maximum a posteriori of the image at each level constrained by the solution at the immediate coarser level. In the Gidas framework, the conditional distributions and temperatures are chosen such that the algorithm finds an approximate maximum a posteriori image at the finest resolution level. Thus, Gidas (1989) defines the model from finer to coarser levels, with finest level dynamics determining coarser levels dynamics. Conversely, Ferreira (2002) and Ferreira et al. (2005) define the model from coarser to finer levels, and in their model, the coarser levels dynamics dominate the finer levels dynamics. Processing of information in the framework of Ferreira et al. (2005) is similar to the processing of information in the Gidas framework, as first the information is sent from finer to coarser levels, processed at the coarsest level, and then sent from coarser to finer levels. A noteworthy distinction is that Ferreira (2002) and Ferreira et al. (2005) simulate realizations of the multiscale field, whereas Gidas obtains maximum a posteriori fields. Thus, while it is not clear how to assess uncertainty in the Gidas framework, the framework of Ferreira et al. (2005) easily provides uncertainty measurements such as posterior variances and credible intervals.

1.5 Inference

Inference is model specific for each of the two classes of models discussed here, spatio-temporal and multiscale models, and for specialized models within each class. Usually, posterior exploration is performed using specialized Markov chain Monte Carlo algorithms. In spatio-temporal modeling, Vivar and Ferreira (2009) develop a forward information filter backward sampler algorithm that samples all the elements of the latent process from their joint full conditional distribution. We refer the reader to the original publications for detailed information on algorithms for exploration of the posterior distribution of the parameters.

In what follows we briefly discuss the specification of priors for the hyperparameters of the model blocks based on proper GMRFs. These priors for the hyperparameters of proper GMRFs have to be carefully specified. For the case when the first specification of the Type I proper GMRF is used as prior, Sun et al. (2000) assume a uniform prior for ρ on the interval $(-1, 1)$. In addition, for the spatio-temporal model discussed in Section 1.3.2 the noninformative priors used by Sun et al. (2000) are $\Pi(\sigma_\epsilon^2) \propto \sigma_\epsilon^{-2}$ for the extra-Poisson variance and $\Pi(\delta) \propto \delta^{-2}$ for the variance parameters of the Type I proper GMRF.

For the case when a Type II proper GMRF is used as a model for the observations, Ferreira and De Oliveira (2007) obtained several non-informative priors for the parameters ϕ and τ. Specifically, Ferreira and De Oliveira (2007) obtained explicit expressions for the Jeffreys-rule, the independence Jeffreys, and the reference priors. They found that the independence Jeffreys prior yields improper posterior distributions, while the reference and Jeffreys-rule priors yield proper posterior distributions. Moreover, they found that the reference-prior-based Bayesian procedure has better frequentist properties than the Jeffreys-prior-based Bayesian procedure. Specifically, when a Type II proper GMRF is used as a model for the observations, the reference prior for (ϕ, τ) is

$$\Pi^r(\phi, \tau) \propto \tau^{-1} \sqrt{\sum_{k=1}^{n-1} \lambda_k^2 (\phi\lambda_k + 1)^{-2} - \frac{1}{n-1} \left[\sum_{k=1}^{n-1} \lambda_k (\phi\lambda_k + 1)^{-1}\right]^2},$$

where $\lambda_1 \geq \ldots \geq \lambda_{n-1} > \lambda_n = 0$ are the ordered eigenvalues of $\boldsymbol{M}$.

Note that reference priors are model-specific. Thus, while the prior above provides good results when the proper GMRF is used as a model for the observations, this prior may not be adequate when proper GMRFs are used as building blocks for more complex models. However, in the case of more complex models, the prior above may serve as a guide for prior specification. For example, for the spatio-temporal models discussed in Section 1.3.3, Vivar and Ferreira (2009) used the knowledge about the tail behavior of the reference prior $\Pi^r(\phi)$ to specify a prior for ϕ. Specifically, Ferreira and De Oliveira (2007) have shown that $\Pi^r(\phi)$ behaves as $O(1)$ as $\phi \to 0$, and as $O(\phi^{-2})$ as $\phi \to \infty$. As a result, the marginal reference prior for ϕ is proper but does not have any positive integer moments. In order to have positive integer moments, Vivar and Ferreira (2009) consider priors for ϕ of the type $\Pi(\phi) \propto 1$ if $0 < \phi < 1$, $\Pi(\phi) \propto \phi^{-2}$ if $1 < \phi < a$, and $\Pi(\phi) \propto 0$ otherwise. In particular, they have found that $a = 20$ works well in their applications.

1.6 Conclusions

Gaussian Markov random fields have been widely used for spatial and spatio-temporal statistical modeling. Most of that work has used improper GMRFs as priors for spatial random effects. In the case of spatio-temporal modeling, the models typically have a common univariate temporal trend for all the regions and time-specific spatial random effects that are modeled with *improper* Gaussian Markov random fields. As a consequence, their scope is limited to modeling processes that have a reasonably simple type of spatio-temporal interaction. As shown by Vivar and Ferreira (2009), the use of *proper* Gaussian Markov random fields in the construction of spatio-temporal models allows for spatio-temporal interaction that can more realistically describe complex spatio-temporal processes such as the spread of contagious diseases and the transport and diffusion of pollutants.

As we discuss in this chapter, because *proper* Gaussian Markov random fields imply proper probability distributions, they may facilitate the construction of highly structured models. In particular, we have discussed here the construction of spatio-temporal and multiscale models. Finally, we note that proper GMRFs may be used as building blocks for other highly structured models.

Bibliography

Assunção, R. M., J. E. Potter, and S. M. Cavenaghi (2002). A Bayesian space varying parameter model applied to estimating fertility schedules. *Statistics in Medicine 21*, 2057–2075.

Banerjee, S., B. P. Carlin, and A. E. Gelfand (2014). *Hierarchical Modeling and Analysis for Spatial Data* (2nd ed.). Boca Raton, FL: CRC Press.

Besag, J. E. (1974). Spatial interaction and the statistical analysis of lattice systems (with discussion). *Journal of the Royal Statistical Society, Series B 36*, 192–236.

Besag, J. E., P. J. Green, D. M. Higdon, and K. L. Mengersen (1995). Bayesian computation and stochastic systems (with discussion). *Statistical Science 10*, 3–66.

Besag, J. E. and C. Kooperberg (1995). On conditional and intrinsic autoregressions. *Biometrika 82*, 733–746.

Besag, J. E., J. York, and A. Mollié (1991). Bayesian image restoration with two applications in spatial statistics (with discussion). *Annals of the Institute of Statistical Mathematics 43*, 1–59.

Boubel, R. W., D. L. Fox, B. Turner, and A. C. Stern (1994). *Fundamentals of Air Pollution* (3rd ed.). Academic Press.

Briggs, W. L., V. E. Henson, and S. F. McCormick (2000). *A Multigrid Tutorial, 2nd edition.* Society for Industrial and Applied Mathematics.

Carlin, B. P. and S. Banerjee (2003). Hierarchical multivariate CAR models for spatio-temporally correlated survival data (with discussion). In J. Bernardo, M. J. Bayarri, J. O. Berger, A. P. Dawid, D. Heckerman, A. F. M. Smith, and M. West (Eds.), *Bayesian Statistics 7*, pp. 45–63. Oxford University Press.

Fedorenko, R. P. (1961). A relaxation method for solving elliptic difference equations. *Zhurnal Vychislitelnoj Matematyki i Matematicheskoj Fiziki 1*, 922–927. Translation appeared in: USSR Computational Mathematics and Mathematical Physics (1962), vol 1, pp 1092–1096.

Ferreira, M. A. R. (2002). *Bayesian Multi-scale Modelling.* Ph. D. thesis, Duke University.

Ferreira, M. A. R. (2019). The limiting distribution of the Gibbs sampler for the intrinsic conditional autoregressive model. *Brazilian Journal of Probability and Statistics 33*(4), 734–744.

Ferreira, M. A. R., Z. Bi, M. West, H. Lee, and D. Higdon (2003). Multi-scale modelling of 1-D permeability fields. In J. Bernardo, M. J. Bayarri, J. O. Berger, A. P. Dawid, D. Heckerman, A. F. M. Smith, and M. West (Eds.), *Bayesian Statistics 7*, pp. 519–527. Oxford University Press.

Ferreira, M. A. R. and V. De Oliveira (2007). Bayesian reference analysis for Gaussian Markov Random Fields. *Journal of Multivariate Analysis 98*, 789–812.

Ferreira, M. A. R., D. Higdon, H. K. H. Lee, and M. West (2005). Multiscale random field models. Technical report, UFRJ - DME.

Ferreira, M. A. R. and H. K. H. Lee (2007). *Multiscale Modeling: A Bayesian Perspective.* Springer Series in Statistics. New York: Springer.

Ferreira, M. A. R., E. M. Porter, and C. T. Franck (2021). Fast and scalable computations for Gaussian hierarchical models with intrinsic conditional autoregressive spatial random effects. *Computational Statistics and Data Analysis 162*, 107264.

Ferreira, M. A. R. and J. C. Vivar (2024). Spatio-temporal models for Poisson areal data. In M. A. R. Ferreira (Ed.), *Modeling Spatio-Temporal Data: Markov Random Fields, Objective Bayes, and Multiscale Models*, pp. 69–89. Chapman & Hall/CRC.

Ferreira, M. A. R., M. West, H. K. H. Lee, and D. Higdon (2006). Multi-scale and hidden resolution time series models. *Bayesian Analysis 1*, 947–968.

Gamerman, D., A. R. B. Moreira, and H. Rue (2003). Space-varying regression models: specifications and simulation. *Computational Statistics and Data Analysis 42*, 513–533.

Gao, L., S. Banerjee, and B. Ritz (2023). Spatial difference boundary detection for multiple outcomes using Bayesian disease mapping. *Biostatistics 24*(4), 922–944.

Garside, L. M. and D. J. Wilkinson (2003). Dynamic lattice-Markov spatio-temporal models for environmental data. In J. Bernardo, J. Berger, A. Dawid, and A. Smith (Eds.), *Bayesian Statistics 7*, pp. 535–542. Oxford: Oxford University Press.

Gelfand, A. E. and P. Vounatsou (2003). Proper multivariate conditional autoregressive models for spatial data analysis. *Biostatistics 4*(1), 11–15.

Gelhar, L. W. (1993). *Stochastic Subsurface Hydrology.* Englewood Cliffs, NJ: Prentice-Hall.

Gidas, B. (1989). A renormalization group approach to image processing problems. *IEEE Transactions on Pattern Analysis and Machine Intelligence 11*, 164–180.

Goodman, J. and A. D. Sokal (1989). Multigrid Monte Carlo method. *Physical Review Letters D 40*, 2035–2072.

Jeffrey, R. C. (1988). Conditioning, kinematics, and exchangeability. In B. Skyrms and W. L. Harper (Eds.), *Causation, Chance, and Credence*, Volume 1, pp. 221–255. Dordrecht: Kluwer.

Kaiser, M. S., M. J. Daniels, K. Furakawa, and P. Dixon (2002). Analysis of particulate matter air pollution using Markov random field models of spatial dependence. *Environmetrics 13*, 615–628.

Keefe, M. J., M. A. R. Ferreira, and C. T. Franck (2018). On the formal specification of sum-zero constrained intrinsic conditional autoregressive models. *Spatial Statistics 24*, 54 – 65.

Keefe, M. J., M. A. R. Ferreira, and C. T. Franck (2019). Objective Bayesian analysis for Gaussian hierarchical models with intrinsic conditional autoregressive priors. *Bayesian Analysis 14*(1), 181–209.

Kim, H., D. Sun, and R. K. Tsutakawa (2001). A bivariate Bayes method for improving the estimates of mortality rates with a twofold conditional autoregressive model. *Journal of the American Statistical Association 96*, pp. 1506–1521.

King, M. J. and A. Datta-Gupta (1998). Streamline simulation: A current perspective. *In Situ 22*, 91–140.

Knorr-Held, L. (2000). Bayesian modelling of inseparable space-time variation in disease risk. *Statistics in Medicine 19*, 2555–2567.

Lee, D. (2013). CARBayes: An R package for Bayesian spatial modeling with conditional autoregressive priors. *Journal of Statistical Software 55*(13), 1–24.

Lee, H. K. H., D. Higdon, Z. Bi, M. A. R. Ferreira, and M. West (2002). Markov random field models for high-dimensional parameters in simulations of fluid flow in porous media. *Technometrics 44*(3), 230–241.

Leroux, B. G., X. Lei, and N. Breslow (2000). Estimation of disease rates in small areas: a new mixed model for spatial dependence. In M. E. Halloran and D. Berry (Eds.), *Statistical Models in Epidemiology, the Environment, and Clinical Trials*, pp. 179–191. Springer.

Liu, J. S. and C. Sabatti (2000). Generalised Gibbs sampler and multigrid Monte Carlo for Bayesian computation. *Biometrika 87*, 353–369.

Mardia, K. V. (1988). Multi-dimensional multivariate Gaussian Markov random fields with application to image processing. *Journal of Multivariate Analysis 24*(2), 265–284.

Min, X., D. Sun, C. Z. He, and M. Schootman (2010). A Bayesian hierarchical model of nontraumatic lower-extremity amputation rates. *Spatial and Spatio-temporal Epidemiology 1*, 169–176.

Pettitt, A. N., I. S. Weir, and A. G. Hart (2002). A conditional autoregressive Gaussian process for irregularly spaced multivariate data with application to modelling large sets of binary data. *Statistics and Computing 12*, 353–367.

Porter, E. M., C. T. Franck, and M. A. R. Ferreira (2023). Objective Bayesian model selection for spatial hierarchical models with intrinsic conditional autoregressive priors. *Bayesian Analysis 19*, accepted for publication.

Porter, E. M., C. T. Franck, and M. A. R. Ferreira (2024). Gaussian spatial hierarchical models with ICAR priors. In M. A. R. Ferreira (Ed.), *Modeling Spatio-Temporal Data: Markov Random Fields, Objective Bayes, and Multiscale Models*, pp. 23–46. Chapman & Hall/CRC.

Porter, E. M., M. J. Keefe, C. T. Franck, and M. A. R. Ferreira (2018). *ref.ICAR: Objective Bayes Intrinsic Conditional Autoregressive Model for Areal Data.* CRAN. R package version 1.0.

R Core Team (2023). *R: A Language and Environment for Statistical Computing.* Vienna, Austria: R Foundation for Statistical Computing.

Rue, H. and L. Held (2005). *Gaussian Markov Random Fields: Theory and Applications.* CRC Press.

Sun, D., R. K. Tsutakawa, H. Kim, and Z. He (2000). Spatio-temporal interaction with disease mapping. *Statistics in Medicine 19*, 2015–2035.

Sun, D., R. K. Tsutakawa, and P. L. Speckman (1999). Posterior distribution of hierarchical models using CAR(1) distributions. *Biometrika 86*, 341–350.

Terzopoulos, D. (1986). Image analysis using multigrid relaxation methods. *IEEE Transactions on Pattern Analysis and Machine Intelligence PAMI-8*, 129–139.

Vasco, D. W. and A. Datta-Gupta (1999). Asymptotic solutions for solute transport: A formalism for tracer tomography. *Water Resources Research 35*, 1–16.

Vivar, J. C. and M. A. R. Ferreira (2009). Spatio-temporal models for Gaussian areal data. *Journal of Computational and Graphical Statistics 18*, 658–674.

Waller, L. A., B. P. Carlin, H. Xia, and A. E. Gelfand (1997). Hierarchical spatio-temporal mapping of disease rates. *Journal of the American Statistical Association 92*, 607–617.

Wikle, C. K., L. M. Berliner, and N. Cressie (1998). Hierarchical Bayesian space-time models. *Environmental and Ecological Statistics 5*, 117–154.

Zhang, S., D. Sun, C. Z. He, and M. Schootman (2006). A Bayesian semiparametric model for colorectal cancer incidences. *Statistics in Medicine 25*, 285–309.

2

Gaussian Spatial Hierarchical Models with ICAR Priors

Erica M. Porter

Clemson University

Christopher T. Franck

Virginia Tech

Marco A. R. Ferreira

Virginia Tech

This chapter summarizes recent developments related to intrinsic conditional autoregressive (ICAR) models for Gaussian data that make inference for these models fast, automatic, and accurate for both model selection and inference on parameters. The chapter culminates with a demonstration of the R package **ref.ICAR** which makes ICAR-based data analysis easily accessible to practitioners. We discuss the use of ICAR models for Gaussian areal data. We explain how to impose the sum-zero constraint on a vector of spatial random

DOI: 10.1201/9781032623443-2

effects when they are assigned an ICAR prior in a hierarchical model. We also describe the automatic reference prior for the parameters of hierarchical models with ICAR spatial random effects, fast computations for these models using spectral decomposition, and model selection based on fractional Bayes factors that accommodates the automatic reference prior and provides consistent model selection. Finally, we demonstrate that many of the techniques for these models can be applied to spatial datasets using the R package **ref.ICAR**.

2.1 Introduction

Gaussian hierarchical models with intrinsic conditional autoregressive (ICAR) spatial random effects are widely used to model spatial areal data. Areal data describes the case where a region of interest is spatially partitioned into smaller subregions whose corresponding data values may inherently rely on those of neighboring subregions (Banerjee et al., 2014). Besag et al. (1991) discussed this idea of intrinsic autoregression for modeling areal data in the context of image restoration, where the spatial subregions consist of pixels of an image and the goal is to estimate the intensity of some measurement in each pixel. For simplicity of exposition, henceforth we refer to Gaussian hierarchical models with ICAR spatial random effects simply as spatial ICAR models. Spatial ICAR models have since been used extensively in many areas, including disease mapping applications (e.g. Best et al., 2005; Reich et al., 2006; Lee, 2011; Goicoa et al., 2016) , ecology (e.g. Ver Hoef et al., 2018), and survey data (e.g. Mercer et al., 2015; Logan et al., 2020).

In contrast to conditional autoregressive (CAR) models that may be used to model observations directly, ICAR priors are not proper densities and can result in identifiability issues if not treated with care. Commonly, computation-based approaches have been used to enforce constraints that make the ICAR prior proper by forcing the spatial random effects to sum to zero. For example, practitioners have often used centering on the fly to make the ICAR model proper. This informal approach uses a Markov chain Monte Carlo (MCMC) algorithm to sample the random effects and re-centers the vector of spatial random effects around zero after each iteration. As an alternative, Rue et al. (2009) proposed the Integrated Nested Laplace Approximation (INLA) approach, which uses a Gaussian approximation to the full conditional of the latent effects to compute the marginal distributions using numerical integration. To make the ICAR model proper, the INLA approach uses standard multivariate Gaussian results to obtain the full conditional distribution of the spatial random effects conditional on their sum being equal to zero (Rue and Held, 2005).

To enable formal Bayesian methodology for spatial ICAR models, Keefe et al. (2018) derived the formal sum-zero constrained ICAR prior, which is a unique, singular Gaussian distribution for the vector of spatial random effects. This explicit density for the ICAR model made it possible to derive objective and automatic priors for the parameters of the spatial ICAR model (Keefe et al., 2019) and then objective Bayesian model selection for ICAR models (Porter et al., 2023). Further, recent developments by Ferreira (2019) and Ferreira et al. (2021) have shown that the full conditional distribution for the spatial random effects with the formal sum-zero constrained ICAR prior and the implied full conditional distribution from centering on the fly are equivalent. Porter et al. (2023) also showed that the formal sum-zero constrained ICAR prior is equivalent to the full conditional distribution of the spatial random effects conditional on their sum being zero, making all three approaches to enforcing the sum-zero constraint for the spatial random effects equivalent.

The remainder of this chapter is organized as follows. Section 2.2 specifies the hierarchical model for Gaussian areal data. Section 2.3 outlines the ICAR prior for a vector of spatial random effects and reviews ways to enforce the sum-zero constraint, i.e. the formal sum-zero constrained ICAR prior and centering on the fly. Section 2.4 describes the reference prior for parameters of the spatial ICAR model. Section 2.5 details fast computations for the spatial ICAR model and reference prior. Section 2.6 describes a posterior maximizer algorithm that uses a large sample asymptotic normality approximation to the joint posterior density of the unknown parameters, which allows us to estimate parameters via maximum a posteriori (MAP) estimation. Section 2.7 presents a model selection approach using the reference prior. Section 2.8 describes the R package **ref.ICAR**, which contains functions for sampling and inference with the reference prior, and Section 2.9 applies these functions to a commonly analyzed dataset. Finally, Section 2.10 provides a summary and discussion of results in this chapter.

2.2 Hierarchical Model for Areal Data

Consider a hierarchical model for areal data, i.e. data measured over a contiguous region that is partitioned into n disjoint subregions indexed by $1, \ldots, n$. The hierarchical model for these data can be written as

$$\boldsymbol{Y} = X\boldsymbol{\beta} + \boldsymbol{\theta} + \boldsymbol{\omega}, \tag{2.1}$$

where $\boldsymbol{Y}$ is a $n \times 1$ response vector, X is a $n \times p$ matrix of covariates, and $\boldsymbol{\beta}$ is a $p \times 1$ vector of regression coefficients for the fixed effects. Additionally, $\boldsymbol{\theta} = (\theta_1, \ldots, \theta_n)^T$ is a $n \times 1$ vector of independent unstructured random effects with distribution $N(\mathbf{0}, \sigma^2 I_n)$ and $\boldsymbol{\omega} = (\omega_1, \ldots, \omega_n)^T$ is a $n \times 1$ vector of spatial random effects that will be assigned an ICAR prior. The first column of the matrix X is assumed to be a vector of ones, denoted by $\mathbf{1}_n$, corresponding to an intercept. The vectors of random effects, $\boldsymbol{\theta}$ and $\boldsymbol{\omega}$, are assumed to be independent *a priori*.

2.3 ICAR Prior

Let $\boldsymbol{\omega}$ be a vector of spatial random effects that *a priori* follows the improper ICAR prior (Besag et al., 1991). Then, the prior density for $\boldsymbol{\omega}$ is given by

$$p(\boldsymbol{\omega}) \propto \exp\left\{ -\frac{\tau_\omega}{2}\boldsymbol{\omega}^T H \boldsymbol{\omega}\right\}, \tag{2.2}$$

where $\tau_\omega > 0$ is a precision parameter, and H is a positive semi-definite matrix such that

$$(H)_{ij} = \begin{cases} h_i, & \text{if } i = j, \\ -g_{ij}, & \text{if } i \in N_j, \\ 0, & \text{otherwise,} \end{cases} \tag{2.3}$$

and $d_1 \geq d_2 \geq \ldots \geq d_{n-1} > d_n = 0$ are the ordered eigenvalues of H. H is a fixed matrix selected by the researcher to define the neighborhood structure of the study region. A common choice for H classifies two subregions as neighbors if they share a border. Then $\{N_j; j = 1 \ldots n\}$ denotes the set of regions that are neighbors to region j, h_i indicates the total number of neighbors of region i, and $g_{ij} = 1$ if regions i and j are neighbors and $g_{ij} = 0$ if regions i and j are not neighbors. All of the subregions are assumed to be connected, so H has rank $n - 1$ and one zero eigenvalue (e.g., see Ferreira and De Oliveira, 2007; De Oliveira and Ferreira, 2011).

There are a few ways to impose the sum-zero constraint on the vector of spatial random effects to make the ICAR prior proper. One approach includes centering on the fly, where the practitioner centers the spatial random effects to sum to zero at the end of each iteration of an MCMC algorithm. A second approach, implemented in INLA, uses the full conditional distribution of the spatial random effects conditional on their sum being equal to zero. Lastly, Keefe et al. (2018) developed a formal sum-zero constrained ICAR prior that has a singular Gaussian distribution with zero mean and a unique covariance matrix.

2.3.1 Formal Sum-Zero Constrained ICAR Prior

Keefe et al. (2018) developed a formal sum-zero constrained ICAR prior for the vector of spatial random effects. The sum-zero constrained ICAR model is proper and well-defined with zero mean vector and a unique covariance matrix, and thus can be used for formal Bayesian model selection and algorithms such as MCMC. This approach uses a limiting argument by imposing the sum-zero constraint on a proper conditional autoregressive (CAR) model assigned to the spatial random effects and taking the limit, which results in a unique ICAR model.

To obtain the sum-zero constrained ICAR, Keefe et al. (2018) considered a proper CAR model for a vector of spatial random effects $\boldsymbol{\phi}^{**}$. For a review of proper CAR models, also known as proper Markov random fields, see Chapter 1 (Ferreira, 2024). Using a signal-to-noise ratio parameterization, the distribution of the proper CAR effects is

$$\boldsymbol{\phi}^{**} \sim N\left(\mathbf{0}, \frac{\sigma^2}{\tau}\Sigma_\lambda\right), \tag{2.4}$$

where σ^2 is the variance of the unstructured random effects, $\tau > 0$ is a precision parameter which controls the strength of spatial dependence, $\lambda > 0$ is a propriety parameter, and $\Sigma_\lambda^{-1} = \lambda I_n + H$ with H as in Equation (2.3). Then, to impose the sum-zero constraint on the CAR prior in Equation (2.4), Keefe et al. (2018) projected $\boldsymbol{\phi}^{**}$ onto the subspace of $\mathbb{R}^n$ that is orthogonal to the subspace spanned by the vector $n^{-1/2}\mathbf{1}_n$ to obtain $\boldsymbol{\phi}^*$, where $\sum_{i=1}^n \phi_i^* = 0$. As a result, $\boldsymbol{\phi}^* = P\boldsymbol{\phi}^{**}$, where $P = I_n - n^{-1}\mathbf{1}_n\mathbf{1}_n^T$ is a projection matrix. Then the distribution of $\boldsymbol{\phi}^*$ follows as

$$\boldsymbol{\phi}^* \sim N\left(\mathbf{0}, \frac{\sigma^2}{\tau}\Sigma_{\phi\lambda}\right), \tag{2.5}$$

where $\Sigma_{\phi\lambda} = P\Sigma_\lambda P^T$.

Finally, to obtain the distribution for the sum-zero constrained ICAR prior, Keefe et al. (2018) took the limit as $\lambda \to 0$. Let the spectral decomposition of H be $H = QDQ^T$, where $Q = (\mathbf{q}_1, \mathbf{q}_2, \ldots, \mathbf{q}_n)$ is a $n \times n$ matrix whose columns are the normalized eigenvectors of H and $D = \text{diag}(d_1, d_2, \ldots, d_{n-1}, 0)$ is a diagonal matrix containing the eigenvalues of H. Letting $\tilde{Q} = (\mathbf{q}_1, \ldots, \mathbf{q}_{n-1})$, the distribution of the spatial random effects $\boldsymbol{\phi}$ can be written as the following singular Gaussian distribution (Keefe et al., 2018, 2019):

$$\boldsymbol{\phi} \sim N\left(\mathbf{0}, \frac{\sigma^2}{\tau}\Sigma_\phi\right), \tag{2.6}$$

where $\Sigma_\phi = \lim_{\lambda\to 0}\Sigma_{\phi\lambda} = \tilde{Q}\text{diag}(d_1^{-1}, \ldots, d_{n-1}^{-1})\tilde{Q}^T$ is the Moore-Penrose pseudoinverse (Penrose, 1955) of the precision matrix H. So the density for the spatial random effects $\boldsymbol{\phi}$ can be written as:

$$\begin{aligned} p(\boldsymbol{\phi}|\sigma^2, \tau) &= (2\pi)^{-(n-1)/2}\left(\frac{\tau}{\sigma^2}\right)^{(n-1)/2}\left(\prod_{i=1}^{n-1} d_i\right)^{1/2} \\ &\quad \times \exp\left\{-\frac{\tau}{2\sigma^2}\boldsymbol{\phi}^T H\boldsymbol{\phi}\right\}\mathbb{1}(\mathbf{1}_n^T\boldsymbol{\phi} = 0). \end{aligned} \tag{2.7}$$

We use $\boldsymbol{\phi}$ to define the formal sum-zero constrained ICAR prior for the vector of spatial random effects, where $\boldsymbol{\phi}$ can be substituted for $\boldsymbol{\omega}$ in Equation (2.1). Upon assigning the distribution in Equation (2.6) to the spatial random effects and integrating $\boldsymbol{\phi}$ out of the hierarchical model in Equation (2.1), the distribution for response variable $\boldsymbol{Y}$ is

$$\boldsymbol{Y}|\boldsymbol{\beta}, \sigma^2, \tau \sim N(X\boldsymbol{\beta}, \sigma^2(I_n + \tau^{-1}\Sigma_\phi)). \tag{2.8}$$

In Section 2.7 we consider model selection both for regressors and for the presence of spatial random effects. In the case of the absence of spatial random effects, the model reduces to an ordinary linear model (OLM) and the distribution of $\boldsymbol{Y}$ becomes

$$\boldsymbol{Y}|\boldsymbol{\beta}, \sigma^2 \sim N(X\boldsymbol{\beta}, \sigma^2 I_n). \tag{2.9}$$

2.3.2 Equivalence Between Formal Sum-Zero ICAR Prior and Centering on the Fly

Prior to the formal sum-zero constrained ICAR prior by Keefe et al. (2018), practitioners would often enforce the sum-zero constraint for the ICAR prior by centering the vector of spatial random effects by subtracting its mean at the end of each iteration of an MCMC algorithm, called centering on the fly. Ferreira et al. (2021) showed that the full conditional distribution for the spatial random effects $\boldsymbol{\phi}$ obtained using the formal sum-zero constrained ICAR prior and the implied full conditional distribution for $\boldsymbol{\phi}$ obtained by using the improper CAR prior with centering on the fly are in fact equivalent.

To see this, first partition the design matrix into $X = [\mathbf{1}_n, F]$ and, similarly, partition the vector of regression coefficients as $\boldsymbol{\beta} = (\alpha, \boldsymbol{\nu}^T)^T$ where α is an intercept. As in Section 2.3.1, write $H = QDQ^T$, where $Q = (\mathbf{q}_1, \mathbf{q}_2, \ldots, \mathbf{q}_n)$ is a $n \times n$ matrix whose columns are the normalized eigenvectors of H, $D = \text{diag}(d_1, d_2, \ldots, d_n)$ is a diagonal matrix containing the eigenvalues $d_1 \geq d_2 \geq \ldots \geq d_{n-1} > d_n = 0$ of H. Additionally, let $\tilde{Q} = (\mathbf{q}_1, \ldots, \mathbf{q}_{n-1})$, so $\Sigma_\phi = \tilde{Q}\text{diag}(d_1^{-1}, \ldots, d_{n-1}^{-1})\tilde{Q}^T$.

Now, assume for the vector of spatial random effects the sum-zero constrained ICAR prior $\boldsymbol{\phi}|\sigma^2, \tau \sim N(\mathbf{0}, \sigma^2\tau^{-1}\Sigma_\phi)$ as in Equation (2.6). Then, the full conditional distribution of $\boldsymbol{\phi}$ is

$$\boldsymbol{\phi}|\tau, \sigma^2, \mathbf{Y}, \boldsymbol{\beta} \sim N(\tilde{Q}\boldsymbol{s}, \sigma^2\tilde{Q}D^*\tilde{Q}^T), \tag{2.10}$$

where $D^* = diag((1 + \tau d_1)^{-1}, \ldots, (1 + \tau d_{n-1})^{-1})$, and $\boldsymbol{s} = D^*\tilde{Q}^T(\mathbf{Y} - F\boldsymbol{\nu})$.

Now, assume for a vector of uncentered spatial random effects $\boldsymbol{\omega}$ the improper ICAR prior given by Equation (2.2). Take the likelihood for the hierarchical model given in Equation (2.1). Let $g = (I_n + \tau H)^{-1}(\mathbf{Y} - X\boldsymbol{\beta}) = Q(I_n + \tau D)^{-1}Q^T(\mathbf{Y} - X\boldsymbol{\beta})$. A vector of centered spatial random effects is obtained as $\boldsymbol{\kappa} = (I_n - n^{-1}\mathbf{1}\mathbf{1}^T)\boldsymbol{\omega}$, and the full conditional distribution for $\boldsymbol{\omega}$ is

$$\boldsymbol{\omega}|\tau, \sigma^2, \mathbf{Y}, \boldsymbol{\beta} \sim N(g, V), \tag{2.11}$$

where $V = \sigma^2(I_n + \tau H)^{-1} = \sigma^2 Q(I_n + \tau D)^{-1}Q^T$. Further, the implied full conditional distribution for the random effects $\boldsymbol{\kappa}$ follows as

$$\boldsymbol{\kappa}|\boldsymbol{Y}, \boldsymbol{\beta}, \sigma^2, \tau \sim N(P\boldsymbol{g}, PVP^T), \tag{2.12}$$

where $P = I_n - n^{-1}\mathbf{1}\mathbf{1}^T$ as in Section 2.3.1. Ferreira et al. (2021) show that $\tilde{Q}\boldsymbol{s} = P\boldsymbol{g}$ and $\sigma^2\tilde{Q}D^*\tilde{Q}^T = PVP^T$. Thus, the full conditional distribution for $\boldsymbol{\phi}$ using the formal sum-zero constrained ICAR prior and the implied full conditional distribution for $\boldsymbol{\omega}$ obtained by using the improper ICAR prior and centering on the fly are equivalent, i.e. the full conditionals in (2.10) and (2.12) are equivalent. See Ferreira et al. (2021) for complete proofs.

2.3.3 Equivalence with the Full Conditional Distribution of Spatial Random Effects Conditional on Their Sum Being Equal to Zero

Another way to enforce the sum-zero constraint for the spatial random effects is by obtaining the full conditional distribution of the spatial random effects, which results in a proper multivariate Gaussian distribution, and then using standard multivariate Gaussian results to obtain the full conditional distribution of the spatial random effects conditional on their sum being equal to zero. Porter et al. (2023) showed that this approach to enforcing the sum-zero constraint is equivalent to using the formal sum-zero constrained ICAR prior.

This equivalence result depends on the full conditional in Equation (2.10), which corresponds to the formal sum-zero ICAR prior, and on the full conditional distribution of $\boldsymbol{\phi}$ conditional on the sum-zero constraint.

To obtain the full conditional distribution conditional on the spatial random effects summing to zero, apply standard multivariate Gaussian results (e.g. see Rue and Held (2005)) to the full conditional given in Equation (2.10). Then, the full conditional distribution of $\boldsymbol{\omega}$ conditional on the sum-zero constraint $\mathbf{1}_n^T\boldsymbol{\omega} = 0$ is

$$\boldsymbol{\omega}|\mathbf{1}_n^T\boldsymbol{\omega} = 0, \tau, \sigma^2, \mathbf{Y}, \boldsymbol{\beta} \sim N(\mu^*, \Sigma^*), \tag{2.13}$$

where $\mu^* = g - V\mathbf{1}_n(\mathbf{1}_n^T V\mathbf{1}_n)^{-1}(\mathbf{1}_n^T g - 0)$ and $\Sigma^* = V - V\mathbf{1}_n(\mathbf{1}_n^T V\mathbf{1}_n)^{-1}\mathbf{1}_n^T V$.

Porter et al. (2023) show that $\mu^* = \tilde{Q}\boldsymbol{s}$ and $\Sigma^* = \sigma^2\tilde{Q}D^*\tilde{Q}^T$. Therefore, the full conditional distributions in Equations (2.10) and (2.13) are the same, and using the formal sum-zero constrained ICAR prior is equivalent to sampling from the full conditional of the spatial random effects conditional on the sum-zero constraint. The results in Sections 2.3.2 and 2.3.3 prove that all three approaches to enforcing the sum-zero constraint are equivalent.

2.4 Reference Prior

In the absence of subjective information for specifying informative priors for the parameters of the ICAR component, practitioners have used vague priors, such as vague Gamma priors for the precisions of the vectors of random effects $\boldsymbol{\theta}$ and $\boldsymbol{\phi}$ (e.g., see Best et al., 1999; Lee, 2013). However, these vague priors may exert undue influence in the analysis. As a safer alternative, Keefe et al. (2019) derived the reference prior for the parameters of the spatial ICAR model. In the absence of subjective information, the reference prior serves as an objective and automatic prior for the parameters of the ICAR component.

The joint reference prior for $\boldsymbol{\beta}, \sigma^2$, and τ follows as

$$\pi(\boldsymbol{\beta}, \sigma^2, \tau) \propto \frac{1}{\sigma^2}\frac{1}{\tau}\left[\sum_{j=1}^{n-p}\left(\frac{\xi_j}{\tau+\xi_j}\right)^2 - \frac{1}{n-p}\left\{\sum_{j=1}^{n-p}\left(\frac{\xi_j}{\tau+\xi_j}\right)\right\}^2\right]^{\frac{1}{2}}, \tag{2.14}$$

where $\xi_1, \ldots, \xi_{n-p}$ are the ordered eigenvalues of $Q^{*T}\Sigma_\phi Q^*$ such that the columns of Q^* are normalized eigenvectors corresponding to the non-zero eigenvalues of the projection matrix $G = I_n - X(X^TX)^{-1}X^T$. The conditional reference prior on the vector of regression coefficients is $\pi(\boldsymbol{\beta}|\sigma^2, \tau) \propto 1$.

When compared to posterior distributions formed with vague Gamma priors, the reference prior exhibits favorable performance with respect to frequentist coverage and average length of credible intervals, as well as mean squared error (MSE) of estimates for all parameters of the ICAR component (Keefe et al., 2019; Ferreira et al., 2021).

2.5 Fast Computations with Spectral Decomposition

Ferreira et al. (2021) used spectral decomposition of the ICAR precision matrix to transform the response variable and the regressors from their respective domains to the spectral domain. Ferreira et al. (2021) also proposed a Spectral Gibbs Sampler (SGS), which uses the spectral transformations to provide remarkable speedups to an MCMC algorithm for sampling the parameters of the spatial ICAR model when assigned the reference prior in Equation (2.14). To understand the computational advantages of this approach, first define the spectral quantities for the response, regressors, and spatial random effects as follows:

$$
\begin{aligned}
\mathcal{Y} &= Q^T \boldsymbol{Y}, \\
\mathcal{X} &= Q^T X, \\
\boldsymbol{\nu} &= Q^T \boldsymbol{\phi}, \\
\boldsymbol{\zeta} &= Q^T \boldsymbol{\theta}.
\end{aligned} \tag{2.15}
$$

Further, let $\boldsymbol{u} = (\mathbf{0}_{n-1}^T, \sqrt{n})^T$. Examining $\boldsymbol{\zeta}$, $\text{Cov}(\boldsymbol{\zeta}) = \text{Cov}(Q^T\boldsymbol{\theta}) = \sigma^2 Q^T Q = \sigma^2 I_n$. Therefore, $\boldsymbol{\zeta}$ is distributed $\boldsymbol{\zeta} \sim N(\mathbf{0}, \sigma^2 I_n)$. Examining $\boldsymbol{\nu}$, $\text{Cov}(\boldsymbol{\nu}) = \text{Cov}(Q^T\boldsymbol{\phi}) = \sigma^2\tau^{-1}Q^T\Sigma_\phi Q = \sigma^2\tau^{-1}D^+ = \sigma^2\tau^{-1}\text{diag}(d_1^{-1}, \ldots, d_{n-1}^{-1}, 0)$. Then the hierarchical model can be written in the spectral domain as:

$$
\begin{aligned}
\mathcal{Y} = \mathcal{X} + \boldsymbol{\nu} + \boldsymbol{\zeta}, \quad \boldsymbol{\zeta} \sim N(\mathbf{0}, \sigma^2 \boldsymbol{I}), \\
\boldsymbol{\nu}|\sigma^2, \tau \sim N(\mathbf{0}, \sigma^2\tau^{-1}D^+).
\end{aligned} \tag{2.16}
$$

Computations using the hierarchical model from Equation (2.16) are much faster than those for the model in Equation (2.1) when the spatial random effects are distributed as in Equation (2.6). This is because computations in the original spatial domain involve matrix multiplications with computational cost that increases quadratically with sample size, whereas in the spectral domain $\boldsymbol{\nu}$ and $\boldsymbol{\zeta}$ both have diagonal covariance matrices, so the computations involving these terms become Hadamard vector and matrix multiplications, which have computational cost that increases linearly with sample size. The SGS algorithm is a Metropolis-within-Gibbs algorithm that uses the hierarchical model in the spectral domain; it contains a Gibbs step for $\boldsymbol{\beta}$ and a joint Metropolis-Hastings step for σ^2 and τ. To obtain the SGS algorithm, first integrate out the spatial random effects $\boldsymbol{\nu}$ to obtain the model:

$$
\mathcal{Y} = \mathcal{X} + \boldsymbol{\delta}, \quad \boldsymbol{\delta} \sim N(\mathbf{0}, \sigma^2\{\boldsymbol{I} + \tau^{-1}D^+\}). \tag{2.17}
$$

Then the basis for the SGS algorithm uses the Hadamard product $\odot$, which returns a matrix of element-wise products and the Kronecker product $\otimes$, which provides an outer product for matrices. Write:

$$
\boldsymbol{b}(\tau) = \left(\frac{\tau d_1}{\tau d_1 + 1}, \ldots, \frac{\tau d_{n-1}}{\tau d_{n-1} + 1}, 1 \right)^T. \tag{2.18}
$$

Then $\mathcal{X}^T(\boldsymbol{I} + \tau^{-1}D^+)^{-1} = \left\{\mathcal{X} \odot (\mathbf{1}_p^T \otimes \boldsymbol{b})\right\}^T$. The operation on the left takes about $2pn^2$ products and $2pn^2$ sums, but the operation on the right uses only $2np$ products. Then the SGS algorithm follows from this result:

1. Sample $\boldsymbol{\beta}$ from its full conditional distribution $N(\mathbf{m}_\beta, \mathbf{C}_\beta)$, where

$$
\mathbf{C}_\beta = \sigma^2\{\mathcal{X}'(\boldsymbol{I} + \tau^{-1}D^+)^{-1}\mathcal{X}\}^{-1} = \sigma^2 \left[\{\mathcal{X} \odot (\mathbf{1}_p' \otimes \boldsymbol{b})\}' \mathcal{X} \right]^{-1}
$$

and

$$\begin{aligned} \mathbf{m}_\beta &= \{\mathcal{X}'(\boldsymbol{I}+\tau^{-1}D^{+})^{-1}\mathcal{X}\}^{-1}\mathcal{X}'(\boldsymbol{I}+\tau^{-1}D^{+})^{-1}\mathcal{Y} && (2.19)\\ &= \left[\{\mathcal{X}\odot(\mathbf{1}_p'\otimes\boldsymbol{b})\}'\mathcal{X}\right]^{-1}\{\mathcal{X}\odot(\mathbf{1}_p'\otimes\boldsymbol{b})\}'\mathcal{Y}. && (2.20)\end{aligned}$$

2. Propose $\sigma^{2(prop)}$ from Lognormal$(\log(\sigma^{2(curr)}), \ell_\sigma)$.
3. Propose $\tau^{(prop)}$ from Lognormal$(\log(\tau^{(curr)}), \ell_T)$.
4. Accept $(\sigma^{2(prop)}, \tau^{(prop)})$ with probability equal to $\min(1, a)$ where

$$\begin{aligned} a &= \frac{p(\tau^{(prop)})}{p(\tau^{(curr)})}\left(\frac{\sigma^{2(prop)}}{\sigma^{2(curr)}}\right)^{-(n+2)/2}\left(\frac{\tau^{(prop)}}{\tau^{(curr)}}\right)^{(n-1)/2}\\ &\quad \left(\frac{\prod_{i=1}^{n-1}(s_i\tau^{(curr)}+1)}{\prod_{i=1}^{n-1}(s_i\tau^{(prop)}+1)}\right)^{1/2}\exp\left[-\frac{1}{2}\Big\{(\mathcal{Y}-\mathcal{X}\beta)\odot\right.\\ &\quad \left.\left(\sigma^{-2(prop)}\boldsymbol{b}^{(prop)}-\sigma^{-2(curr)}\boldsymbol{b}^{(curr)}\right)\Big\}'(\mathcal{Y}-\mathcal{X}\beta)\right].\end{aligned}$$

Finally, samples for the spectral random effects $\boldsymbol{\zeta}$ can be obtained through composite sampling. Usually, matrix inversions require $O(n^3)$ operations. Keefe et al. (2019) perform spectral decomposition based on the eigenvalues and eigenvectors of Σ_ϕ prior to the MCMC algorithm to achieve matrix inversions costing $O(n^2)$ operations. By using spectral transformation of the dependent variable and the regressors, Ferreira et al. (2021) are able to achieve matrix inversions requiring only $O(n)$ operations.

2.6 Spectral Posterior Maximizer

When performing estimation and model selection for large datasets, large sample approximations (i.e. asymptotic normality) can be used to estimate the parameters of the ICAR model via maximum a posteriori (MAP) estimates. The MAP can be computed by optimizing the posterior density, and then credible intervals can be obtained using the MAP and the information matrix. Ferreira et al. (2021) called this the spectral posterior maximizer (SPM) approach.

As Ferreira et al. (2021) showed, once the spectral decomposition of the neighborhood matrix H is performed and the data are transformed onto the spectral domain, computations for every model can be performed in the spectral domain, and no further matrix decompositions are required, allowing for efficient computations in the case of many data points or a large number of possible models.

Additional speed-ups in computation can be achieved by using the approximate reference prior proposed by Ferreira et al. (2021):

$$p^{(a)}(\boldsymbol{\beta}, \sigma^2, \tau) \propto \frac{1}{\sigma^2(a_\tau + \tau)^2}, \tag{2.21}$$

where a_τ is a hyperparameter. An effective choice for a_τ is 0.5. Keefe et al. (2019) showed that the reference prior for τ behaves as $O(1)$ as $\tau \to 0$ and as $O(\tau^{-2})$ as $\tau \to \infty$, so the approximate reference prior has the same tail behavior as the reference prior. Use of the approximate reference prior obviates the need to calculate the eigenvalues $\xi_1, \ldots \xi_{n-p}$, which can reduce computing times for large sample sizes.

Assuming the approximate reference prior for the parameters of the ICAR model, the joint posterior density of $\boldsymbol{\beta}, \sigma^2$, and τ is $p(\boldsymbol{\beta}, \sigma^2, \tau | \mathcal{Y}, \mathcal{X}) \propto p^{(a)}(\boldsymbol{\beta}, \sigma^2, \tau) p(\mathcal{Y} | \mathcal{X}, \boldsymbol{\beta}, \sigma^2, \tau)$. Since σ^2 and τ are positive quantities, reparameterizing to the logarithmic scale with $\gamma = \log(\sigma^2)$ and $\psi = \log(\tau)$ will speed up convergence to normality at large sample sizes. The updated approximate reference prior after multiplying by the Jacobian for the transformation to γ and ψ is:

$$p^{(a)}(\boldsymbol{\beta}, \gamma, \psi) \propto \frac{e^\psi}{(a_\tau + e^\psi)^2}. \tag{2.22}$$

Then the estimate for the MAP (posterior mode) follows as:

$$\begin{aligned} (\widehat{\boldsymbol{\beta}}, \widehat{\gamma}, \widehat{\psi}) &= \max_{(\boldsymbol{\beta}, \gamma, \psi)} \log[p^{(a)}(\boldsymbol{\beta}, \gamma, \psi) p(\boldsymbol{y} | \boldsymbol{X}, \boldsymbol{\beta}, \gamma, \psi)] \\ &= \max_{(\boldsymbol{\beta}, \gamma, \psi)} \log[p^{(a)}(\boldsymbol{\beta}, \gamma, \psi) p(\mathcal{Y} | \mathcal{X}, \boldsymbol{\beta}, \gamma, \psi)]. \end{aligned} \tag{2.23}$$

Given a particular value of ψ, the posterior mode of $\boldsymbol{\beta}$ is

$$\widehat{\boldsymbol{\beta}}(\psi) = \left[\left\{ \mathcal{X} \odot (\mathbf{1}_p^T \otimes \boldsymbol{b}(e^\psi)) \right\}^T \mathcal{X} \right]^{-1} \left\{ \mathcal{X} \odot (\mathbf{1}_p^T \otimes \boldsymbol{b}(e^\psi)) \right\}^T \mathcal{Y}, \tag{2.24}$$

where $\boldsymbol{b}(\tau)$ is defined as

$$\boldsymbol{b}(\tau) = \left(\frac{\tau d_1}{\tau d_1 + 1}, \ldots, \frac{\tau d_{n-1}}{\tau d_{n-1} + 1}, 1 \right)^T, \tag{2.25}$$

and for fixed $\boldsymbol{\beta}$ and τ the posterior mode of γ is

$$\widehat{\gamma}(\boldsymbol{\beta}, \psi) = -\log(n) + \log\left\{ \sum_{i=1}^{n-1} \frac{(\mathcal{Y}_i - \mathcal{X}_i^T \boldsymbol{\beta})^2}{1 + e^{-\psi} d_i^{-1}} + (\mathcal{Y}_n - \mathcal{X}_n^T \boldsymbol{\beta})^2 \right\}. \tag{2.26}$$

Ferreira et al. (2021) express the joint posterior density of (β, γ, ψ) as a function of ψ. A numerical maximizer gives the posterior mode $\hat{\psi}$, and substituting $\hat{\psi}$ into Equations (2.24) and (2.26) gives $\hat{\beta}$ and $\hat{\gamma}$. Then the estimates of τ and σ^2 are found by $\hat{\tau} = e^{\hat{\psi}}$ and $\hat{\sigma}^2 = e^{\hat{\gamma}}$.

2.7 Model Selection

The reference prior in Equation (2.14) enables automatic, objective Bayesian inference for the ICAR model and can easily be used for estimation with an MCMC algorithm. However, care must be taken when performing model selection, as the marginal reference priors for $\boldsymbol{\beta}$ and σ^2 are improper. Improper priors on model parameters contain undefined normalizing constants that cause model selection via standard Bayes factors to be invalid (O'Hagan, 1995). To allow for objective Bayesian model selection that takes advantage of the properties and performance of the reference prior, Porter et al. (2023) developed model selection via fractional Bayes factors (FBF) for the spatial ICAR model using the reference prior on model parameters (O'Hagan, 1995). The fractional Bayes factor uses a portion of the likelihood to update the priors on parameters, leading undefined normalizing constants to cancel and resulting in a trained prior for selection that is proper. To understand how the FBF works to allow objective Bayesian model selection with the reference prior, first consider the model space $\mathcal{M} = \{M_c,\ c = 1, \dots C\}$ of size $C = 2^p$, consisting of spatial ICAR models and OLMs with all combinations of the candidate predictors. Let $\boldsymbol{\eta}$ denote the vector of model parameters, containing $\boldsymbol{\beta}$ and σ^2 for OLMs and containing $\boldsymbol{\beta}$, σ^2, and τ for spatial ICAR models.

In general, Bayesian model selection can be performed with the use of the integrated likelihoods for each model M_c:

$$p(\mathbf{Y}|M_c) = \int p(\mathbf{Y}|\boldsymbol{\eta}_c, M_c)\pi(\boldsymbol{\eta}_c|M_c)d\boldsymbol{\eta}_c, \tag{2.27}$$

where model M_c has corresponding parameter vector $\boldsymbol{\eta}_c$. Then two models, say M_1 and M_2, are often compared using the corresponding standard Bayes factor, BF_{12}, defined as the ratio of the two models' integrated likelihoods,

$$BF_{12} = \frac{p(\mathbf{Y}|M_1)}{p(\mathbf{Y}|M_2)}. \tag{2.28}$$

Porter et al. (2023) form all Bayes factors with respect to a single baseline model, M_l. The baseline model has the largest integrated likelihood of all models in the model space and using this baseline model for all Bayes factors helps to alleviate concerns around numerical underflow when computing integrated likelihoods. With all Bayes factors $\{BF_{1l}, \dots, BF_{Cl}\}$ defined, the posterior probability of a single model M_r in the model space can be found using Bayes' Rule:

$$P(M_r|\mathbf{Y}) = \frac{p(\mathbf{Y}|M_r)P(M_r)}{\sum_{c=1}^{C} p(\mathbf{Y}|M_c)P(M_c)} = \left(\sum_{c=1}^{C} BF_{cl}P(M_c)\right)^{-1} \times BF_{rl} \times P(M_r). \tag{2.29}$$

For a given model, the FBF approach updates the prior on model parameters using a fraction $b = m/n$ of the likelihood. This fraction of the likelihood is used to train $\pi(\boldsymbol{\eta}_c)$, resulting in an updated/trained prior $\pi^*(\boldsymbol{\eta}_c)$ for $\boldsymbol{\eta}_c$:

$$\pi^*(\boldsymbol{\eta}_c) = \frac{\pi(\boldsymbol{\eta}_c)\{p(\mathbf{Y}|\boldsymbol{\eta}_c, M_c)\}^b}{\int \pi(\boldsymbol{\eta}_c)\{p(\mathbf{Y}|\boldsymbol{\eta}_c, M_c)\}^b d\boldsymbol{\eta}_c}. \tag{2.30}$$

The trained prior is used to form the fractional integrated likelihood, $q_c(b, \mathbf{Y})$, as:

$$\begin{aligned} q_c(b, \mathbf{Y}) &= \int \pi^*(\boldsymbol{\eta}_c)\{p(\mathbf{Y}|\boldsymbol{\eta}_c, M_c)\}^{1-b} d\boldsymbol{\eta}_c \qquad (2.31)\\ &= \int \frac{\pi(\boldsymbol{\eta}_c)p^b(\mathbf{Y}|\boldsymbol{\eta}_c, M_c)\{p^{1-b}(\mathbf{Y}|\boldsymbol{\eta}_c, M_c)\}d\boldsymbol{\eta}_c}{\int p^b(\mathbf{Y}|\boldsymbol{\eta}_c, M_c)\pi(\boldsymbol{\eta}_c)d\boldsymbol{\eta}_c} \\ &= \frac{\int p(\mathbf{Y}|\boldsymbol{\eta}_c, M_c)\pi(\boldsymbol{\eta}_c)d\boldsymbol{\eta}_c}{\int p^b(\mathbf{Y}|\boldsymbol{\eta}_c, M_c)\pi(\boldsymbol{\eta}_c)d\boldsymbol{\eta}_c}, \end{aligned}$$

where $\{p^{1-b}(\mathbf{Y}|\boldsymbol{\eta}_c, M_c)\}$ is the fraction of the likelihood remaining after training the prior for $\boldsymbol{\eta}_c$, to be used in the posterior. Then the FBF, denoted BF_{12}^b for comparing models M_1 and M_2, follows as

$$BF_{12}^b = \frac{q_1(b, \mathbf{Y})}{q_2(b, \mathbf{Y})}. \tag{2.32}$$

The set of FBFs can be used in place of the standard Bayes factors to calculate posterior model probabilities as in Equation (2.29) for each model in the model space $\mathcal{M}$. Since the same $\pi(\boldsymbol{\eta}_c)$ appears in both the numerator and denominator of the trained prior as seen in Equation (2.30), all undefined normalizing constants cancel and thus the posterior model probabilities resulting from the FBF approach are valid.

One necessary consideration when using the FBF approach is the choice of training fraction b, or the corresponding training size m, where $b = m/n$. Porter et al. (2023) derived the minimal training size for the OLM and the ICAR model when the reference prior from Equation (2.14) is assigned to the model parameters. The minimal training size is defined as the smallest integer value for m such that $\int p^b(\boldsymbol{Y}|\boldsymbol{\eta}_c, M_c)$ is finite for all models in the model space, and thus the trained prior for each vector of parameters is proper. The minimal training size for both the OLM and the ICAR model with the reference prior is $m = p + 1$, where p here indicates the length of the largest $\boldsymbol{\beta}$ vector in the model space. A training size less than $p + 1$ may result in invalid selection, as some of the trained priors may be improper. Increasing m past the minimal training size may forfeit statistical power by over-training the prior, lowering the ability of the FBF to detect signal.

2.8 ref.ICAR Package

Porter et al. (2018) released version 2.0 of the open-source R package **ref.ICAR**, which allows researchers to implement an automatic, objective Bayesian analysis for areal data. **ref.ICAR** is publicly available on the Comprehensive R Archive Network (CRAN). **ref.ICAR** uses the formal sum-zero constrained ICAR prior for the spatial random effects (Keefe et al., 2018) and reference prior for model parameters (Keefe et al., 2019), so users do not need to specify any hyperparameters or constraints. **ref.ICAR** provides functions to construct the neighborhood matrix from Equation (2.3), fit the spatial ICAR model with the reference prior, perform MCMC sampling for model parameters and spatial random effects, and perform the FBF-based model selection proposed by Porter et al. (2023) for covariates and spatial model structure. The functions return useful parameter estimates and intervals for model parameters, fitted values for each spatial subregion, and posterior model probabilities in the case of model selection.

The sampling function in **ref.ICAR** takes either a shapefile or a prespecified neighborhood matrix. The user can specify a different neighborhood matrix than (2.3) using packages such as **spdep** (Bivand and Wong, 2018). **ref.ICAR** obtains MCMC samples for model parameters $\boldsymbol{\beta}$, σ^2, and τ. Posterior medians are returned for σ^2 and τ, and posterior means are returned for the vector of regression coefficients $\boldsymbol{\beta}$, along with 95% credible intervals for all parameters. All MCMC chains are returned in a matrix, so the user can obtain other desired estimates, diagnostics, and plots. If a shapefile is available for the spatial region, the output from **ref.ICAR** can be used to create spatial maps by subregion using functions such as **ggplot** (Wickham, 2016). The model selection function outputs a list of posterior model probabilities for all candidate models and posterior inclusion probabilities for each candidate covariate. An accompanying vignette gives a detailed description of each function in **ref.ICAR** and example code to demonstrate an analysis of verbal SAT scores over the contiguous United States. This dataset is available in **ref.ICAR** for users to practice analyzing spatially correlated areal data using an ICAR component on the vector of spatial random effects. The vignette can be accessed by running `browseVignettes(package="ref.ICAR")` in R once the **ref.ICAR** package is installed.

2.9 Applications

The methods described in this chapter can be applied to any spatial areal dataset with a connected graph (i.e. no islands). To illustrate this, we apply some of the sampling and inference methods discussed in this chapter for the formal sum-zero constrained ICAR model to a widely used dataset.

One common dataset that has been analyzed by Cressie and Read (1985), Banerjee et al. (2014), and others is available in the **spdep** package. The data consists of aggregated counts of SIDS (sudden infant death syndrome) deaths in North Carolina in 1974 and 1979. The SIDS counts and rates are recorded for each of North Carolina's 100 counties. To load the data, first create a `SpatialPolygonDataFrame` object called `nc.sids`.

```
> library(spdep)
> nc.sids <- st_read(system.file("shapes/sids.shp",
                    package="spData")[1], quiet=TRUE)
> row.names(nc.sids) <- as.character(nc.sids$FIPSNO)
```

We can calculate the raw SIDS rates for the year 1979 by using the `probmap()` function to divide all of the SIDS counts by the corresponding total births in that year.

```
> nc.sids$raw <- probmap(nc.sids$SID79, nc.sids$BIR79)$raw
```

Then we can use the ggplot2 package to plot the raw SIDS rates.

```
ggplot() + geom_sf(data = nc.sids, aes(fill = raw)) +
    theme_bw() +
    scale_fill_distiller(palette = "Greys") +
    theme(legend.key.size = unit(1, 'cm'),
          legend.title = element_text(size=25),
          legend.text = element_text(size=20),
          axis.text.y=element_text(size=15),
          axis.text.x = element_text(size=15))
```

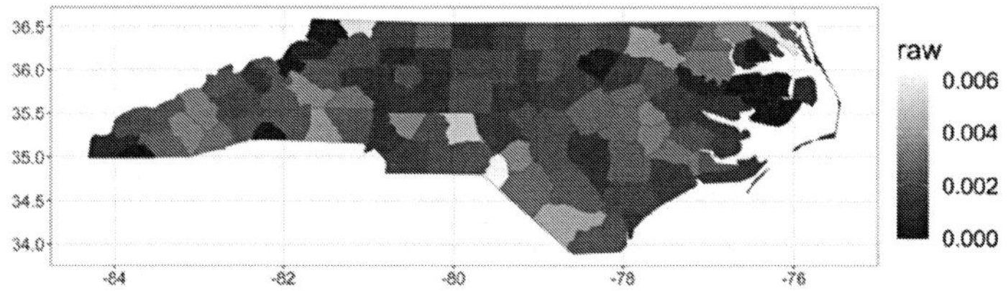

FIGURE 2.1
Raw SIDS rates for 1979.

Alternatively, we can create and plot the SIDS rates using a Freeman-Tukey transformation to help with variance stabilizing.

```
> nc.sids$FT <- sqrt(1000) * (sqrt(nc.sids$SID79/nc.sids$BIR79)
  + sqrt((nc.sids$SID79 + 1)/nc.sids$BIR79))
> ggplot() + geom_sf(data = nc.sids, aes(fill = FT)) +
      theme_bw() +
      scale_fill_distiller(palette = "Greys") +
      theme(legend.key.size = unit(1, 'cm'),
            legend.title = element_text(size=25),
            legend.text = element_text(size=20),
            axis.text.y=element_text(size=15),
            axis.text.x = element_text(size=15))
```

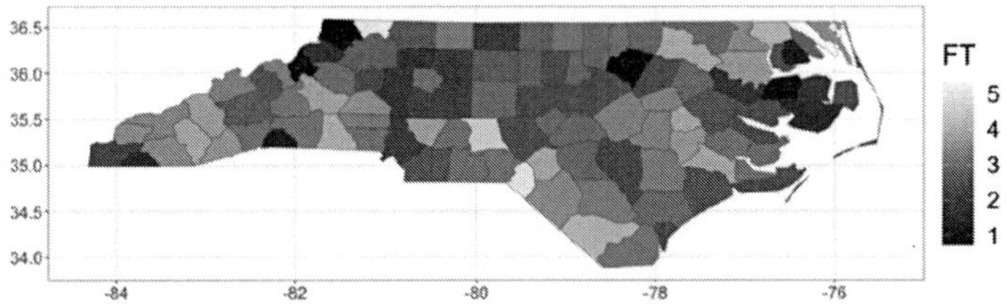

FIGURE 2.2
Freeman-Tukey transformed SIDS rates for 1979.

We can create a covariate from the non-white birth rates, also during 1979.

```
> nc.sids.nwbir.FT = sqrt(1000) * (sqrt(nc.sids$NWBIR79/
   nc.sids$BIR79) + sqrt((nc.sids$NWBIR79 + 1)/nc.sids$BIR79))
> nc.sids$nwbir.FT = nc.sids.nwbir.FT
```

Then the response variable and the X matrix can be defined as:

```
> Y <- nc.sids$FT
> X <- cbind(1, nc.sids$nwbir.FT)
```

Next, we can define the adjacency matrix, W, and obtain a summary of the neighborhood object.

```
> nc.listw <- poly2nb(nc.sids)
> summary(nc.listw)
  Neighbour list object:
  Number of regions: 100
  Number of nonzero links: 490
  Percentage nonzero weights: 4.9
  Average number of links: 4.9
  Link number distribution:
```

```
    2  3  4  5  6  7  8  9
    8 15 17 23 19 14  2  2
  8 least connected regions:
  37053 37041 37177 37055 37149 37137 37043 37129 with 2 links
  2 most connected regions:
  37097 37125 with 9 links
> W <- nb2mat(nc.listw, style="B")
```

We can construct the neighborhood matrix H manually as follows.

```
> Dmat <- diag(apply(W,1,sum))
> H <- Dmat - W
> H <- (H+t(H))/2
> rownames(H) <- NULL
> isSymmetric(H)
```

Now we are ready to fit the formal sum-zero constrained ICAR model (2.8) with reference prior (2.14).

```
> nc.sids.ref <- ref.MCMC(Y,X,H, iters=15000, burnin=5000,
                                 verbose=TRUE)
```

Now we can obtain parameter estimates for β, σ^2, and τ, using the `ref.summary()` function from **ref.ICAR**.

```
> ref.summary(MCMCchain=nc.sids.ref$MCMCchain,
  tauc.MCMC=nc.sids.ref$tauc.MCMC,
  sigma2.MCMC=nc.sids.ref$sigma2.MCMC,
  beta.MCMC=nc.sids.ref$beta.MCMC,
  phi.MCMC=nc.sids.ref$phi.MCMC,
  accept.phi = nc.sids.ref$accept.phi,
  accept.sigma2 = nc.sids.ref$accept.sigma2,
  accept.tauc = nc.sids.ref$accept.tauc)
  $beta.median

[1] 2.33602285 0.01765517

$beta.hpd
           lower      upper
var1  1.67189157 3.01757894
var2 -0.00107504 0.03888486

$tauc.median
[1] 0.8842467

$tauc.hpd
     lower      upper
0.02405608 7.89078477
```

```
$sigma2.median
[1] 0.4996799

$sigma2.hpd
    lower     upper
0.1102295 0.7444294

$tauc.accept
[1] 0.312

$sigma2.accept
[1] 0.312
```

The user may also obtain these and other desired estimates manually by using the MCMC output from the `ref.MCMC` function in the appropriate R functions.

Note that the estimated τ value is 0.8843, which indicates some level of spatial dependence in the Freeman-Tukey transformed SIDS rates. The above output shows that the data have a positive intercept around 2.34 for the transformed data, but the effect of the non-white birth rate covariate is small and does not appear to be significant after accounting for spatial correlation through the spatial ICAR model.

From the MCMC chains, we can obtain the fitted values for the transformed SIDS rates by region.

```
> summary.region <- reg.summary(nc.sids.ref$MCMCchain, X, Y,
                                burnin = 5000)
> nc.sids$reg.medians <- summary.region$reg.medians
> ggplot() + geom_sf(data = nc.sids, aes(fill = reg.medians))
      + theme_bw() +
      scale_fill_distiller(palette = "Greys") +
      theme(legend.key.size = unit(1, 'cm'),
            legend.title = element_text(size=25),
            legend.text = element_text(size=20),
            axis.text.y=element_text(size=15),
            axis.text.x = element_text(size=15))
```

To demonstrate the model selection capabilities of **ref.ICAR**, consider a spatial dataset previously analyzed by Anselin (1988), Banerjee et al. (2014), and Porter et al. (2023). This data is also available in the **spdep** package. The response variable is defined as the residential crime rate across the 49 neighborhoods of Columbus, Ohio. The data also includes five candidate predictors: average housing value, average household income, the amount of open space in each neighborhood, the number of housing units without available plumbing, and distance from the Columbus business district. After loading **ref.ICAR**, which imports **spdep**, the user can read in the data as follows:

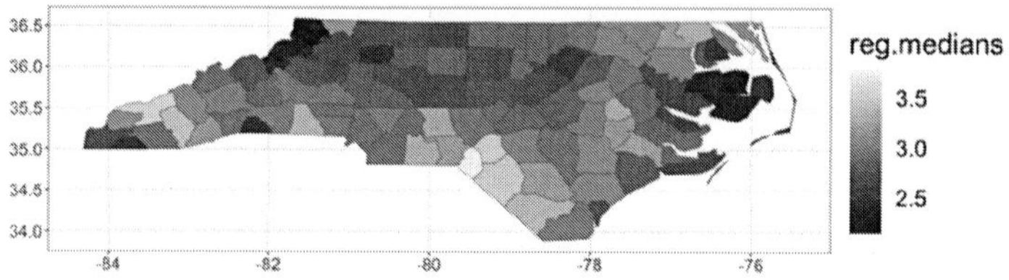

FIGURE 2.3
Fitted posterior medians for Freeman-Tukey transformed 1979 SIDS rates.

```
> library(ref.ICAR)

> # read in the data as contained in the spdep package
> columbus <- st_read(system.file("shapes/columbus.shp",
                      package="spData")[1], quiet=TRUE)
```

The model selection function in **ref.ICAR** requires the user to input the neighborhood matrix H from Equation (2.3).

```
> # create neighborhood matrix
> columbus.listw <- poly2nb(columbus)
> summary(columbus.listw)
> W <- nb2mat(columbus.listw, style="B")
> Dmat <- diag(apply(W,1,sum))
> num.reg <- length(columbus$CRIME)

> H <- Dmat - W
> H <- (H+t(H))/2
> rownames(H) <- NULL
> isSymmetric(H)  # check that neighborhood matrix is
                  # symmetric before proceeding
```

probs.icar(), the model selection function in **ref.ICAR**, allows the user to take advantage of the spectral decomposition of the spatial covariance matrix.

```
> # spectral quantities for use in model selection
> H.spectral <- eigen(H, symmetric=TRUE)
> Q <- H.spectral$vectors
> eigH <- H.spectral$values
> phimat <- diag(1/sqrt(eigH[1:(num.reg-1)]))
> Sig_phi <- matrix(0,num.reg, num.reg) #initialize
   for(i in 1:(num.reg-1)){
       total <- (1/(eigH[i]))*Q[,i]%*%t(Q[,i])
       Sig_phi <- Sig_phi + total
   }
```

probs.icar() also requires as input the response vector and design matrix for the full candidate model.

```
> # define response and design matrix
> Y <- columbus$CRIME
> X <- cbind(1, columbus$HOVAL, columbus$INC, columbus$OPEN,
             columbus$PLUMB, columbus$DISCBD)
```

We can plot the response variable to visualize the neighborhood crime rates prior to selecting a model.

```
> ggplot() + geom_sf(data = columbus, aes(fill = CRIME)) +
    scale_fill_distiller(palette = "Greys") +
    theme_bw() +
    theme(axis.text.y=element_blank(),
          axis.text.x = element_blank(),
          axis.ticks.x=element_blank(),
          axis.ticks.y =element_blank(),
          legend.key.size = unit(1,'cm'),
          legend.title = element_text(size=25),
          legend.text = element_text(size=20))
```

FIGURE 2.4
Columbus, OH residential crime rates by neighborhood.

Lastly, the **probs.icar()** function requires the user to specify the training fraction b to use with the fractional Bayes factor approach. Porter et al. (2023) recommend using the minimal training fraction as follows:

```
> b <- (ncol(X)+1)/num.reg
 # specify the minimal training size
```

Then model selection can be performed for this data using **probs.icar()**.

```
> columbus.select <- probs.icar(Y=Y,X=X,H=H,
                                 H.spectral=H.spectral,
                                 Sig_phi=Sig_phi,
                                 b=b,verbose=FALSE)
```

The **probs.icar()** function outputs a list containing: a data frame of all candidate models and their corresponding posterior model probabilities, a vector of the priors used on the model space, the log integrated likelihoods for all candidate models, the largest integrated likelihood value, and a vector of all the fractional Bayes factor values used to calculate the posterior model probabilities.

```
> names(columbus.select)

  [1] "probs.mat"     "mod.prior"      "logmargin.all"
  [4] "base.model"     "BF.vec"
```

The first element of the list, **probs.mat**, is a data frame that can be used to find the selected model among all possible OLM and spatial ICAR models for the given candidate predictors. The data frame output lists information for one candidate model per row. The columns indicate the posterior model probability, model type (independent if OLM and spatial if ICAR), and the model form indicating the predictors contained in the model. All models include an intercept.

```
> names(columbus.select$probs.mat)
[1] "model prob"    "model type"    "model form"

> columbus.select$probs.mat[
              which.max(columbus.select$probs.mat[,1]),]

     model prob  model type    model form
  19  0.119395    Independent   Y ~ Intercept + X1 + X2 + X5
```

The OLM with the first, second, and fifth predictors was selected with highest posterior model probability, i.e. average housing value, average household income, and distance from the Columbus business district with posterior model probability 0.119. The corresponding model prior for the prior probability for an OLM with three covariates was 0.0083, so the posterior has moved off this prior considering that the sample size is small. The total posterior probability of OLMs was 0.569, moving only slightly away from the 1/2 prior probability of selecting an OLM.

2.10 Conclusions

This chapter has traced some of the recent developments in Gaussian spatial ICAR models for areal data. The use of the ICAR prior for a vector of spatial random effects has been widely used in applications such as image restoration and disease mapping. The spatial ICAR model differs from the CAR model applied directly to the observations because the spatial ICAR model requires a sum-zero constraint to result in a proper density. This constraint has often been enforced in practice through the use of computational tools such as MCMC with centering on the fly, but these informal approaches can obscure the joint prior density. We discussed the key development of the formal sum-zero constrained ICAR prior, which clearly specifies a proper joint density for the spatial random effects. The availability of the proper density provided by the formal sum-zero constrained prior allowed the derivation of the reference prior for the parameters of the ICAR model. The reference prior enables objective, automatic Bayesian analysis without the need to specify hyperparameters for priors. Notably, objective Bayesian model selection using the reference prior performs well, and fast and scalable computations were developed by studying the hierarchical model with sum-zero constrained ICAR prior in the spectral domain.

This chapter also demonstrated the current capabilities of **ref.ICAR**, which performs estimation and Bayesian model selection using the formal sum-zero constrained ICAR prior with the reference prior on model parameters. **ref.ICAR**'s estimation capabilities were illustrated using a common dataset consisting of SIDS rates in North Carolina and model selection was demonstrated using data on crime rates in Columbus, OH neighborhoods.

ICAR models are a useful tool for Gaussian areal data that have seen further development and options for implementation. Future extensions include objective ICAR models for data that can be modeled by generalized linear mixed models (Xu et al., 2023), and the generality of intrinsic autoregressions may allow for the study of related data types such as network data.

Bibliography

Anselin, L. (1988). *Spatial Econometrics: Methods and Models.* Springer Science & Business Media.

Banerjee, S., B. P. Carlin, and A. E. Gelfand (2014). *Hierarchical Modeling and Analysis for Spatial Data* (2nd ed.). Boca Raton, FL: CRC Press.

Besag, J. E., J. York, and A. Mollié (1991). Bayesian image restoration with two applications in spatial statistics (with discussion). *Annals of the Institute of Statistical Mathematics 43*, 1–59.

Best, N., R. Arnold, A. Thomas, L. A. Waller, and E. M. Conlon (1999). Bayesian models for spatially correlated disease and exposure data. *Bayesian Statistics 6*, 65–82.

Best, N., S. Richardson, and A. Thomson (2005). A comparison of Bayesian spatial models for disease mapping. *Statistical Methods in Medical Research 14*(1), 35–59.

Bivand, R. and D. W. S. Wong (2018). Comparing implementations of global and local indicators of spatial association. *TEST 27*(3), 716–748.

Cressie, N. and T. Read (1985). Do sudden infant deaths come in clusters? *Statistics and Decisions Supplement Issue 2*, 333–349.

De Oliveira, V. and M. A. R. Ferreira (2011). Maximum likelihood and restricted maximum likelihood estimation for a class of Gaussian Markov random fields. *Metrika 74*, 167–183.

Ferreira, M. A. R. (2019). The limiting distribution of the Gibbs sampler for the intrinsic conditional autoregressive model. *Brazilian Journal of Probability and Statistics 33*(4), 734–744.

Ferreira, M. A. R. (2024). Proper Gaussian Markov random fields. In M. A. R. Ferreira (Ed.), *Modeling Spatio-Temporal Data: Markov Random Fields, Objective Bayes, and Multiscale Models*, pp. 1–22. Chapman & Hall/CRC.

Ferreira, M. A. R. and V. De Oliveira (2007). Bayesian reference analysis for Gaussian Markov Random Fields. *Journal of Multivariate Analysis 98*, 789–812.

Ferreira, M. A. R., E. M. Porter, and C. T. Franck (2021). Fast and scalable computations for Gaussian hierarchical models with intrinsic conditional autoregressive spatial random effects. *Computational Statistics and Data Analysis 162*, 107264.

Goicoa, T., M. D. Ugarte, J. Etxeberria, and A. F. Militino (2016). Age–space–time CAR models in Bayesian disease mapping. *Statistics in Medicine 35*(14), 2391–2405.

Keefe, M. J., M. A. R. Ferreira, and C. T. Franck (2018). On the formal specification of sum-zero constrained intrinsic conditional autoregressive models. *Spatial Statistics 24*, 54–65.

Keefe, M. J., M. A. R. Ferreira, and C. T. Franck (2019). Objective Bayesian analysis for Gaussian hierarchical models with intrinsic conditional autoregressive priors. *Bayesian Analysis 14*(1), 181–209.

Lee, D. (2011). A comparison of conditional autoregressive models used in Bayesian disease mapping. *Spatial and Spatio-temporal Epidemiology 2*(2), 79–89.

Lee, D. (2013). CARBayes: An R package for Bayesian spatial modeling with conditional autoregressive priors. *Journal of Statistical Software 55*(13), 1–24.

Logan, J. R., C. Bauer, J. Ke, H. Xu, and F. Li (2020). Models for small area estimation for census tracts. *Geographical Analysis 52*(3), 325–350.

Mercer, L. D., J. Wakefield, A. Pantazis, A. M. Lutambi, H. Masanja, and S. Clark (2015). Space-time smoothing of complex survey data: Small area estimation for child mortality. *The Annals of Applied Statistics 9*(4), 1889.

O'Hagan, A. (1995). Fractional Bayes factors for model comparison. *Journal of the Royal Statistical Society. Series B (Methodological) 57*(1), 99–138.

Penrose, R. (1955). A generalized inverse for matrices. *Mathematical Proceedings of the Cambridge Philosophical Society 51*(3), 406–413.

Porter, E. M., C. T. Franck, and M. A. R. Ferreira (2023). Objective Bayesian model selection for spatial hierarchical models with intrinsic conditional autoregressive priors. *Bayesian Analysis 19*, accepted for publication.

Porter, E. M., M. J. Keefe, C. T. Franck, and M. A. R. Ferreira (2018). *ref.ICAR: Objective Bayes Intrinsic Conditional Autoregressive Model for Areal Data.* CRAN. R package version 1.0.

Reich, B. J., J. S. Hodges, and V. Zadnik (2006). Effects of residual smoothing on the posterior of the fixed effects in disease-mapping models. *Biometrics 62*(4), 1197–1206.

Rue, H. and L. Held (2005). *Gaussian Markov Random Fields: Theory and Applications.* CRC Press.

Rue, H., S. Martino, and N. Chopin (2009). Approximate Bayesian inference for latent Gaussian models by using integrated nested Laplace approximations. *Journal of the Royal Statistical Society: Series B (Statistical Methodology) 71*(2), 319–392.

Ver Hoef, J. M., E. E. Peterson, M. B. Hooten, E. M. Hanks, and M.-J. Fortin (2018). Spatial autoregressive models for statistical inference from ecological data. *Ecological Monographs 88*(1), 36–59.

Wickham, H. (2016). *ggplot2: Elegant Graphics for Data Analysis.* Springer-Verlag New York.

Xu, S., M. A. R. Ferreira, E. M. Porter, and C. T. Franck (2023). Bayesian model selection for generalized linear mixed models. *Biometrics 79*(4), 3266–3278.

3

Objective Priors for Spatio-Temporal Models

Marco A. R. Ferreira

Virginia Tech

One important aspect in the practical implementation of spatio-temporal models is the choice of priors for the hyperparameters. Because many spatio-temporal models are built as highly structured multivariate state-space models, results for state-space models may inform us about appropriate priors for large classes of spatio-temporal models. Unfortunately, the literature on objective priors for state-space models is sparse. This chapter starts by reviewing this literature. We then explore the relationship between state-space models and hierarchical models with ICAR random effects to motivate the specification of objective approximate reference priors for spatio-temporal models. Using these ideas, we develop an objective approximate reference prior for a class of spatio-temporal models that includes a time varying overall mean level and dynamic spatio-temporal random effects. Of importance, the spatio-temporal random effects evolve through time with a singular evolution equation such that at each time point, the spatio-temporal random effects satisfy a sum-zero constraint. Further, we develop an efficient Markov chain Monte Carlo algorithm to explore the posterior distribution of the unknown quantities of the model. Finally, applications to simulated data and to an unemployment dataset illustrate the usefulness of our spatio-temporal framework.

DOI: 10.1201/9781032623443-3

3.1 Introduction

Spatio-temporal data are ubiquitous in science and society (Cressie and Wikle, 2011). Examples of spatio-temporal data include temperature and humidity measured at several monitoring stations through time, the monthly number of cases of flu per county, and yearly economic variables per state through several years. Here, we focus on spatio-temporal dynamical models that describe the spatio-temporal evolution of the process under study with multivariate linear state-space equations.

The literature on objective Bayesian methods for state-space models is sparse. To the best of our knowledge, Pole and West (1989) is the only scientific article published to date on the subject. Specifically, Pole and West (1989) developed Bayesian analysis for univariate state space models with an objective flat prior for the latent state at time $t = 1$ and an objective reciprocal prior for the observational variance. These are excellent objective prior choices for the latent state at time $t = 1$ and for the observational variance. However, Pole and West (1989) assumed the covariance matrix of the system equation to be equal to the observational variance times a known positive definite symmetric matrix. This latter assumption implies knowledge about the ratio between the variability due to the signal contained in the latent state process and the variability due to measurement error – the signal-to-noise ratio. When dealing with univariate time series, the lack of knowledge about the signal-to-noise ratio is often dealt with in a subjective manner using discount factor implementations (West and Harrison, 1997; Prado et al., 2021). Unfortunately, the lack of objective Bayes priors for the signal-to-noise ratio parameter in state-space models implies an important gap in the knowledge about objective Bayes methods for spatio-temporal models that are based on state-space models.

To reduce this gap in the literature, in this chapter we introduce objective approximate reference priors for spatio-temporal models. We focus on models for areal data, where the region of interest is partitioned into n subregions and at each time point there is one observation per subregion. Examples of such data include socio-economic and epidemiological data by state in the United States. To obtain approximate reference priors, we use the fact that the latent states in linear state-space models are Markov random fields. Thus, there are similarities between the linear state-space models considered in this chapter and the hierarchical models with intrinsic conditional autoregressive priors for Gaussian spatial data presented in Chapter 2 (Porter et al., 2024). Therefore, here we explore these similarities and use reference prior results reviewed in Chapter 2 to obtain approximate reference priors for spatio-temporal models.

Specifically, we consider a spatio-temporal model that includes a time varying overall mean level and dynamic spatio-temporal random effects. These features are very useful in the analysis of spatio-temporal data. For example, in one of the applications in this chapter, we consider the logarithm of unemployment rate per state in the United States over time. In this example, the spatio-temporal framework we propose provides an estimate of the overall mean level of the logarithm of unemployment rate over time, and estimates of the spatio-temporal random effects that inform about time varying underlying differences between the mean logarithm of unemployment rate per state and the overall mean level.

The spatio-temporal model we consider is similar to one of the models proposed by Vivar and Ferreira (2009) except that, instead of proper Markov random field innovations, here we consider sum-zero constrained intrinsic conditional autoregressive innovations. As a consequence, the covariance matrix in the system evolution equation is singular. To deal with this singularity, we develop a Markov chain Monte Carlo (MCMC) algorithm (Gelfand and Smith, 1990) that includes a singular forward filter backward sampler for the latent states. In addition, the approximate reference prior for the signal-to-noise ratio parameters is not conjugate. To overcome this difficulty, we simulate the signal-to-noise ratio parameters with efficient Metropolis-Hastings steps.

We present two applications of our spatio-temporal framework. First, we present an application to simulated data that demonstrates that our inferential approach based on the approximate reference prior and MCMC algorithm works well at quantifying the uncertainty in the estimation of the parameters and latent states. Second, we present an application to a dataset on logarithm of unemployment rate that illustrates the relevance of the information provided by our spatio-temporal framework.

The remainder of this chapter is organized as follows. Section 3.2 presents the spatio-temporal model that we consider. Section 3.3 presents the approximate reference prior that we propose. Section 3.4 presents the MCMC algorithm for posterior exploration. Section 3.5 presents the application of our spatio-temporal framework to two datasets. Finally, Section 3.6 presents a short discussion and conclusions.

3.2 A Spatio-Temporal Model

Here we consider univariate spatio-temporal areal data, where the region of interest is partitioned into n subregions and at each time point there is one observation per subregion. Examples of such data include gross state product, inflation, and unemployment rate by state in the United States. Let $\boldsymbol{y}_t$ be the n-dimensional vector of observations of the n subregions at time t.

Let N_i be the set of subregions that are neighbors of subregion i. Following Section 2.3.1 in Chapter 2 (Porter et al., 2024), let $\boldsymbol{H}$ be a positive semidefinite matrix that reflects the neighborhood and dependence structures of the subregions under study. Specifically, let $\boldsymbol{H}$ be such that

$$(\boldsymbol{H})_{ij} = \begin{cases} h_i, & \text{if } i = j, \\ -g_{ij}, & \text{if } j \in N_i, \\ 0, & \text{otherwise,} \end{cases} \tag{3.1}$$

where g_{ij} reflects the strength of association between subregions i and j conditional on all the other subregions. Further, $h_i = \sum_{j \in N_i} g_{ij}$. A common choice for neighborhood structure defines two subregions as neighbors if they share a border and sets $g_{ij} = 1$. Conversely, if subregions i and j do not share a border then $g_{ij} = 0$. In this case, h_i is the number of neighbors of subregion i. Here we assume that all subregions are connected, which implies that H has rank $n-1$ and one zero eigenvalue (Ferreira and De Oliveira, 2007; De Oliveira and Ferreira, 2011).

To specify the spatio-temporal model that we consider, let $\boldsymbol{\phi}_t$ be an n-dimensional vector of spatio-temporal random effects for n subregions at time t, μ_t be a scalar that is the overall mean level at time t, $\mathbf{1}_n$ be an n-dimensional vector of ones, $\mathbf{0}_n$ be an n-dimensional vector of zeros, and $\boldsymbol{I}_n$ be an n by n identity matrix. Then, the model we consider is

$$\boldsymbol{y}_t = \mu_t \mathbf{1}_n + \boldsymbol{\phi}_t + \boldsymbol{\epsilon}_t, \quad \boldsymbol{\epsilon}_t \sim N(\mathbf{0}, \sigma^2 \boldsymbol{I}_n), \tag{3.2}$$

$$\boldsymbol{\phi}_t = \boldsymbol{\phi}_{t-1} + \boldsymbol{\omega}_t^{\phi}, \quad \boldsymbol{\omega}_t^{\phi} \sim N(\mathbf{0}_n, \sigma^2 \xi^{\phi} \boldsymbol{W}^{\phi}), \tag{3.3}$$

$$\mu_t = \mu_{t-1} + \omega_t^{\mu}, \quad \omega_t^{\mu} \sim N(0, \sigma^2 \xi^{\mu}), \tag{3.4}$$

$$\boldsymbol{\phi}_0 \sim N(\mathbf{0}_n, C_0^{\phi} \boldsymbol{W}^{\phi}), \tag{3.5}$$

$$\mu_0 \sim N(0, C_0^{\mu}), \tag{3.6}$$

where σ^2 is the variance of each element of the observational error vector $\boldsymbol{\epsilon}_t$. In addition, ξ^{ϕ} is a signal-to-noise ratio parameter of the evolution of the spatio-temporal random effects. Further, ξ^{μ} is a signal-to-noise ratio parameter of the evolution of the overall mean. Furthermore, C_0^{ϕ} and C_0^{μ} are known multiplicative factors in the prior variances of $\boldsymbol{\phi}_0$ and μ_0 at time $t = 0$.

The innovation vector $\boldsymbol{\omega}_t^{\phi}$ of the evolution of the spatio-temporal random effects $\boldsymbol{\phi}_t$ follows a sum-zero constrained intrinsic conditional autoregressive (ICAR) distribution (Keefe et al., 2018, 2019) that is reviewed in Section 2.3.1 of Chapter 2 (Porter et al., 2024). Thus, $\boldsymbol{W}^{\phi} = \boldsymbol{H}^{+}$ is a singular covariance matrix that is the Moore-Penrose inverse of the matrix $\boldsymbol{H}$ defined in Equation (3.1). In addition, $\boldsymbol{\phi}_0$ also follows a sum-zero constrained ICAR distribution. As a consequence, for any time t with probability one the elements of $\boldsymbol{\phi}_t$ sum to zero. Therefore, for a subregion i at time t the corresponding spatio-temporal random effect $\boldsymbol{\phi}_{ti}$ represents the departure of the mean level of subregion i with respect to the overall mean level μ_t.

We may rewrite the spatio-temporal model given in Equations (3.2)–(3.6) in matrix notation. Let $\boldsymbol{\theta}_t = (\mu_t, \boldsymbol{\phi}_t')'$, $\boldsymbol{F} = (\mathbf{1}_n | \boldsymbol{I}_n)$, $\boldsymbol{G} = \boldsymbol{I}_{n+1}$, $\boldsymbol{W}^{\theta} = \text{blockdiag}(\xi^{\mu}, \xi^{\phi} \boldsymbol{W}^{\phi})$, $\boldsymbol{m}_0 = \mathbf{0}_{n+1}$, and $\boldsymbol{C}_0 = \text{blockdiag}(C_0^{\mu}, C_0^{\phi} \boldsymbol{W}^{\phi})$. Then, the spatio-temporal model can be written as

$$\begin{aligned} \boldsymbol{y}_t &= \boldsymbol{F}\boldsymbol{\theta}_t + \boldsymbol{\epsilon}_t, & \boldsymbol{\epsilon}_t &\sim N(\mathbf{0}, \sigma^2 \boldsymbol{I}_n), & (3.7)\\ \boldsymbol{\theta}_t &= \boldsymbol{G}\boldsymbol{\theta}_{t-1} + \boldsymbol{\omega}_t^{\theta}, & \boldsymbol{\omega}_t^{\theta} &\sim N(\mathbf{0}_n, \sigma^2 \boldsymbol{W}^{\theta}), & (3.8)\\ \boldsymbol{\theta}_0 &\sim N(\boldsymbol{m}_0, \boldsymbol{C}_0). & & & (3.9) \end{aligned}$$

This alternative parameterization of the model facilitates joint simulation of the latent states $\boldsymbol{\mu}_t$ and $\boldsymbol{\phi}_t$, $t = 1, \ldots, T$, as we explain in Section 3.4.

3.3 Approximate Reference Prior

From a philosophical point of view, a prior distribution should be a mathematical representation of the information that we have about the parameters of a model before we look at the data. However, in practice, the elicitation of prior distributions is far from trivial. In particular, for many models, such as the spatio-temporal model given in Equations (3.2)–(3.6), the parameters control the behavior of the process in an intricate manner, making prior elicitation even more challenging. As a safe alternative, practitioners often use non-informative objective priors when they are known for the models at hand.

For the spatio-temporal model given in Equations (3.2)–(3.6), safe non-informative priors are not known. In addition, the naïve use of priors that would be non-informative for other models may have undesirable consequences. For example, a well-known non-informative objective prior for the variance in a random sample from the Gaussian distribution is proportional to the reciprocal of the variance. However, the use of such prior for ξ^{μ} and ξ^{ϕ} in the context of the spatio-temporal model we consider may lead to a useless improper posterior distribution.

Alternative objective Bayes priors include Jeffreys priors (Jeffreys, 1961; Bernardo and Smith, 2000; Fonseca et al., 2012) and those developed with the reference prior approach (Bernardo, 1979). While Jeffreys priors work well for regular models with one-dimensional parameters, for multi-parameter models they often lead to Bayesian procedures with undesirable frequentist properties such as frequentist coverage of credible intervals far from the nominal level. In contrast, priors obtained with the reference prior approach (Bernardo, 1979; Berger and Bernardo, 1992a,b,c) usually lead to Bayesian procedures with favorable frequentist properties that can be safely used as default priors in software for general use by practitioners.

Bayesian reference priors have been developed for many models such as Gaussian random fields (Berger et al., 2001), Markov random fields (Ferreira and De Oliveira, 2007), multivariate normal models (Sun and Berger, 2007), elapsed times in continuous-time Markov chains (Ferreira and Suchard, 2008), regression models with exponential power (Ferreira and Salazar, 2014) and

Student-t errors (He et al., 2021), and hierarchical models with spatial intrinsic conditional autoregressive random effects (Keefe et al., 2019). For these models, reference priors lead to Bayesian procedures with favorable frequentist properties such as better-behaved mean squared error than competing procedures and frequentist coverage close to nominal. Unfortunately, to the best of our knowledge, currently there are no known reference priors for the parameters of the spatio-temporal model given by Equations (3.2)–(3.6).

Ferreira and Suchard (2008) introduced the concept of approximate reference priors in the context of continuous-time Markov chains. Specifically, they obtained the reference prior for elapsed times and noted that such prior has a complicated functional form. Then, they proposed the use of an approximate reference prior for elapsed times with the same tail behavior of the reference prior. In their case, the approximate reference prior is a proper gamma distribution. After that, they showed that the frequentist properties of Bayesian procedures based on the reference prior and approximate reference prior are very similar. Finally, when using standard software that allows only standard distributions, in the context of continuous-time Markov chains, Ferreira and Suchard (2008) recommended the use of the approximate reference prior.

More recently, Ferreira et al. (2021) proposed an approximate reference prior for the analysis of hierarchical models with ICAR spatial random effects. This approximate reference prior has the same tail behavior of the exact reference prior for such a model obtained by Keefe et al. (2019). In addition, computation implementations of the approximate reference prior are much faster than those of the exact reference prior. Importantly, Bayesian procedures based on the reference prior and on the approximate reference prior have similar frequentist properties. In addition, Xu et al. (2023) used an approximate reference prior similar to that proposed by Ferreira et al. (2021) for the analysis of generalized linear mixed models. In this later case, the reference prior is not known. Of importance, simulation studies presented by Xu et al. (2023) show that when compared to widely used model selection criteria such as the deviance information criterion (DIC) (Spiegelhalter et al., 2002) and the Watanabe-Akaike information criterion (WAIC) (Watanabe, 2010), Bayesian model selection based on the approximate reference prior has better performance.

Reference priors are model-specific. That is, if we extend a model and the extension includes an additional parameter, the reference prior for the parameters which were in the original model may change. However, to obtain approximate reference priors for ξ^{μ} and ξ^{ϕ} it may be helpful to consider models that include only one of these parameters at a time. This is an idea reminiscent of the algorithm to obtain independence Jeffreys priors, which obtains the Jeffreys prior for each parameter keeping the other parameters fixed, and then multiplies the resulting priors to obtain the independence Jeffreys prior. Independence Jeffreys priors have been proposed for many models, and often coincide with reference priors such as in the case of regression models with exponential power errors (see Salazar et al., 2012; Ferreira and Salazar, 2014) and Student-t errors (see Fonseca et al., 2008; He et al., 2021). Thus, in what follows we first obtain an approximate reference prior for ξ^μ, after that we obtain an approximate reference prior for ξ^ϕ, and finally, we use the product of these two priors as the joint prior for ξ^μ and ξ^ϕ.

To obtain an approximate reference prior for ξ^μ, let us consider a one-dimensional dynamic linear model (West and Harrison, 1997; Migon et al., 2005; Prado et al., 2021) with a random walk evolution equation. Specifically, let x_t be the one-dimensional observation at time t and μ_t be its level at time t. Then, the model we consider is

$$x_t = \mu_t + \epsilon_t, \qquad \epsilon_t \sim N(0, \sigma^2), \tag{3.10}$$

$$\mu_t = \mu_{t-1} + \omega_t^\mu, \qquad \omega_t^\mu \sim N(0, \sigma^2 \xi^\mu). \tag{3.11}$$

We note that the μ_t process can be thought of as following a Gaussian Markov random field (Rue and Held, 2005) and, thus, the model given by Equations (3.10) and (3.11) has the same probabilistic structure as the hierarchical model that has observations with distribution given in Equation (2.8). Thus, the reference prior for the parameters of the model given by Equations (3.10) and (3.11) is going to be of the form given in Equation (2.14) with $\tau = (\xi^\mu)^{-1}$. Using results from Keefe et al. (2019) and performing a transformation of variables, we can show that the corresponding reference prior for ξ^μ behaves as $O(1)$ as $\xi^\mu \to 0$ and as $O((\xi^\mu)^{-2})$ as $\xi^\mu \to \infty$. This then justifies the use of the approximate reference prior

$$p(\sigma^2, \xi^\mu) \propto \frac{1}{\sigma^2 (a + \xi^\mu)^2}, \tag{3.12}$$

where a is a hyperparameter. Approximate reference priors of this same functional form have been proposed by Ferreira et al. (2021) for the spatial hierarchical models reviewed in Chapter 2 and by Xu et al. (2023) for generalized linear mixed models. Here we set $a = 2$ which is equivalent to the values used by Ferreira et al. (2021) and Xu et al. (2023).

To obtain an approximate reference prior for ξ^ϕ, let us consider an n-dimensional dynamic linear model with a multivariate random walk evolution equation for spatio-temporal random effects with innovations that follow a sum-zero constrained ICAR distribution. Let $\boldsymbol{x}_t$ be the n-dimensional vector of observations at time t and $\boldsymbol{\phi}_t$ be the vector of spatio-temporal random effects at time t. Specifically, the model we consider is

$$\boldsymbol{x}_t = \boldsymbol{\phi}_t + \boldsymbol{\epsilon}_t, \qquad \boldsymbol{\epsilon}_t \sim N(\boldsymbol{0}, \sigma^2 \boldsymbol{I}_n), \tag{3.13}$$

$$\boldsymbol{\phi}_t = \boldsymbol{\phi}_{t-1} + \boldsymbol{\omega}_t^\phi, \qquad \boldsymbol{\omega}_t^\phi \sim N(\boldsymbol{0}_n, \sigma^2 \xi^\phi \boldsymbol{W}^\phi). \tag{3.14}$$

Similarly to the μ_t process in Equation (3.11), the $\boldsymbol{\phi}_t$ process in Equation (3.14) can be thought of as following a multivariate Gaussian Markov random field. As such, the $\boldsymbol{\phi}_t$ process follows a similar probabilistic structure as the hierarchical model that has observations with distribution given in

Equation (2.8). Thus, we propose the use of the approximate reference prior

$$p(\sigma^2, \xi^\phi) \propto \frac{1}{\sigma^2(a+\xi^\phi)^2}, \tag{3.15}$$

where a is a hyperparameter that, following Ferreira et al. (2021) and Xu et al. (2023), we set to 2.

Based on these developments, we propose for the parameters of the spatio-temporal model given in Equations (3.2)–(3.6) the approximate reference prior

$$p(\sigma^2, \xi^\mu, \xi^\phi) \propto \frac{1}{\sigma^2(a+\xi^\mu)^2(a+\xi^\phi)^2}, \tag{3.16}$$

where, following the rationality above, we set $a = 2$. This approximate reference prior performs well in an application to simulated data shown in Section 3.5.1. The development of formal reference priors for $(\sigma^2, \xi^\mu, \xi^\phi)$ remains an open research problem.

3.4 Posterior Exploration

We have developed a hybrid Gibbs sampler (Robert and Casella, 2004; Gamerman and Lopes, 2006) to explore the posterior distribution of the spatio-temporal model given in Equations (3.2)–(3.6). Specifically, we simulate σ^2 from its full conditional distribution. In addition, because the approximate reference prior for ξ^μ and ξ^ϕ is not conjugate, we simulate these parameters using efficient Metropolis-Hastings steps. Finally, we simulate $\boldsymbol{\theta}_1, \ldots, \boldsymbol{\theta}_T$ jointly using a singular forward filter backward sampler algorithm. In what follows, we provide details of the hybrid Gibbs sampler.

Simulation of σ^2

We simulate σ^2 from its full conditional distribution $\sigma^2|- \sim IG(n_\sigma/2, n_\sigma s_\sigma/2)$, where $n_\sigma = 2Tn$ and

$$n_\sigma s_\sigma = \sum_{t=1}^{T} \left[(\boldsymbol{y} - \boldsymbol{F}\boldsymbol{\theta}_t)'(\boldsymbol{y} - \boldsymbol{F}\boldsymbol{\theta}_t) + (\boldsymbol{\theta}_t - \boldsymbol{\theta}_{t-1})'(\boldsymbol{W}^\theta)^-(\boldsymbol{\theta}_t - \boldsymbol{\theta}_{t-1}) \right].$$

Simulation of ξ^μ

The full conditional density of ξ^μ is

$$p(\xi^\mu|-) \propto p(\xi^\mu)(\xi^\mu)^{-0.5T+1-1} \exp\left\{ -\frac{1}{2\sigma^2\xi^\mu} \sum_{t=1}^{T} (\mu_t - \mu_{t-1})^2 \right\},$$

which is not available for sampling. However, this density is the product of the prior $p(\xi^{\mu})$ and the kernel of an inverse gamma density. Thus, we use a Metropolis-Hastings step that simulates a proposal $\xi^{\mu(prop)}$ from $IG(0.5(T-2), (2\sigma^2)^{-1}\sum_{t=1}^{T}(\mu_t-\mu_{t-1})^2)$ and accepts $\xi^{\mu(prop)}$ with probability $\min(1,a)$ where

$$a = \frac{(\xi^{\mu(old)}+2)^2}{(\xi^{\mu(prop)}+2)^2}.$$

Simulation of ξ^{ϕ}

The full conditional density of ξ^{ϕ} is

$$\begin{aligned} p(\xi^{\phi}|-) \quad &\propto \quad p(\xi^{\phi})(\xi^{\phi})^{-0.5T(n-1)+1-1} \\ &\times \exp\left\{-\frac{1}{2\sigma^2\xi^{\phi}}\sum_{t=1}^{T}(\boldsymbol{\phi}_t-\boldsymbol{\phi}_{t-1})'(\boldsymbol{W}^{\phi})^{-}(\boldsymbol{\phi}_t-\boldsymbol{\phi}_{t-1})\right\}, \end{aligned}$$

which is not available for sampling. Similarly to ξ^{μ}, the full conditional density of ξ^{ϕ} is the product of the prior $p(\xi^{\phi})$ and the kernel of an inverse gamma density. Thus, we use a Metropolis-Hastings step that simulates a proposal $\xi^{\phi(prop)}$ from $IG(0.5(T(n-1)-2), (2\sigma^2)^{-1}\sum_{t=1}^{T}(\boldsymbol{\phi}_t-\boldsymbol{\phi}_{t-1})'(\boldsymbol{W}^{\phi})^{-}(\boldsymbol{\phi}_t-\boldsymbol{\phi}_{t-1}))$ and accepts $\xi^{\phi(prop)}$ with probability $\min(1,a)$ where

$$a = \frac{(\xi^{\phi(old)}+2)^2}{(\xi^{\phi(prop)}+2)^2}.$$

Simulation of $\boldsymbol{\theta}_1,\ldots,\boldsymbol{\theta}_T$

To simulate $\boldsymbol{\theta}_1,\ldots,\boldsymbol{\theta}_T$, we use a variation of the forward filter backward sampler algorithm (FFBS) (Carter and Kohn, 1994; Frühwirth-Schnatter, 1994) that takes into account the fact that the covariance matrix of the system evolution equation $\boldsymbol{W}^{\theta}$ is singular.

Specifically, the singular forward filter is as follows:

(1) Prior at $t=0$: $\boldsymbol{\theta}_0|D_0 \sim N(\boldsymbol{m}_0, \boldsymbol{C}_0)$.

(2) Posterior at $t-1$: $\boldsymbol{\theta}_{t-1}|D_{t-1} \sim N(\boldsymbol{m}_{t-1}, \boldsymbol{C}_{t-1})$.

(3) Prior at t: $\boldsymbol{\theta}_t|D_{t-1} \sim N(\boldsymbol{a}_t, \boldsymbol{R}_t)$, where $\boldsymbol{a}_t = \boldsymbol{m}_{t-1}$ and $\boldsymbol{R}_t = \boldsymbol{C}_{t-1} + \sigma^2\boldsymbol{W}^{\theta}$.

(4) Posterior at t: $\boldsymbol{\theta}_t|D_t \sim N(\boldsymbol{m}_t, \boldsymbol{C}_t)$, where $\boldsymbol{C}_t = (\boldsymbol{R}_t^{-} + \sigma^{-2}\boldsymbol{F}'\boldsymbol{F})^{-}$ and $\boldsymbol{m}_t = \boldsymbol{C}_t(\boldsymbol{R}_t^{-}\boldsymbol{a}_t + \sigma^{-2}\boldsymbol{F}'\boldsymbol{y}_t)$.

After we run the singular forward filter, the singular backward sampler proceeds as follows:

(1) At time T, sample $\boldsymbol{\theta}_T$ from the distribution $\boldsymbol{\theta}_T | D_T \sim N(\boldsymbol{m}_T, \boldsymbol{C}_T)$.

(2) Backward in time, for $t = T-1, \ldots, 0$ sample $\boldsymbol{\theta}_t$ from the distribution $\boldsymbol{\theta}_t | \boldsymbol{\theta}_{t+1}, D_t \sim N(\boldsymbol{h}_t, \boldsymbol{H}_t)$, where $\boldsymbol{H}_t = (\boldsymbol{C}_t^- + \sigma^{-2}(\boldsymbol{W}^\theta)^-)^-$ and $\boldsymbol{h}_t = \boldsymbol{H}_t(\boldsymbol{C}_t^- \boldsymbol{m}_t + \sigma^{-2}(\boldsymbol{W}^\theta)^- \boldsymbol{\theta}_{t+1})$.

As we show with an application to simulated data in Section 3.5.1, the hybrid Gibbs sampler algorithm presented above combined with the conditional reference prior presented in Section 3.3 provides excellent quantification of the uncertainty in the analysis of spatio-temporal data with the model given in Equations (3.2)–(3.6).

3.5 Applications

We present two applications of our spatio-temporal model. The first application considers simulated data and demonstrates that our inferential approach based on the approximate reference prior and MCMC algorithm works well at quantifying the uncertainty in the estimation of the parameters and latent states. The second application analyzes the unemployment rate and illustrates the relevance of the information provided by our spatio-temporal framework.

3.5.1 Application: Simulated Data

We consider spatio-temporal data simulated with the model given in Equations (3.2)–(3.6). Specifically, we simulate data on a 10×10 regular grid in a total of $n = 100$ subregions through $T = 30$ time points. The true values of the parameters are $\xi^\phi = 0.15$, $\xi^\mu = 0.2$, and $\sigma^2 = 5$. We assume the approximate reference prior for ξ^μ, ξ^ϕ, and σ^2, as well as $C_0^\mu = C_0^\phi = 10^{10}$ which implies vague priors for the latent states at time $t = 0$.

We explore the posterior distribution with the hybrid Gibbs sampler presented in Section 3.4 with 5,000 iterations. Figure 3.1 presents trace plots (left column) and posterior densities (right column) of ξ^μ, ξ^ϕ, and σ^2. The trace plots indicate convergence of the MCMC algorithm in less than 1,000 iterations. Thus, we discard the first 1,000 iterations as burnin and use the last 4,000 iterations to compute the posterior densities and posterior summaries. The dashed vertical lines in the plots of posterior densities indicate the true values of the respective parameters, and show that the posterior densities are concentrating near the true values. Specifically, the posterior median of ξ^μ is 0.26 and a 95% credible interval for ξ^μ is (0.15, 0.48), which includes the true value 0.2. In addition, the posterior median of ξ^ϕ is 0.15 and a 95% credible interval for ξ^ϕ is (0.12, 0.20), which includes the true value 0.15. Finally, the

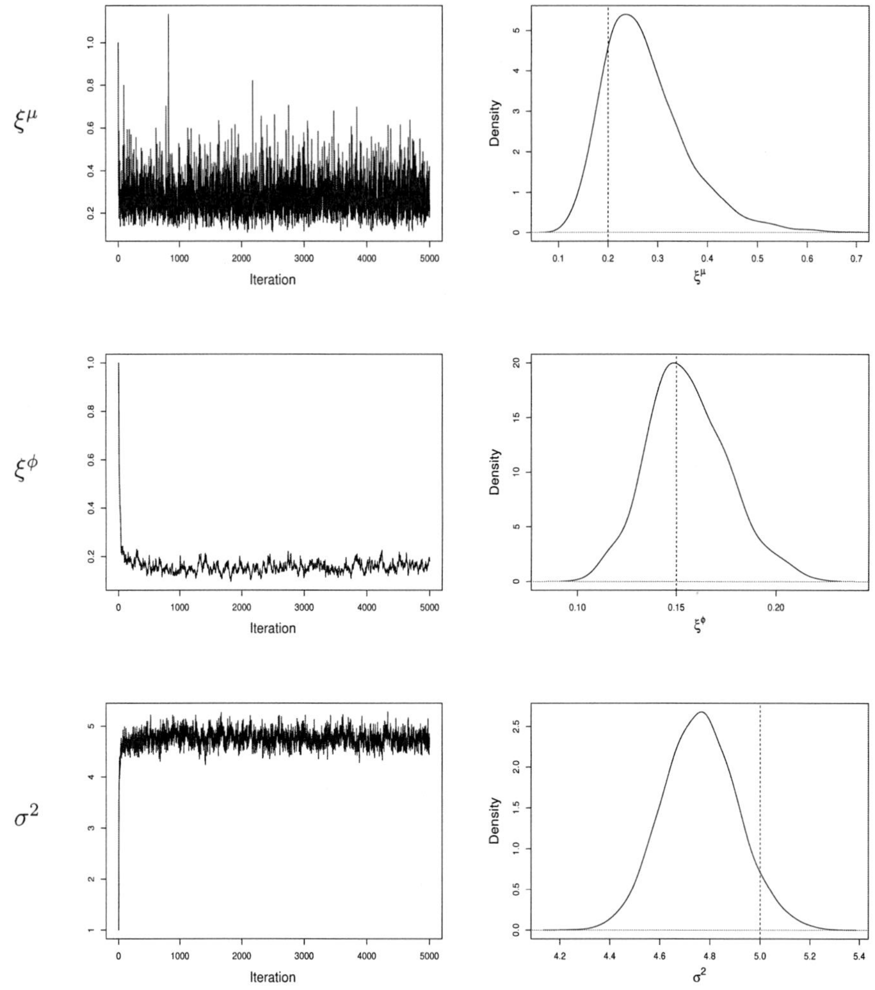

FIGURE 3.1
Analysis of simulated spatio-temporal data. Trace plots (left column) and posterior densities (right column) of ξ^μ, ξ^ϕ, and σ^2. The trace plots indicate that the MCMC chain converged quickly. The dashed vertical lines in the plots of posterior densities indicate the true values of the respective parameters, and show that the posterior densities are concentrating near the true values.

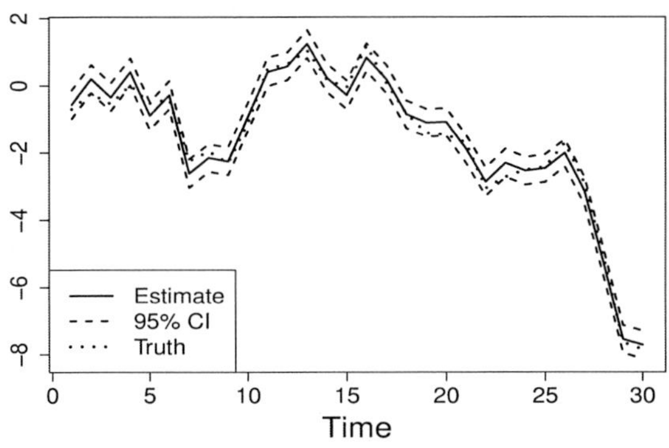

FIGURE 3.2
Analysis of simulated spatio-temporal data. Overall mean level μ_t over time.

posterior median of σ^2 is 4.76 and a 95% credible interval for σ^2 is (4.48, 5.06), which includes the true value 5.

Figure 3.2 presents the mean level μ_t over time, where the solid line represents the posterior mean, the dashed line represents pointwise 95% credible intervals, and the dotted line represents the true mean level simulated during the simulation of the data. Our inferential framework captures the true μ_t, $t = 1, \ldots, 30$, very precisely. In addition, the 95% credible intervals include the true values of μ_t for 96.7% of the time points and, thus, provide an excellent quantification of the uncertainty involved in the estimation of μ_t.

Figure 3.3 presents time series plots of the spatio-temporal random effects $\phi_{t,10}$, $\phi_{t,50}$, and $\phi_{t,100}$ for subregions 10, 50, and 100, respectively. Specifically, the solid line represents the posterior mean, the dashed line represents pointwise 95% credible intervals, and the dotted line represents the true spatio-temporal random effects simulated during the simulation of the data. The posterior means of these ϕ_t's follow very closely the true values. In addition, for these three subregions, the 95% credible intervals include the true values 96.7% of the time. Moreover, for all 100 subregions, the 95% credible intervals include the true values 94.6% of the time, which is very close to the nominal level and reflects excellent quantification of the uncertainty involved in the estimation of ϕ_t.

In summary, this application to simulated data demonstrates that our inferential approach based on the approximate reference prior and MCMC algorithm works well at estimating unknown quantities and quantifying the related uncertainty.

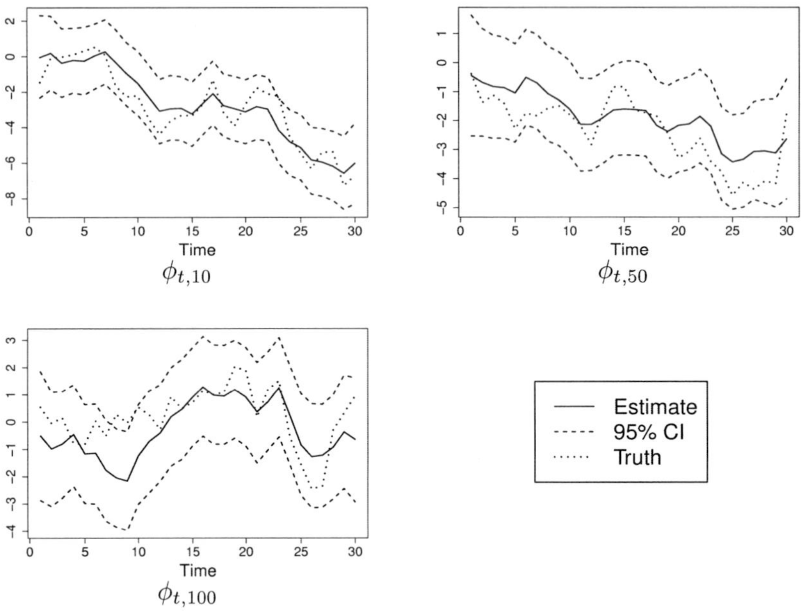

FIGURE 3.3
Analysis of simulated spatio-temporal data. Spatio-temporal random effects ϕ_t for subregions 10, 50, and 100.

3.5.2 Application: Unemployment Rate

We analyze here the unemployment rate in the contiguous United States. Specifically, we apply our spatio-temporal framework to analyze the logarithm of unemployment rate by state from 1970 to 1986. This dataset is available from the R package plm (Croissant and Millo, 2008). Figure 3.4, prepared with the function stplot from the package plm, presents maps of the logarithm of unemployment rate by state from 1970 to 1986. We are interested in estimating the overall mean level of the logarithm of unemployment rate over time, and the spatio-temporal random effects that provide the underlying latent differences between the mean logarithm of unemployment rate per state and the overall mean level.

We apply to the logarithm of unemployment data the hybrid Gibbs sampler presented in Section 3.4. We run the algorithm for a total of 5,000 iterations. In addition to the approximate reference prior for ξ^μ, ξ^ϕ, and σ^2, we assign for the latent states at time $t = 0$ prior variances equal to $C_0^\mu = C_0^\phi = 10^{10}$. These choices of prior variances imply vague priors for the latent states at time $t = 0$. Similarly to the simulated example presented in Section 3.5.1, trace plots (not shown) of the MCMC output for ξ^μ, ξ^ϕ, and σ^2 show fast convergence of

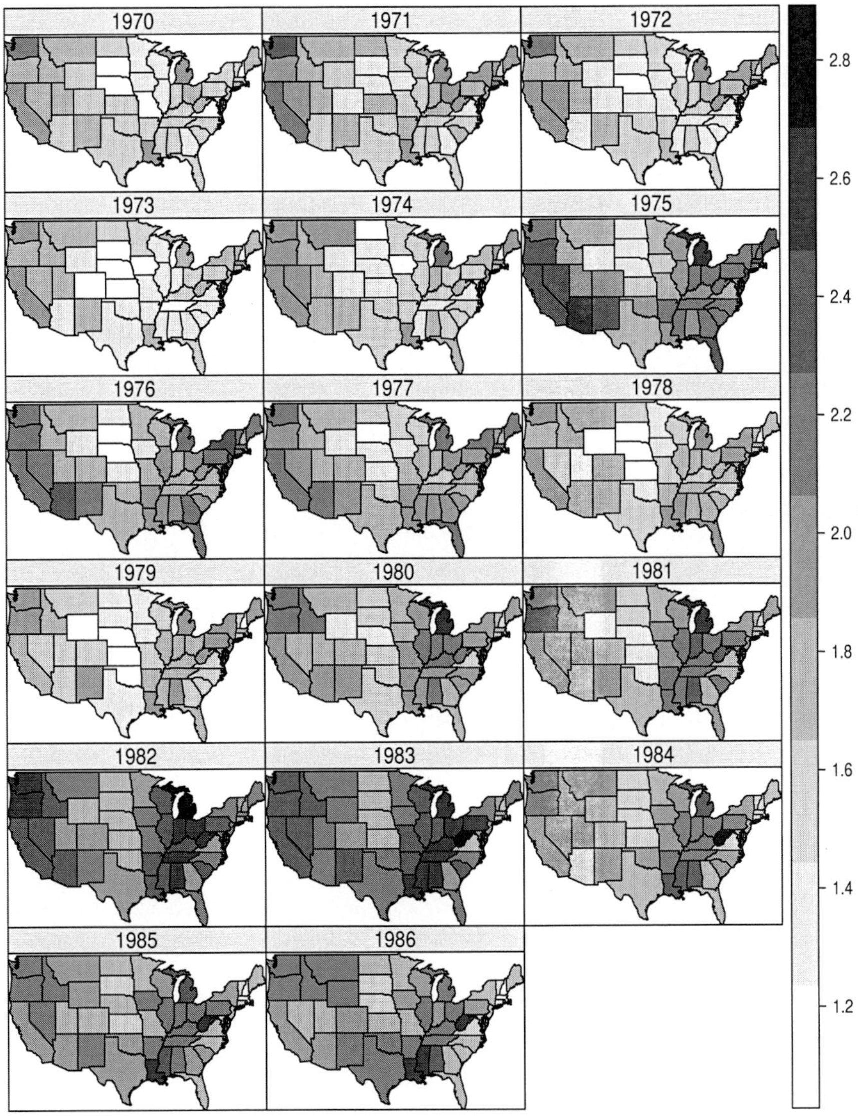

FIGURE 3.4
Maps of logarithm of unemployment per state in the United States from 1970 to 1986.

the algorithm in less than 1,000 iterations. This provides another example of the efficiency of the singular forward filter backward sampler presented in Section 3.4. Thus, we discard the first 1,000 iterations as burnin and use the last 4,000 iterations to perform the posterior analysis.

Figure 3.5 presents posterior densities of ξ^{μ}, ξ^{ϕ}, and σ^2. The signal-to-noise ratio parameter ξ^{μ} for the temporal evolution of the mean level is large, with a posterior median of 8.57 and an equal-tailed 95% credible interval of (4.32, 19.37). Similarly, the signal-to-noise ratio parameter ξ^{ϕ} for the temporal evolution of the spatio-temporal random effects is also large, with a posterior median equal to 7.99 and an equal-tailed 95% credible interval equal to (5.19, 12.69). In contrast, the posterior median of the variance σ^2 of the observational errors is small equal to 0.0032 and an equal-tailed 95% credible interval for σ^2 is (0.0024, 0.0042). These point estimates and interval estimates of ξ^{μ}, ξ^{ϕ}, and σ^2 indicate that the variance of the observational errors is much smaller than the variances of the innovations of the system evolution equations (3.3) and (3.4). Therefore, most of the variability in this spatio-temporal unemployment dataset is explained by the time varying mean level μ_t and the spatio-temporal random effects $\boldsymbol{\phi}_t$.

Figure 3.6 presents the posterior mean (solid line) and 95% credible intervals (dashed lines) of the overall mean level μ_t for the logarithm of unemployment over time from 1970 to 1986 in the contiguous United States. The 95% credible intervals are tight, indicating high precision in the estimation of the overall mean level μ_t. From an applied point of view, two main features emerge from this plot. First, unemployment had an increasing trend during the period under study with average unemployment around 4.9% in 1970 and 6.9% in 1986. Second, in addition to this increasing trend, the overall mean level of logarithm of unemployment exhibits what seem to be economic cycles. While out of the scope of this work, it would be interesting to investigate how these cyclical behaviors in the logarithm of unemployment relate to other economic variables such as price of commodities, gross domestic product, and interest rates.

Finally, Figure 3.7 presents maps of the posterior means of the spatio-temporal random effects per state through time. Several features can be observed in this figure. First, during the period of study, the states of Kansas, Nebraska, South Dakota, and North Dakota had consistently lower unemployment rates than the other states under study. Second, at the beginning of the study period the states on the Pacific Coast, and in particular Washington State, tended to have higher unemployment rates than the other states; this feature somehow attenuated through time. Third, the state of Michigan had a much higher level of unemployment rate than other states from 1980 to 1982. Finally, the state of West Virginia had much higher unemployment rate than other states from 1983 to 1986.

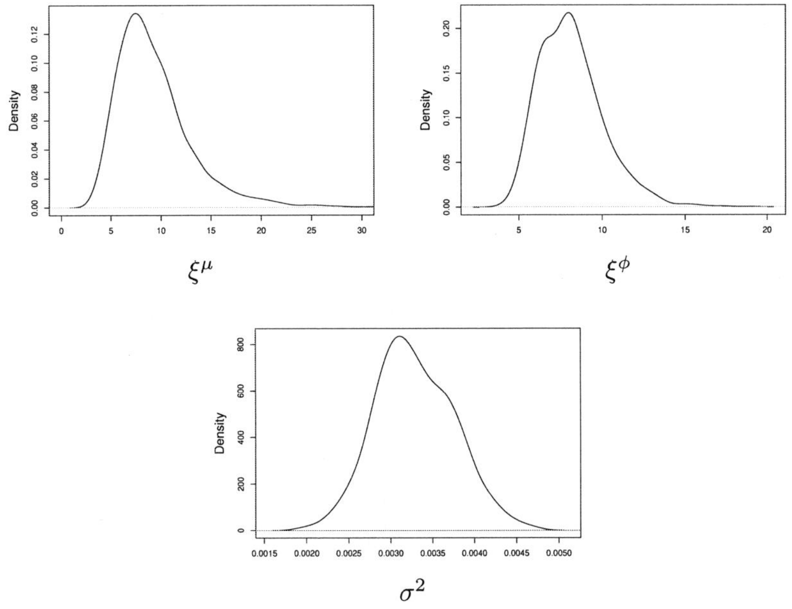

FIGURE 3.5
Logarithm of unemployment data. Posterior densities of ξ^{μ}, ξ^{ϕ}, and σ^2.

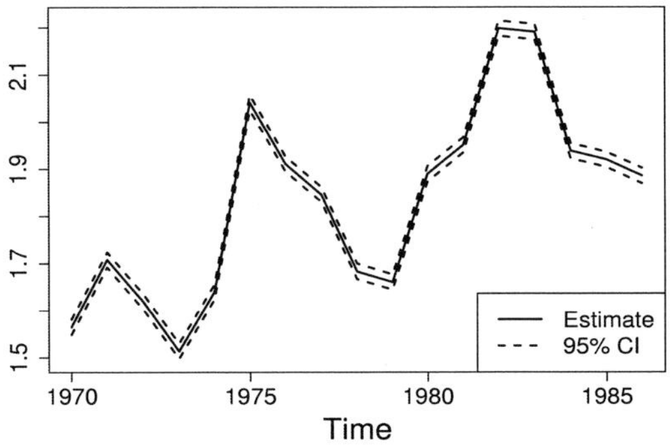

FIGURE 3.6
Logarithm of unemployment data. Overall mean level μ_t over time. Posterior mean (solid line) and 95% credible intervals (dashed lines).

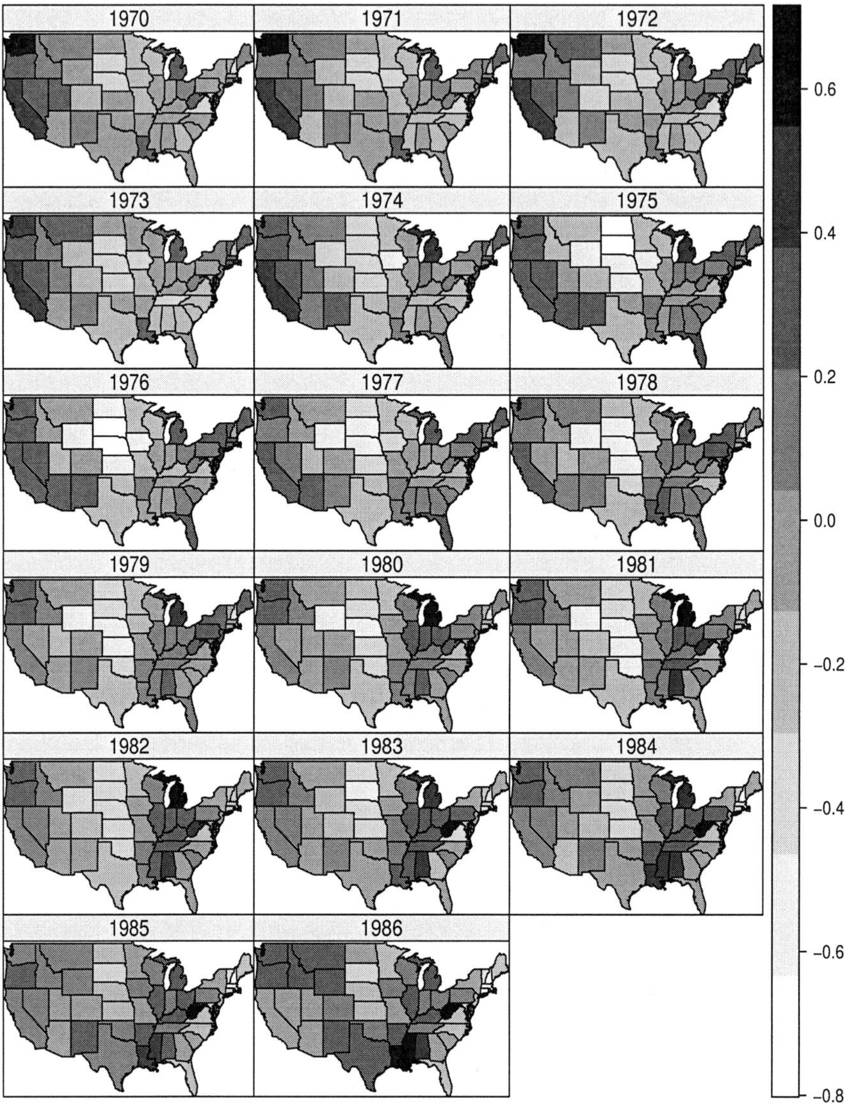

FIGURE 3.7
Posterior mean of spatio-temporal random effects for the logarithm of unemployment.

3.6 Conclusions

We have proposed an approximate reference prior for spatio-temporal models. Specifically, the spatio-temporal model we have considered includes an overall mean level and spatio-temporal random effects. The spatio-temporal random effects evolve through time with a multivariate random walk process where the innovations follow a sum-zero constrained ICAR process. In addition, we have presented an efficient MCMC algorithm to explore the posterior distribution. An application to simulated data shows that our proposed framework works well for the estimation of the parameters that govern the considered spatio-temporal process. An application to spatio-temporal unemployment data illustrates the usefulness of the resulting spatio-temporal analysis.

There are many worthwhile avenues for research related to objective Bayesian analysis of spatio-temporal data. In that regard, we are currently pursuing two research topics. The first is the development of reference priors for general linear state-space models. The second is the development of reference priors for the spatio-temporal model considered here. In particular, a joint reference prior for the signal-to-noise ratio parameters of the temporal evolution of the overall mean level and spatio-temporal random effects may be particularly useful. Research on these topics is progressing and will be reported elsewhere.

Bibliography

Berger, J. O. and J. M. Bernardo (1992a). On the development of reference priors (Disc: p49-60). In J. M. Bernardo, J. O. Berger, A. P. Dawid, and A. F. M. Smith (Eds.), *Bayesian Statistics 4*, pp. 35–49. Oxford University Press.

Berger, J. O. and J. M. Bernardo (1992b). Ordered group reference priors with application to the multinomial problem. *Biometrika 25*, 25–37.

Berger, J. O. and J. M. Bernardo (1992c). Reference priors in a variance components problem. In P. K. Goel and N. S. Iyengar (Eds.), *Bayesian Analysis in Statistics and Econometrics*, pp. 323–340. Berlin: Springer.

Berger, J. O., V. De Oliveira, and B. Sansó (2001). Objective Bayesian analysis of spatially correlated data. *Journal of the American Statistical Association 96*, 1361–1374.

Bernardo, J. M. (1979). Reference posterior distributions for Bayesian inference (with discussion). *Journal of the Royal Statistical Society, Series B, Methodological 41*, 113–147.

Bernardo, J. M. and A. F. M. Smith (2000). *Bayesian Theory.* John Wiley & Sons.

Carter, C. K. and R. Kohn (1994). On Gibbs sampling for state space models. *Biometrika 81*, 541–553.

Cressie, N. and C. K. Wikle (2011). *Statistics for Spatio-Temporal Data.* New York: Wiley.

Croissant, Y. and G. Millo (2008). Panel data econometrics in R: The plm package. *Journal of Statistical Software 27*(2), 1–43.

De Oliveira, V. and M. A. R. Ferreira (2011). Maximum likelihood and restricted maximum likelihood estimation for a class of Gaussian Markov random fields. *Metrika 74*, 167–183.

Ferreira, M. A. R. and V. De Oliveira (2007). Bayesian reference analysis for Gaussian Markov Random Fields. *Journal of Multivariate Analysis 98*, 789–812.

Ferreira, M. A. R., E. M. Porter, and C. T. Franck (2021). Fast and scalable computations for Gaussian hierarchical models with intrinsic conditional autoregressive spatial random effects. *Computational Statistics and Data Analysis 162*, 107264.

Ferreira, M. A. R. and E. Salazar (2014). Bayesian reference analysis for exponential power regression models. *Journal of Statistical Distributions and Applications 1*(1), 1–12.

Ferreira, M. A. R. and M. A. Suchard (2008). Bayesian analysis of elapsed times in continuous-time Markov chains. *Canadian Journal of Statistics 36*, 355–368.

Fonseca, T. C. O., M. A. R. Ferreira, and H. S. Migon (2008). Objective Bayesian analysis for the Student-t regression model. *Biometrika 95*, 325–333.

Fonseca, T. C. O., H. S. Migon, and M. A. R. Ferreira (2012). Bayesian analysis based on the Jeffreys prior for the hyperbolic distribution. *Brazilian Journal of Probability and Statistics 26*, 327–343.

Frühwirth-Schnatter, S. (1994). Data augmentation and dynamic linear models. *Journal of Time Series Analysis 15*(2), 183–202.

Gamerman, D. and H. F. Lopes (2006). *Markov Chain Monte Carlo: Stochastic Simulation for Bayesian Inference* (2nd ed.). Boca Raton, FL: Chapman and Hall/CRC.

Gelfand, A. E. and A. F. M. Smith (1990). Sampling-based approaches to calculating marginal densities. *Journal of the American Statistical Association 85*(410), 398–409.

He, D., D. Sun, and L. He (2021). Objective Bayesian analysis for the Student-t linear regression. *Bayesian Analysis 16*, 129–145.

Jeffreys, H. (1961). *Theory of Probability.* London: Oxford University Press.

Keefe, M. J., M. A. R. Ferreira, and C. T. Franck (2018). On the formal specification of sum-zero constrained intrinsic conditional autoregressive models. *Spatial Statistics 24*, 54–65.

Keefe, M. J., M. A. R. Ferreira, and C. T. Franck (2019). Objective Bayesian analysis for Gaussian hierarchical models with intrinsic conditional autoregressive priors. *Bayesian Analysis 14*(1), 181–209.

Migon, H. S., D. Gamerman, H. F. Lopes, and M. A. R. Ferreira (2005). Dynamic models. In D. Dey and C. R. Rao (Eds.), *Bayesian Thinking, Modeling and Computation*, Handbook of Statistics, pp. 553–588. North Holland.

Pole, A. and M. West (1989). Reference analysis of the dynamic linear model. *Journal of Time Series Analysis 10*, 131–147.

Porter, E. M., C. T. Franck, and M. A. R. Ferreira (2024). Gaussian spatial hierarchical models with ICAR priors. In M. A. R. Ferreira (Ed.), *Modeling Spatio-Temporal Data: Markov Random Fields, Objective Bayes, and Multiscale Models*, pp. 23–46. Chapman & Hall/CRC.

Prado, R., M. A. R. Ferreira, and M. West (2021). *Time Series: Modeling, Computation, and Inference* (2nd ed.). Boca Raton: CRC Press.

Robert, C. P. and G. Casella (2004). *Monte Carlo Statistical Methods* (2nd ed.). Springer-Verlag.

Rue, H. and L. Held (2005). *Gaussian Markov Random Fields: Theory and Applications.* CRC Press.

Salazar, E., M. A. R. Ferreira, and H. S. Migon (2012). Objective Bayesian analysis for exponential power regression models. *Sankhya - Series B 74*, 107–125.

Spiegelhalter, D. J., N. G. Best, B. P. Carlin, and A. Van Der Linde (2002). Bayesian measures of model complexity and fit. *Journal of the Royal Statistical Society: Series B (Statistical Methodology) 64*(4), 583–639.

Sun, D. and J. O. Berger (2007). Objective Bayesian analysis for the multivariate normal model. In J. M. Bernardo, M. Bayarri, J. Berger, A. Dawid, D. Heckerman, A. Smith, and M. West (Eds.), *Bayesian Statistics 8*, pp. 525–547. Oxford: Oxford University Press.

Vivar, J. C. and M. A. R. Ferreira (2009). Spatio-temporal models for Gaussian areal data. *Journal of Computational and Graphical Statistics 18*, 658–674.

Watanabe, S. (2010). Asymptotic equivalence of Bayes cross validation and widely applicable information criterion in singular learning theory. *Journal of Machine Learning Research 11*(12), 3571–3594.

West, M. and P. J. Harrison (1997). *Bayesian Forecasting and Dynamic Models* (2nd. ed.). New York: Springer.

Xu, S., M. A. R. Ferreira, E. M. Porter, and C. T. Franck (2023). Bayesian model selection for generalized linear mixed models. *Biometrics 79*(4), 3266–3278.

4

Spatio-Temporal Models for Poisson Areal Data

Marco A. R. Ferreira

Virginia Tech

Juan C. Vivar

U.S. Food and Drug Administration

We present a class of spatio-temporal models for Poisson areal data suitable for the analysis of emerging infectious diseases. These models assume Poisson observations are related through a link equation to a latent random field process. This latent random field process evolves through time with proper Gaussian Markov random field convolutions. Our approach naturally accommodates flexible structures such as distinct but interacting temporal trends for each region and across-time contamination among neighboring regions. We develop a Bayesian analysis approach with a simulation-based procedure: specifically, we construct a Markov chain Monte Carlo algorithm based on the extended Kalman filter to obtain samples from an approximate posterior distribution. Finally, for the comparison of Poisson spatio-temporal models, we develop a simulation-based conditional Bayes factor. We illustrate the utility and flexibility of our Poisson spatio-temporal framework with an application to the number of acquired immunodeficiency syndrome (AIDS) cases during the period 1982–2007 in Rio de Janeiro.

DOI: 10.1201/9781032623443-4

4.1 Introduction

As the recent COVID pandemic has demonstrated, there is an increasing need for models for the spatio-temporal spread of emerging infectious diseases. This increasing need arises from several factors such as the impact of global warming on the spread of insect vectors (Lafferty, 2009; Epstein, 2010) and the higher global mobility of goods and people (Apostolopoulos and Sönmez, 2007). These factors have substantially increased the risk of epidemics of new, emerging, and re-emerging infectious diseases (Brower and Chalk, 2003). In terms of publicly available information, datasets on diseases are usually in the form of areal count data; that is, the total number of cases of the disease available in a partition of the geographical domain of interest (Banerjee et al., 2014). These areal count data are usually modeled using the Poisson distribution.

Often based on dynamic generalized linear models (West et al., 1985; Ferreira and Gamerman, 2000; Alves et al., 2010), several spatio-temporal models for Poisson areal data have been developed in the disease mapping literature (e.g., Bernardinelli et al., 1995; Waller et al., 1997; Knorr-Held, 2000; Schmid and Held, 2004; Tzala and Best, 2008; Lawson, 2018). Usually, these models assume a common univariate temporal trend for all the regions and time-specific spatial random effects. When modeling non-infectious diseases such as cancer, these models are usually effective to explain the remaining spatio-temporal noise after covariates are taken into account. However, these models are not appropriate for modeling the spatio-temporal dynamics of infectious diseases. Chapter 14 of Lawson (2018) provides a thorough discussion of spatio-temporal models for infectious diseases, including the models proposed by Mugglin et al. (2002) and Knorr-Held and Richardson (2003).

Here, we are interested in spatio-temporal evolutions similar to those proposed by Vivar and Ferreira (2009) for a general class of linear Gaussian spatio-temporal models for areal data that allows for more complex spatio-temporal processes. Even though the Gaussian model framework of Vivar and Ferreira (2009) may be applied to transformed count data when the counts are large, their framework is inadequate for dealing with counts that are small. Hence, their framework would not be appropriate for modeling the early epidemic expansion stages of an emerging disease. Here we develop a methodology that can be directly applied to areal count data. Specifically, we propose a spatio-temporal class of models that allows for more complex spatio-temporal processes for Poisson areal data. In particular, in our framework, each subregion may have a distinct stochastic temporal trend.

There has been some previous literature where each subregion of the geographical domain of interest may have a different temporal trend (Sun et al., 2000; Assunção et al., 2001; MacNab and Dean, 2001). Typically, these models assume time-specific spatial random effects that follow a Gaussian Markov random field. Furthermore, they assume region-specific temporal trends that are deterministic functions of time and may be linear (Sun et al., 2000), quadratic (Assunção et al., 2001), or piecewise cubic (MacNab and Dean, 2001). Finally, Sun et al. (2000), Assunção et al. (2001), and MacNab and Dean (2001) assume that the temporal trend coefficients follow Gaussian Markov random fields. Even though these previous works allow for similar deterministic temporal trends for neighboring regions, they do not allow for more flexible temporal trends. Conversely, as we detail in Section 4.2.1, our novel class of models allows for more flexible stochastic temporal trends and for spatio-temporal interaction among the different regions.

There are two main modeling approaches to analyze epidemic data. One of these approaches is through compartmental susceptible/exposed/infectious/removed (SEIR) models (e.g., see Anderson and May, 1992). Based on the local behavior of individuals, these SEIR models assume that the expected epidemic dynamics in the population can be represented by a set of partial differential equations. Because SEIR models are based on the local behavior of individuals, Mugglin et al. (2002) say that SEIR models provide a "particle" view of the spread of an epidemic. The other modeling approach is what Mugglin et al. (2002) call a "field" view. This field view approach pioneered by Mugglin et al. (2002) is particularly adequate for spatio-temporal areal data and models the stochastic geographic spread of the epidemic through time using a multivariate dynamic generalized linear model. Specifically, Mugglin et al. (2002) model the spatio-temporal dynamics of an influenza epidemic with what they call a spatially descriptive temporally dynamic hierarchical model. Their model is a particular case of our class of models where they use what we call a contamination model. Similarly to Mugglin et al. (2002), here we take a field view of the spatio-temporal epidemic process.

To perform Bayesian statistical analysis for our proposed class of spatio-temporal models, here we develop a simulation-based procedure. Specifically, we construct a Markov chain Monte Carlo (MCMC) algorithm (Gelfand and Smith, 1990; Ferreira et al., 1997; Robert and Casella, 2004; Gamerman and Lopes, 2006) to obtain samples from an approximate posterior distribution. The MCMC algorithm we propose combines the extended Kalman filter (p. 352, Fahrmeir and Tutz, 2001) and the forward filter backward sampler (FFBS) (Frühwirth-Schnatter, 1994; Carter and Kohn, 1994) for the simulation of the latent fields. Our computational methodology is general enough to be applied to observations with any distribution in the regular exponential family. However, here we focus on Poisson observations.

With the ability to fit several nonnested spatio-temporal models comes the need to compare those models. In addition, when analyzing a particular dataset, each of the fitted models corresponds to a distinct scientific hypothesis. As such, comparison of the different models becomes paramount for understanding the nature of the process underlying the observed data. For the comparison of the resulting nonnested Poisson spatio-temporal models, we develop a conditional Bayes factor (p. 190, Ghosh et al., 2006) computed by using the output of our MCMC algorithm. As we discuss in Section 4.3.2, this conditional Bayes factor provides a model comparison criterion that is justified from a Bayesian decision-theoretic point of view (Berger, 1985). In addition, this conditional Bayes factor is computed based on the one-step ahead predictive densities of the competing models. Therefore, as an important practical consequence, this conditional Bayes factor favors models that provide better probabilistic predictions.

We illustrate the utility and flexibility of our Poisson spatio-temporal framework with an application to the number of acquired immunodeficiency syndrome (AIDS) cases in Rio de Janeiro State, Brazil. Specifically, we consider the annual number of cases of AIDS from 1982 to 2007 for each of the 92 counties in Rio de Janeiro State. These data are publicly available from the Ministry of Health of Brazil and may be downloaded from the webpage https://datasus.saude.gov.br/. AIDS is caused by the human immunodeficiency virus (HIV), which has a long incubation period with a median of about 10 years in young adults (Bacchetti and Moss, 1989). Because the incubation period is much longer than the observational time unit of one year, we may expect the observed number of cases in one year to be a convolution of infections that have occurred in several previous years. As a result, the change from year to year in the underlying latent expected number of cases may have a strong spatial dependence. Finally, given the time scale of the incubation period of about one decade and the fact that in Rio de Janeiro the first identified AIDS cases appeared at the beginning of the 1980s, the AIDS epidemic provides an excellent case study for what may happen in terms of spatio-temporal dynamics when a contagious disease emerges in a previously unaffected region.

This chapter is organized as follows: Section 4.2 presents the proposed general class of spatio-temporal Poisson models for epidemic areal data, as well as several useful specific models. Section 4.3 develops simulation-based Bayesian estimation and model selection for these spatio-temporal models. Section 4.4 illustrates our spatio-temporal framework with an application to the number of AIDS cases in the State of Rio de Janeiro, Brazil. Conclusions and possible extensions are presented in Section 4.5.

4.2 Spatio-Temporal Models for Poisson Areal Data

Consider a geographical domain of interest partitioned in a collection of S regions indexed by the integers $1, \ldots, S$ forming a regular or irregular grid. Assume that this grid is endowed with a neighborhood system $\{N_s, s = 1, \ldots, S\}$, where N_s is the set of regions that are neighbors of region s. Let y_{ts} be the number of cases observed at time t in region s, $t = 1, \ldots, T$, $s = 1, \ldots, S$. As usual for count data, assume that the observation y_{ts} follows a Poisson distribution. Specifically, assume that $y_{ts}|\lambda_{ts} \sim Po(n_{ts}\lambda_{ts})$, where n_{ts} denotes the population size and λ_{ts} is the underlying risk at time t in region s, $t = 1, \ldots, T, s = 1, \ldots, S$. Thus, the probability function of y_{ts} conditional on λ_{ts} is

$$p(y_{ts}|\lambda_{ts}) = \frac{(n_{ts}\lambda_{ts})^{y_{ts}} e^{-n_{ts}\lambda_{ts}}}{y_{ts}!}. \tag{4.1}$$

Let $\eta_{ts} = \log(n_{ts}\lambda_{ts})$, $a = 1$, $b(\eta_{ts}) = n_{ts}\lambda_{ts} = \exp(\eta_{ts})$, and $c(y_{ts}) = -\log(y_{ts}!)$. Then, the Poisson probability function can be rewritten as (Casella and Berger, 2001)

$$p(y_{ts}|\eta_{ts}, \upsilon) \quad \propto \quad \exp\{(y_{ts}\eta_{ts} - b(\eta_{ts}))/a + c(y_{ts})\}, \tag{4.2}$$

and thus belongs to the regular exponential family of distributions. Here, the natural parameter is $\eta_{ts} = \log(\mu_{ts})$, where $\mu_{ts} = b'(\eta_{ts}) = exp(\eta_{ts}) = n_{ts}\lambda_{ts}$ is the mean of y_{ts}. Hence, we may rewrite the natural parameter as $\eta_{ts} = \log n_{ts} + \log \lambda_{ts}$.

To model the mean level, generalized linear models assume a monotone differentiable link function $g(.)$ that transforms the mean μ_{ts} from its restricted range to the unrestricted real line (McCullagh and Nelder, 1989). For example, in the case of Poisson data with mean μ a typical choice is the canonical link function $g(\mu) = \log \mu$. In the spatio-temporal case we consider, the canonical link function becomes $g(\mu_{ts}) = \log(\mu_{ts}) = \log n_{ts} + \log \lambda_{ts}$, where $\log n_{ts}$ is a known offset. In this case, we define $\theta_{ts} = \log \lambda_{ts}$ as the value of the latent log-risk random field at time t in region s. Then, we may directly relate θ_{ts} to the natural parameter η_{ts} through the function $\delta_{ts}(.)$ defined as

$$\eta_{ts} \quad = \quad \delta_{ts}(\theta_{ts}) \quad = \quad \theta_{ts} + \log n_{ts}. \tag{4.3}$$

Finally, assuming the canonical link function $g(\mu_{ts}) = \log(\mu_{ts}) = \log n_{ts} + \log \lambda_{ts}$ implies that the so-called response function, i.e. the inverse of the link function, is $f(\theta_{ts}) = n_{ts} \exp(\theta_{ts})$. The response function plays an important role in the construction of the extended forward filter detailed in Algorithm 3.1.

Let $\boldsymbol{\theta}_t = (\theta_{t1}, \ldots, \theta_{tS})'$ be the vectorized latent log-risk random field at time t. We relate $\boldsymbol{\theta}_t$ to a dynamic latent vector $\boldsymbol{\beta}_t$ through the linear equation

$$\boldsymbol{\theta}_t \quad = \quad \boldsymbol{F}_t'\boldsymbol{\beta}_t, \tag{4.4}$$

where $\boldsymbol{F}_t$ is typically known up to some few unknown hyperparameters. Finally, we assume that the latent vector $\boldsymbol{\beta}_t$ evolves through time following the system (also known as evolution or state-space) equation

$$\boldsymbol{\beta}_t \quad = \quad \boldsymbol{G}_t\boldsymbol{\beta}_{t-1} + \boldsymbol{\omega}_t, \qquad \boldsymbol{\omega}_t \sim PGMRF(\mathbf{0}, \boldsymbol{W}_t^{-1}), \tag{4.5}$$

where the evolution matrix $\boldsymbol{G}_t$ is usually known up to a few unknown hyperparameters. Equation (4.5) extends the dynamic generalized linear model system equation of West et al. (1985) to the multivariate spatio-temporal case. Further, $\boldsymbol{\omega}_t$ is a vector of evolution innovations that follows the proper Gaussian Markov random field (PGMRF) process discussed in Section 1.2.3 of Chapter 1 (Ferreira, 2024) with zero mean vector and precision matrix $\boldsymbol{W}_t^{-1}$, that is, the density of $\boldsymbol{\omega}_t$ is proportional to

$$|\boldsymbol{W}_t|^{-0.5} \exp(-0.5\boldsymbol{\omega}_t' \boldsymbol{W}_t^{-1} \boldsymbol{\omega}_t),$$

where the precision matrix is $\boldsymbol{W}_t^{-1} = \tau(\boldsymbol{I} + \phi \boldsymbol{M})$. Here, $\tau > 0$ is a scale parameter, $\phi > 0$ controls the amount of spatial dependence, and $\boldsymbol{I}$ is the identity matrix. In addition, the neighborhood matrix $\boldsymbol{M}$ is such that $(\boldsymbol{M})_{k,l} = m_k$ if $k = l$, $(\boldsymbol{M})_{k,l} = -g_{kl}$ if $k \in N_l$, and $(\boldsymbol{M})_{k,l} = 0$ otherwise, where $g_{kl} > 0$ is a measure of similarity between regions k and l, and the diagonal elements are defined as $m_k = \sum_{l \in N_k} g_{kl}, k = 1, \ldots, S$. For additional details on this type of PGMRFs, see Ferreira and De Oliveira (2007).

4.2.1 Spatio-Temporal Models for Epidemics

This section discusses several important specific models within the general class of models defined by Equations (4.1), (4.3), (4.4), and (4.5) that are useful for spatio-temporal modeling of epidemics. Epidemic processes of emerging diseases, as is the case of the AIDS data that we consider, are usually nonstationary at the start of the epidemic. Thus, here we consider five distinct nonstationary state-space equation specifications. Further, for each of these five specifications, we consider two distinct state-space covariance matrices, comprising a total of ten models. The first covariance matrix is diagonal, and the second covariance matrix is defined (as discussed above) as the covariance matrix of a PGMRF.

The first state-space equation specification we consider is a first-order temporal trend evolution. Specifically, the first-order temporal trend evolution assumes $\boldsymbol{F}_t' = \boldsymbol{I}_S$ and $\boldsymbol{G}_t = \boldsymbol{I}_S$. This matrix $\boldsymbol{F}_t$ implies that β_{ts} is the log of the risk at time t for each person in subregion s. Thus, in this specification $\boldsymbol{\beta}_t = (\beta_{t1}, \ldots, \beta_{tS})'$ is the log-risk level field at time t. When the state-space precision matrix is the PGMRF matrix $\boldsymbol{W}_t^{-1} = \tau(\boldsymbol{I}_S + \phi \boldsymbol{M})$, the implied expected value of β_{ts}, $t = 2, \ldots, T-1$, conditional on the entire latent process is $0.5(\beta_{t-1,s} + \beta_{t+1,s}) + m_s^{-1} \sum_{k \in N_s} \{\beta_{tk} - 0.5(\beta_{t-1,k} + \beta_{t+1,k})\}$. Hence, the conditional mean of β_{ts} depends on the latent process in region s at times $t-1$ and $t+1$, as well as the latent process in the spatial neighbors of region s at times $t-1$, t, and $t+1$. Thus, the first-order temporal trend evolution may be used for spatio-temporal smoothing. Therefore, this state-space specification may be useful after the epidemic ends its expansion phase and the disease becomes endemic.

The second state-space specification incorporates spatio-temporal contamination. Specifically, evolution with contamination assumes $\boldsymbol{F}_t' = \boldsymbol{I}_S$, and a contamination matrix $\boldsymbol{G}_t = (1 + \kappa h)^{-1} \boldsymbol{H}$, with $\{\boldsymbol{H}\}_{kl} = 1$ for $k = l$,

$\{\boldsymbol{H}\}_{kl} = \kappa$ for $k \in C_l$, and $\{\boldsymbol{H}\}_{kl} = 0$ otherwise. Here, C_l is the set of across-time neighbors of subregion l, $h = \max_k \mathbf{1}(k \in C_l)$ is the maximum number of across-time neighbors, and $\kappa > 0$ is a contamination coefficient. Note that for full generality the set of across-time neighbors C_l may differ from the set of time-specific neighbors N_l. This contamination specification allows for abnormal increases in a region at a given time point to spill over to neighboring regions at subsequent time points. Thus, we expect this contamination state-space specification to be useful during the initial periods of a disease emerging in a previously unaffected region.

The third state-space specification assumes a second-order temporal trend. The second-order temporal trend model assumes that $\boldsymbol{\beta}_t$ is comprised of two latent fields at time t, $\boldsymbol{\beta}_{1t} = (\beta_{1t1}, \ldots, \beta_{1tS})'$ and $\boldsymbol{\beta}_{2t} = (\beta_{2t1}, \ldots, \beta_{2tS})'$. Hence, this model assumes $\boldsymbol{\beta}_t = (\boldsymbol{\beta}_{1t}', \boldsymbol{\beta}_{2t}')'$. In this specification, $\boldsymbol{\beta}_{1t}$ is equal to $\boldsymbol{\theta}_t$, the log-risk level field at time t. Furthermore, $\boldsymbol{\beta}_{2t}$ is the gradient field, that is, the expected increase in the log-risk level field from time t to time $t+1$. Specifically, the second-order temporal trend model assumes $\boldsymbol{F}_t' = (\boldsymbol{I}_S, \mathbf{0}_{SS})$, and evolution matrix

$$\boldsymbol{G}_t = \begin{pmatrix} \boldsymbol{G}_{1t} & \boldsymbol{G}_{1t} \\ \mathbf{0}_{SS} & \boldsymbol{G}_{2t} \end{pmatrix},$$

where $\boldsymbol{G}_{it} = \boldsymbol{I}_S$, $i = 1, 2$, and $\mathbf{0}_{SS}$ is an $S \times S$ matrix of zeros. Further, in the case of spatially correlated state-space innovations, we assume that $\boldsymbol{W}_t^{-1}$ is block diagonal, that is, $\boldsymbol{W}_t^{-1} = \text{blockdiag}(\boldsymbol{W}_{1t}^{-1}, \boldsymbol{W}_{2t}^{-1})$, where $\boldsymbol{W}_{it}^{-1} = \tau_i(\boldsymbol{I}_S + \phi_i \boldsymbol{M})$, $i = 1, 2$. The second-order temporal trend model is appropriate when the temporal trend for each region can be approximated by a local linear trend that changes smoothly over time. Therefore, this model may be useful for describing the spatio-temporal evolution of the early stages of an emerging disease.

The fourth state-space specification assumes a log-risk level field and a common gradient for all counties at each time point. The common gradient at time t is a univariate parameter β_{2t} and evolves through time according to the evolution equation $\beta_{2t} = \beta_{2,t-1} + \omega_{2t}$, where $\omega_{2t} \sim N(0, \psi^{-1})$ and ψ is the evolution precision for the common gradient. For this state-space specification, $\boldsymbol{\beta}_t = (\boldsymbol{\beta}_{1t}', \beta_{2t})'$, $\boldsymbol{F}_t' = (\boldsymbol{I}_S, \mathbf{0}_S)$, and the evolution matrix is

$$\boldsymbol{G}_t = \begin{pmatrix} \boldsymbol{I}_S & \mathbf{1}_S \\ \mathbf{0}_S' & 1 \end{pmatrix},$$

where $\mathbf{0}_S$ and $\mathbf{1}_S$ are S-dimensional vectors of zeros and ones, respectively. For the fourth state-space specification, the state-space precision matrix is

$$\boldsymbol{W}_t^{-1} = \begin{pmatrix} \boldsymbol{W}_{1t}^{-1} & \mathbf{0}_S \\ \mathbf{0}_S' & \psi^{-1} \end{pmatrix}.$$

Finally, the fifth state-space specification assumes a contamination field model for the log-risk combined with a common gradient for all counties at each time point. Hence, almost all the components of the fifth state-space specification are the same as the components of the fourth state-space specification, except that the evolution matrix includes a contamination matrix and is given by

$$\boldsymbol{G}_t = \begin{pmatrix} (1+\kappa h)^{-1}\boldsymbol{H} & \mathbf{1}_S \\ \mathbf{0}'_S & 1 \end{pmatrix},$$

where κ, h, and $\boldsymbol{H}$ are defined as in the second state-space specification described above.

The models we describe above by no means exhaust the general class of models defined by Equations (4.1), (4.3), (4.4), and (4.5). In particular, many other models may be obtained as stochastic discretized versions of deterministic continuous time-space mathematical models. For example, these deterministic continuous time-space mathematical models may be based on partial differential equations and on integro-difference equations. For details on several possible deterministic continuous time-space mathematical models and how to discretize them, see Wikle and Hooten (2010), Cressie and Wikle (2011), and references therein.

4.3 Bayesian Inference

This section develops simulation-based methods to perform full Bayesian statistical analysis for the class of spatio-temporal models defined by Equations (4.1), (4.3), (4.4), and (4.5). Specifically, we construct a Markov chain Monte Carlo (MCMC) algorithm (Gelfand and Smith, 1990; Robert and Casella, 2004; Gamerman and Lopes, 2006) to obtain samples from an approximate posterior distribution. Our computational methodology is general enough to be applied to observations with any distribution in the regular exponential family. However, here we focus on Poisson observations.

Let $\boldsymbol{\beta}_{0:T} = (\boldsymbol{\beta}'_0, \ldots, \boldsymbol{\beta}'_T)'$ be the vectorized latent process collected through time, with similar notation for other quantities of the model. In addition, let $\boldsymbol{\psi}$ be the vector of unknown hyperparameters. Further, let $\boldsymbol{D}_t$, $t = 1, \ldots, T$, represent all the information up to time t. Thus, $\boldsymbol{D}_t$ is recursively defined as $\boldsymbol{D}_t = \boldsymbol{D}_{t-1} \cup \{\boldsymbol{y}_t\}$. Then, by Bayes Theorem the exact posterior density for $(\boldsymbol{\beta}_{0:T}, \boldsymbol{\psi})$ is proportional to

$$p(\boldsymbol{\beta}_{0:T}, \boldsymbol{\psi}|\boldsymbol{D}_T) \propto \left\{\prod_{t=1}^{T} p\left(\boldsymbol{y}_t|\boldsymbol{\theta}_t, \boldsymbol{\eta}_t\right)\right\} \left\{\prod_{t=1}^{T} p\left(\boldsymbol{\beta}_t|\boldsymbol{\beta}_{t-1}, \boldsymbol{\psi}\right)\right\} p\left(\boldsymbol{\beta}_0|\boldsymbol{\psi}\right) p(\boldsymbol{\psi}).$$

The Markov chain used in our procedure has to be tailored to each specific spatio-temporal model and will depend on how the hyperparameter vector $\boldsymbol{\psi}$ appears in the matrices $\boldsymbol{F}_t, \boldsymbol{G}_t$ and $\boldsymbol{W}_t$. However, the Markov chain may be partitioned in two blocks: simulation of $\boldsymbol{\psi}$, which is model specific, and simulation of the latent process $(\boldsymbol{\beta}_0, \ldots, \boldsymbol{\beta}_T)$. For the simulation of the latent process, we propose in Section 4.3.1 a novel extended forward filter backward sampler (EFFBS).

4.3.1 Extended Forward Filter Backward Sampler

This section introduces a novel extended forward filter backward sampler for the simulation of the latent process $(\boldsymbol{\beta}_0, \ldots, \boldsymbol{\beta}_T)$. The EFFBS we propose combines the extended Kalman filter (p. 352, Fahrmeir and Tutz, 2001) and the forward filter backward sampler (Frühwirth-Schnatter, 1994; Carter and Kohn, 1994). Hence, the EFFBS is composed of two subsequent steps, that we describe below as the extended forward filter in Algorithm 3.1 and the backward sampler in Algorithm 3.2.

Let $\widehat{\boldsymbol{\theta}}_t = (\widehat{\theta}_{t1}, \ldots, \widehat{\theta}_{tS})'$ be a preliminary estimate of $\boldsymbol{\theta}_t$. This preliminary estimate may, for example, be the posterior mode obtained by the extended Kalman filter. In addition, let $f'(\theta_{ts}) = n_{ts}e^{\theta_{ts}}$ denote the first derivative of the response function $f(\theta_{ts})$. Furthermore, let $\Sigma_{ts} = Var(y_{ts}) = n_{ts}e^{\theta_{ts}}$ be the variance of the number of disease cases at time t and region s. Then, the extended forward filter proceeds as follows.

Algorithm 4.3.1 (Extended forward filter) *Initiate the algorithm at time $t = 0$ with the distribution $\boldsymbol{\beta}_0|\boldsymbol{\psi}, \boldsymbol{D}_0 \sim N(\boldsymbol{m}_0, \boldsymbol{C}_0)$. Then, for $t = 1, \ldots, T$, do:*

1. *Prior at t: $\boldsymbol{\beta}_t|\boldsymbol{\psi}, \boldsymbol{D}_{t-1} \sim N(\boldsymbol{a}_t, \boldsymbol{R}_t)$, with*

$$\begin{aligned} \boldsymbol{a}_t &= \boldsymbol{G}_t \boldsymbol{m}_{t-1}, \\ \boldsymbol{R}_t &= \boldsymbol{G}_t \boldsymbol{C}_{t-1} \boldsymbol{G}_t' + \boldsymbol{W}_t. \end{aligned}$$

2. *Based on the first-order Taylor expansion of the response function f around $\widehat{\boldsymbol{\theta}}_t = \boldsymbol{F}_t' \boldsymbol{a}_t$, compute the artificial observation vector $\widehat{\boldsymbol{y}}_t = (\widehat{y}_{t1}, \ldots, \widehat{y}_{tS})$ with*

$$\widehat{y}_{ts} = \left\{ f'(\widehat{\theta}_{ts}) \right\}^{-1} \left\{ y_{ts} - f(\widehat{\theta}_{ts}) \right\} + \widehat{\theta}_{ts}, \tag{4.6}$$

and the approximate precision matrix

$$\widehat{\boldsymbol{V}}_t^{-1} = \operatorname{diag} \left[\frac{\left\{ f'\left(\widehat{\theta}_{ts}\right) \right\}^2}{\widehat{\Sigma}_{ts}} \right]. \tag{4.7}$$

3. *Posterior at t: $\boldsymbol{\beta}_t|\boldsymbol{\psi}, \boldsymbol{D}_t \sim N(\boldsymbol{m}_t, \boldsymbol{C}_t)$, with*

$$\boldsymbol{C}_t = \left(\boldsymbol{R}_t^{-1} + \boldsymbol{F}_t \widehat{\boldsymbol{V}}_t^{-1} \boldsymbol{F}_t' \right)^{-1}, \tag{4.8}$$

$$\boldsymbol{m}_t = \boldsymbol{C}_t \left(\boldsymbol{F}_t \widehat{\boldsymbol{V}}_t^{-1} \widehat{\boldsymbol{y}}_t + \boldsymbol{R}_t^{-1} \boldsymbol{a}_t \right). \tag{4.9}$$

The extended forward filter is an approximate method to calculate the mean vectors and covariance matrices of the posterior distributions $p(\boldsymbol{\beta}_t|\boldsymbol{\psi}, \boldsymbol{D}_t)$, $t = 1, \ldots, T$. At the end of the extended forward filter, we have an approximation for the posterior distribution of the latent process at the last time point T given all the available information as well as the hyperparameter vector $\boldsymbol{\psi}$, that is, $p(\boldsymbol{\beta}_T|\boldsymbol{\psi}, \boldsymbol{D}_T)$. Then, the backward sampler proceeds as follows.

Algorithm 4.3.2 (Backward sampler) *Sampling from $\boldsymbol{\beta}_{1:T}|\boldsymbol{\psi}, \boldsymbol{D}_t$:*

1. Sample $\boldsymbol{\beta}_T^*$ from the distribution $N(\boldsymbol{m}_T, \boldsymbol{C}_T)$, obtained from the extended forward filter.

2. For $t = T-1, \ldots, 0$, sample backwards $\boldsymbol{\beta}_t^*$ from the conditional distribution $N(\boldsymbol{b}_t, \boldsymbol{B}_t)$, where

$$\begin{aligned} \boldsymbol{B}_t &= \left(\boldsymbol{C}_t^{-1} + \boldsymbol{G}_{t+1}' \boldsymbol{W}_{t+1}^{-1} \boldsymbol{G}_{t+1}\right)^{-1}, \quad (4.10) \\ \boldsymbol{b}_t &= \boldsymbol{B}_t \left(\boldsymbol{G}_{t+1}' \boldsymbol{W}_{t+1}^{-1} \boldsymbol{\beta}_{t+1}^* + \boldsymbol{C}_t^{-1} \boldsymbol{m}_t\right). \quad (4.11) \end{aligned}$$

The EFFBS assumes an approximate Gaussian observational equation with artificial observation vector defined in (4.6) and approximate precision matrix given by (4.7). This implies that the EFFBS implicitly assumes the approximate likelihood function

$$k\left(\boldsymbol{y}_{1:T}|\boldsymbol{\theta}_{1:T}, \boldsymbol{\psi}\right) = \prod_{t=1}^{T} k\left(\boldsymbol{y}_t|\boldsymbol{\theta}_t, \boldsymbol{\psi}\right), \quad (4.12)$$

where $k\left(\boldsymbol{y}_t|\boldsymbol{\theta}_t, \boldsymbol{\psi}\right)$ is proportional to

$$\left(\prod_{s=1}^{S} \widehat{\Sigma}_{ts}\right)^{-1/2} \exp\left[-0.5\sum_{s=1}^{S}(n_{ts}e^{\widehat{\theta}_{ts}})^{-1}\left\{y_{ts} - n_{ts}e^{\widehat{\theta}_{ts}} + n_{ts}e^{\widehat{\theta}_{ts}}\left(\widehat{\theta}_{ts} - (\boldsymbol{F}_t'\boldsymbol{\beta}_t)_s\right)\right\}^2\right].$$

Therefore, the MCMC algorithm with the embedded EFFBS defines a Markov chain that converges to an approximate posterior distribution with density

$$k(\boldsymbol{\beta}_{0:T}, \boldsymbol{\psi}|\boldsymbol{D}_T) \propto \left\{\prod_{t=1}^{T} k\left(\boldsymbol{y}_t|\boldsymbol{\theta}_t, \boldsymbol{\psi}\right)\right\}\left\{\prod_{t=1}^{T} p\left(\boldsymbol{\beta}_t|\boldsymbol{\beta}_{t-1}, \boldsymbol{\psi}\right)\right\} p\left(\boldsymbol{\beta}_0|\boldsymbol{\psi}\right) p(\boldsymbol{\psi}).$$

Finally, at the end of the MCMC algorithm we have a sample $\left(\boldsymbol{\psi}^{(1)}, \boldsymbol{\beta}_{0:T}^{(1)}\right), \ldots, \left(\boldsymbol{\psi}^{(G)}, \boldsymbol{\beta}_{0:T}^{(G)}\right)$ from this approximate posterior distribution.

4.3.2 Model Selection

With the ability to fit several nonnested spatio-temporal models comes the need to select among these models. In addition, when analyzing a particular dataset, each of the possible models corresponds to a distinct scientific hypothesis. As such, comparison of the different models becomes paramount for understanding the nature of the process underlying the observed data. For the comparison of the nonnested spatio-temporal models introduced in Section 4.2, we develop a conditional Bayes factor (p. 190, Ghosh et al., 2006) that uses the output of the MCMC algorithm for the approximate posterior distribution. As we discuss below, this conditional Bayes factor provides a model selection criterion that is justified from a Bayesian decision-theoretic point of view (Berger, 1985; Wu et al., 2016, 2020; Williams et al., 2023).

Bayesian model selection is usually performed by comparing the posterior probabilities of the competing models. When the competing models have equal prior probabilities, their posterior probabilities are proportional to their respective predictive densities (also known as integrated likelihoods). A potential difficulty in Bayesian model selection is that these predictive densities will be sensitive to the specification of the prior distribution for the hyperparameter vector $\boldsymbol{\psi}$. To overcome this difficulty, we use here a training sample approach (Frühwirth-Schnatter, 1995) of the first t^* time observations; this results in calibrated priors for the parameters of each model. Then, Monte Carlo integration is used to compute the predictive distribution under each model for the remaining $T - t^*$ time observations.

Suppose there are Q competing spatio-temporal models denoted by $M_1, \ldots, M_Q$. The qth model has observational density $p_q(\boldsymbol{y}_t|\boldsymbol{\beta}_t, \boldsymbol{\Psi})$ and evolution density $p_q(\boldsymbol{\beta}_t|\boldsymbol{\beta}_{t-1}, \boldsymbol{\Psi})$. Let $p_q(\boldsymbol{\beta}_{0:(t-1)}, \boldsymbol{\Psi}|\boldsymbol{D}_{t-1})$ denote the joint posterior distribution of $\boldsymbol{\Psi}$ and $\beta_0, \ldots, \boldsymbol{\beta}_{t-1}$ under the qth model given the information up to time $t-1$. Note that the definitions of the hyperparameter vector $\boldsymbol{\Psi}$ and the latent vector $\boldsymbol{\beta}_t$ may be (and in general will be) different under each model, but this distinction is omitted here to keep the notation simple. Then, under model q the one-step ahead predictive density of $\boldsymbol{y}_t$ given the information up to time $t-1$ is

$$p_q(\boldsymbol{y}_t|\boldsymbol{D}_{t-1}) = \int p_q(\boldsymbol{y}_t|\boldsymbol{\beta}_t, \boldsymbol{\Psi}) p_q(\boldsymbol{\beta}_t|\boldsymbol{\beta}_{t-1}, \boldsymbol{\Psi}) \times p_q(\boldsymbol{\beta}_{0:(t-1)}, \boldsymbol{\Psi}|\boldsymbol{D}_{t-1}) d\boldsymbol{\beta}_{0:(t-1)} d\boldsymbol{\beta}_t d\boldsymbol{\Psi}. \quad (4.13)$$

We apply the simulation scheme outlined in Section 4.3 to obtain G posterior draws

$$\left[\left(\boldsymbol{\beta}_{0:(t-1)}^{(1)}, \boldsymbol{\Psi}^{(1)}\right), \left(\boldsymbol{\beta}_{0:(t-1)}^{(2)}, \boldsymbol{\Psi}^{(2)}\right), \ldots, \left(\boldsymbol{\beta}_{0:(t-1)}^{(G)}, \boldsymbol{\Psi}^{(G)}\right)\right].$$

Next, for $g = 1, \ldots, G$, we simulate $\boldsymbol{\beta}_t^{(g)}$ given $(\boldsymbol{\beta}_{0:(t-1)}^{(g)}, \boldsymbol{\Psi}^{(g)})$ from the evolution distribution with density $p_q(\boldsymbol{\beta}_t|\boldsymbol{\beta}_{0:(t-1)}^{(g)}, \boldsymbol{\Psi}^{(g)})$. This yields the posterior sample $[(\boldsymbol{\beta}_{0:t}^{(1)}, \boldsymbol{\Psi}^{(1)}), \ldots, (\boldsymbol{\beta}_{0:t}^{(G)}, \boldsymbol{\Psi}^{(G)})]$.

Now rewrite Equation (4.13) as

$$p_q(\boldsymbol{y}_t|\boldsymbol{D}_{t-1}) = \int p_q(\boldsymbol{y}_t|\boldsymbol{\beta}_t, \boldsymbol{\Psi}) p_q(\boldsymbol{\beta}_{0:t}, \boldsymbol{\Psi}|\boldsymbol{D}_{t-1}) d\boldsymbol{\beta}_{0:t} d\boldsymbol{\Psi}. \tag{4.14}$$

Equation (4.14) shows that $p_q(\boldsymbol{y}_t|\boldsymbol{D}_{t-1})$ is the expectation of $p_q(\boldsymbol{y}_t|\boldsymbol{\beta}_t, \boldsymbol{\Psi})$ with respect to $p_q(\boldsymbol{\beta}_{0:t}, \boldsymbol{\Psi}|\boldsymbol{D}_{t-1})$. Therefore, we can estimate the one-step ahead predictive density $p_q(\boldsymbol{y}_t|\boldsymbol{D}_{t-1})$ with

$$\widehat{p}_q(\boldsymbol{y}_t|\boldsymbol{D}_{t-1}) = \frac{1}{G}\sum_{g=1}^{G} p_q(\boldsymbol{y}_t|\boldsymbol{\beta}_t^{(g)}, \boldsymbol{\Psi}^{(g)}). \tag{4.15}$$

Because the joint predictive density of $\boldsymbol{y}_{t^*+1}, \ldots, \boldsymbol{y}_T$ can be written as $p_q(\boldsymbol{y}_{t^*+1}, \ldots, \boldsymbol{y}_T|\boldsymbol{D}_{t^*}) = \prod_{t=t^*+1}^{T} p_q(\mathbf{y}_t|\boldsymbol{D}_{t-1})$, and as $\boldsymbol{D}_t = \boldsymbol{D}_{t-1} \cup \{\boldsymbol{y}_t\}$, an estimate of this density under model q is (Vivar and Ferreira, 2009)

$$\widehat{p}_q(\boldsymbol{y}_{t^*+1}, \ldots, \boldsymbol{y}_T|\boldsymbol{D}_{t^*}) = \prod_{t=t^*+1}^{T} \widehat{p}_q(\boldsymbol{y}_t|\boldsymbol{D}_{t-1}), \tag{4.16}$$

where t^* is such that $p_q(\boldsymbol{\psi}|\boldsymbol{D}_{t^*})$ is proper for all $q = 1, \ldots, Q$. For a detailed discussion on the choice of t^*, see Vivar and Ferreira (2009). Then, based on the joint predictive density (4.16), for each pair of models we compute the conditional Bayes factor of model m against model n as

$$B_{mn} = \frac{\hat{p}_m(\boldsymbol{y}_{t^*+1}, \ldots, \boldsymbol{y}_T|\boldsymbol{D}_{t^*})}{\hat{p}_n(\boldsymbol{y}_{t^*+1}, \ldots, \boldsymbol{y}_T|\boldsymbol{D}_{t^*})}.$$

Finally, we use the conditional Bayes factors to decide what is the best model. In addition to being well justified theoretically, these conditional Bayes factors favor models that provide better probabilistic predictions.

4.4 Application

We illustrate the utility and flexibility of our Poisson spatio-temporal framework with an application to the number of acquired immunodeficiency syndrome (AIDS) cases in Rio de Janeiro State, Brazil. Specifically, we consider the annual number of new cases of AIDS from 1982 to 2007 for each of the 92 counties in Rio de Janeiro State. Data on the annual number of new cases of AIDS and on population size are publicly available from the Ministry of Health of Brazil and may be downloaded from the webpage https://datasus.saude.gov.br/. From 1982 to 2007 new counties were created in Rio de Janeiro State. The data available on https://datasus.saude.gov.br/ about the number of new cases already use the geopolitical organization of

Rio de Janeiro State as of 2007. However, for previous years the data on population sizes are available based on each year's geopolitical organization. We performed backcasting to obtain estimates of annual population sizes per county using the geopolitical organization of Rio de Janeiro State as of 2007. The analysis below is conditional on these county population size estimates.

AIDS is caused by the human immunodeficiency virus (HIV), which has a long incubation period with a median of about 10 years in young adults (Bacchetti and Moss, 1989). Because the incubation period is much longer than the observational time unit of one year, we may expect the observed number of cases in one year to be a convolution of infections that have occurred in several previous years. As a result, the change from year to year in the underlying risk of observing new cases may exhibit a strong spatial dependence. This spatial dependence is accounted for in our modeling framework through the system equation innovation $\boldsymbol{\omega}_t$ and its covariance matrix $\boldsymbol{W}_t$. Finally, given the time scale of the incubation period of about one decade and the fact that the first cases of AIDS in Rio de Janeiro were identified at the beginning of the 1980s, the AIDS epidemic provides an excellent case study of what may happen in terms of spatio-temporal dynamics when a previously unaffected region becomes infected by an emerging infectious disease.

We have implemented the Bayesian analysis methodology developed in Section 4.3 for each of the five spatio-temporal models described in Section 4.2.1. Specifically, we consider the following models with spatially dependent innovations: I – First-order polynomial field model; II – Contamination field model; III - Second-order polynomial field model; IV - Model with a log-risk level field and a common gradient for all counties at each time point; V - Contamination field model for the log-risk combined with a common gradient for all counties at each time point. Further, we have considered for each of these five models two specifications for the system equation innovations: spatial dependence, and spatial independence.

A Bayesian analysis requires the specification of priors both for the hyperparameters and for the latent process $\boldsymbol{\beta}_0$ at time $t = 0$ given the information up to time 0 denoted by $\boldsymbol{D}_0$. For the specification of the prior for $\boldsymbol{\beta}_0$ for an emerging disease, we may use the information about the type of disease that we anticipate to observe. In the case of slowly progressing emerging epidemics such as in the case of AIDS in the beginning of the 1980s, we expect the standardized risk for each county of the first case to appear at a given time to be about the reciprocal of the state population size. In the case of Rio de Janeiro State, the population size in 1981 (t=0) was about 11.5 million people. Hence, this corresponds to an expected logarithm of the risk equal to $E(\boldsymbol{F}'\boldsymbol{\beta}_0|\boldsymbol{D}_0) = \mathbf{1}_S \; log(1/11,500,000) \approx -16.25 \; \mathbf{1}_S$. In addition, for an emerging disease we believe that the risk of the first case to appear at a given time to be within 20-fold of its expectation with probability 0.997. Further, we assume that before the disease emerges the standardized risk is stochastically independent between counties. These assumptions correspond in the logarithm scale to assuming $Var(\boldsymbol{F}'\boldsymbol{\beta}_0|\boldsymbol{D}_0) = \boldsymbol{I}_S$.

When assigning priors for the hyperparameters, we follow closely the recommendations of Vivar and Ferreira (2009). Specifically, we assume for the scale parameter τ_i a weakly informative $Gamma(1,1)$ prior; this prior imparts little information and concurrently guarantees posterior propriety. In addition, for the contamination coefficient κ we assume a noninformative uniform prior on the interval $(0,1)$. Further, for the spatial dependence parameter ϕ_i we assign a prior of the type $\Pi(\phi_i) \propto 1$ if $0 < \phi_i < 1$, $\Pi(\phi_i) \propto \phi_i^{-2}$ if $1 < \phi_i < a$, and $\Pi(\phi_i) \propto 0$ otherwise. The tail behavior of this prior is the same as the tail behavior of the marginal reference prior for ϕ_i proposed by Ferreira and De Oliveira (2007) for PGMRFs, except that we truncate the prior at a so that the posterior mean and variance are finite. In particular, we have found that $a = 100$ works well in our current application.

Finally, for Models IV and V that include a common gradient, we assign a conditionally conjugate prior $Gamma(a_\psi, b_\psi)$ for the log-gradient evolution precision ψ. We choose a_ψ and b_ψ in an informative manner using the following rationality. First, to obtain a reasonably vague prior we assume $b_\psi = 0.1$ that implies the prior variance for ψ to be 10 times larger than the prior mean. Moreover, we expect the common gradient of the risk to be reasonably stable at subsequent time points. In particular, we expect the gradient to do not vary by much more than 20% between subsequent time points. This informs us that in the logarithm scale the evolution standard deviation $\psi^{-1/2}$ is probably less than 0.1; hence, we assign for this event a prior probability of about 0.95. This probability, together with assuming $b_\psi = 0.1$ implies $a_\psi = 16$.

With respect to the MCMC algorithm, we ran two parallel chains, with distant initial starting points, for a total of 20,000 iterations for each chain, discarding the first 10,000 iterations for burn-in. Thus, inference was based on the last 10,000 iterations from each chain, for a total of 20,000 iterations. Convergence has been verified through the Gelman and Rubin convergence diagnostics (Gelman and Rubin, 1992), implemented in the CODA package of the R statistical software. Gelman and Rubin convergence diagnostics indicated convergence for each of the hyperparameters and for a sample of the latent process parameters.

We have applied the model selection approach developed in Section 4.3.2 to compare the five models. Here we use the observations at the first $t^* = 9$ time points as training sample for the computation of the conditional Bayes factors. Table 4.1 presents the conditional Bayes factors taking Model I as the baseline. A first feature clearly shown by this table is that, when compared with their spatially independent innovations counterparts, models with spatially dependent innovations (SDI) are better supported by the data. This supports the claim that the long incubation period of the HIV (Bacchetti and Moss, 1989) together with the annual frequency of the data would lead to spatially correlated field innovations. Thus, the data provide evidence of the usefulness of spatially dependent innovations to account for the fact that an increase in the risk of new cases of AIDS in a given year may reflect the convolution of HIV infections in several previous years.

TABLE 4.1
AIDS Data – Logarithm of conditional Bayes factor (BF) (Baseline: Model I with spatially dependent innovations). SDI: spatially dependent innovations.

SDI	Model				
	I	II	III	IV	V
Yes	0	44.7	-1474.7	79.3	**84.7**
No	-3932.0	-1681.2	-2809.0	26.1	26.0

Thus, we now focus on the comparison of models with spatially dependent innovations. With a conditional Bayes factor equal to 84.7, the best is Model V that includes a contamination field together with a common gradient for all counties. In addition, Model IV has a conditional Bayes factor somehow lower equal to 79.3. Model IV includes a first-order level field and a common gradient for all counties. Furthermore, Model III that is a second-order field model with a gradient field, that is a gradient for each county, has a very small Bayes factor equal to -1474.7. Thus, the data does not support the hypothesis of a distinct gradient for each county. Therefore, the data provides evidence of the usefulness of the inclusion of both a contamination evolution for the log-risk level and a common gradient for all counties.

Table 4.2 provides posterior means and 95% credible intervals for the hyperparameters of the best model, Model V. A first feature that emerges from the table is that the contamination coefficient κ is most probably small with a posterior mean equal to 0.001 and a 95% credible interval given by (0.0005, 0.0014). Even though κ seems to be small, the logarithm of the conditional Bayes factor of Model V against the corresponding model without contamination, that is Model IV, is equal to $84.7 - 79.3 = 5.4$ and gives evidence in favor of including the contamination component in the model. Another feature that emerges is that ϕ, the parameter that controls the degree of spatial dependence of the innovations, is also small with a posterior mean equal to 0.50 and a 95% credible interval given by (0.15, 1.14). This small value for ϕ has to be

TABLE 4.2
Posterior summaries for the parameters of Model V that includes spatially dependent innovations with contamination field and a common gradient for all counties.

Parameter	Mean	95% Credible Interval
κ	0.001	(0.0005, 0.0014)
τ	7.16	(3.67, 11.92)
ϕ	0.50	(0.15, 1.14)
ψ	136.7	(79.2, 211.2)

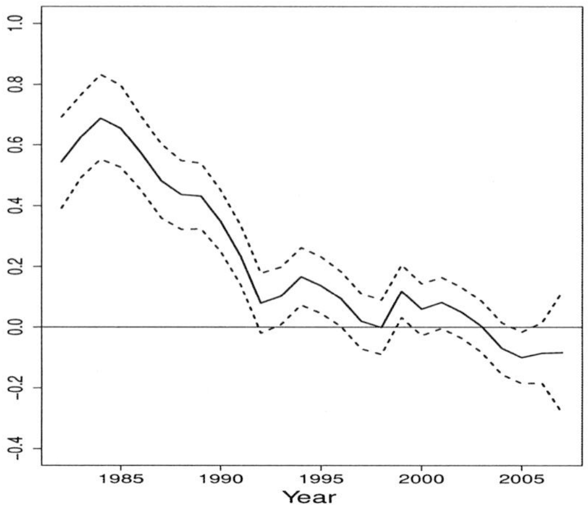

FIGURE 4.1
AIDS data – Common gradient for all counties for Model V. Posterior mean (solid line) and 95% credible interval (dashed lines).

put in perspective by computing the conditional Bayes factor of Model V with spatially dependent against spatially independent innovations, that is equal to $84.7 - 26.0 = 58.7$. Hence, the data supports the hypothesis of spatially dependent innovations. Finally, the precision parameter ψ for the evolution of the common gradient has a posterior mean equal to 136.7 and a 95% credible interval given by (79.2, 211.2), implying that the common gradient evolves through time with small changes between subsequent time points.

Figure 4.1 presents the plot of the common gradient of the log-risk level for all counties through time for Model V. Specifically, the figure presents the posterior mean (solid line) and 95% credible interval (dashed lines). This figure is quite informative and clearly shows that there was a rapid expansion of the epidemic during the 1980s. Even though this rapid expansion was followed by a sharp reduction in the gradient in the beginning of the 1990s, the gradient continued to be small but mostly positive until about 2001. Finally, from 2003 to 2007 the gradient was mostly negative.

Figure 4.2 shows for Model V the observed standardized ratio of new cases (left) and the posterior mean of the risk (right) per 100,000 inhabitants for the years 1982, 1988, 1994, 2000, and 2006. The maps show the level of risk in a scale from white (low risk) to dark gray (high risk). From a methodological perspective, our procedure accomplishes a nice balance between spatial and temporal smoothing resulting from the fact that Model V takes into account the information from spatial neighbors and from across-time neighbors. From an epidemiological perspective, two relevant features emerge from Figure 4.2.

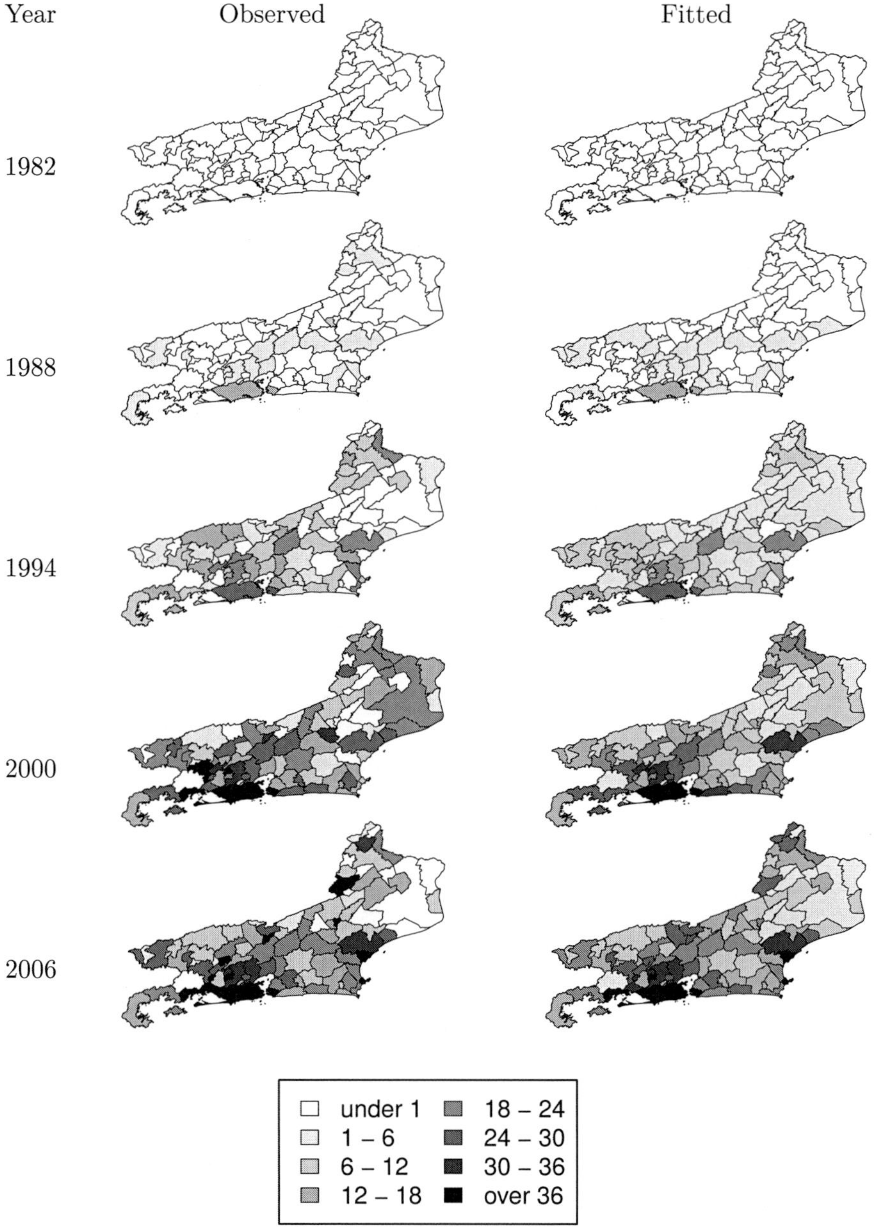

FIGURE 4.2
AIDS data – Model V. Standardized ratio of new cases per 100,000 inhabitants: observed (left panels) and fitted (right panels).

First, corroborating the findings from Figure 4.1, the rate of new cases increased substantially from the beginning of the 1980s to the end of the 1990s, and somehow stabilized from 2000 to 2006. Second, the rate of new cases is substantially higher in the metropolitan region of the City of Rio de Janeiro, located in the center-south of Rio de Janeiro State. Findings such as these may be used by departments of health to decide how to allocate resources both for the treatment and prevention of AIDS.

4.5 Conclusions

We have presented a class of spatio-temporal models for Poisson areal data that are useful for the statistical modeling of epidemics. Further, we have developed several specific models for count data with several distinct specifications of the evolution system equation to allow different spatio-temporal process behaviors with time-specific spatially correlated innovations. In addition, we have developed simulation-based full Bayesian analysis for both parameter estimation and model selection.

We have applied our spatio-temporal framework to analyze the number of new cases of AIDS per county in the State of Rio de Janeiro. Our spatio-temporal modeling framework has achieved a nice balance between spatial and temporal smoothing. In particular, our analysis makes it clear that in the period of study, the rate of new cases of AIDS was substantially higher in the metropolitan region of the City of Rio de Janeiro, and that the rate of new cases seems to have stabilized between 2003 and 2006.

From a modeling point of view, our general class of models may be tailored to specific applications. For example, when monitoring a particular disease, information on the natural history of the disease such as incubation time and rate of infection could potentially be incorporated into the spatio-temporal model. In one possible specification, parameters of the spatio-temporal model could be assumed to be functions of parameters of the natural history. In an alternative specification, the natural history parameters could be incorporated in the priors for the spatio-temporal model parameters. Finally, even though we have focused here on the analysis of Poisson data, our methodology is general and can be applied to observations with any distribution in the regular exponential family. This may be useful for the analysis of spatio-temporal gamma-distributed survival data, and binomial-distributed survey data.

DISCLAIMER: Although Juan C. Vivar is an FDA/CDER employee, this work was not done as part of his official duties. This publication reflects the views of the authors and should not be construed to reflect the FDA/CDER's views or policies.

Bibliography

Alves, M. B., D. Gamerman, and M. A. R. Ferreira (2010). Transfer functions in dynamic generalized linear models. *Statistical Modelling 10*, 03–40.

Anderson, R. M. and R. M. May (1992). *Infectious Diseases of Humans: Dynamics and Control.* New York: Oxford University Press.

Apostolopoulos, Y. and S. Sönmez (2007). Demographic and epidemiological perspectives of human movement. In Y. Apostolopoulos and S. Sönmez (Eds.), *Population Mobility and Infectious Disease*, pp. 1–16. New York: Springer.

Assunção, R. M., I. A. Reis, and C. D. L. Oliveira (2001). Diffusion and prediction of Leishmaniasis in a large metropolitan area in Brazil with a Bayesian space-time model. *Statistics in Medicine 20*, 2319–2335.

Bacchetti, P. and A. R. Moss (1989). Incubation period of AIDS in San Francisco. *Nature 338*(6212), 251–253.

Banerjee, S., B. P. Carlin, and A. E. Gelfand (2014). *Hierarchical Modeling and Analysis for Spatial Data* (2nd ed.). Boca Raton, FL: CRC Press.

Berger, J. O. (1985). *Statistical Decision Theory and Bayesian Analysis, Second Ed.* New York: Springer.

Bernardinelli, L., D. Clayton, C. Pascutto, C. Montomoli, M. Ghislandi, and M. Songini (1995). Bayesian analysis of space-time variation in disease risk. *Statistics in Medicine 14*, 2433–2443.

Brower, J. and P. Chalk (2003). The global threat of new and reemerging infectious diseases: Reconciling U.S. national security and public health policy. Technical report, RAND Corporation, Santa Monica, CA.

Carter, C. K. and R. Kohn (1994). On Gibbs sampling for state space models. *Biometrika 81*, 541–553.

Casella, G. and R. L. Berger (2001). *Statistical Inference* (2nd ed.). Duxbury Press.

Cressie, N. and C. K. Wikle (2011). *Statistics for Spatio-Temporal Data.* New York: Wiley.

Epstein, P. (2010). The ecology of climate change and infectious diseases: comment. *Ecology 91*(3), 925–928.

Fahrmeir, L. and G. Tutz (2001). *Multivariate Statistical Modelling Based on Generalized Linear Models* (2nd ed.). New York: Springer.

Ferreira, M. A. R. (2024). Proper Gaussian Markov random fields. In M. A. R. Ferreira (Ed.), *Modeling Spatio-Temporal Data: Markov Random Fields, Objective Bayes, and Multiscale Models*, pp. 1–22. Chapman & Hall/CRC.

Ferreira, M. A. R. and V. De Oliveira (2007). Bayesian reference analysis for Gaussian Markov Random Fields. *Journal of Multivariate Analysis 98*, 789–812.

Ferreira, M. A. R. and D. Gamerman (2000). Dynamic generalized linear models. In G. Dey and Mallick (Eds.), *Generalized Linear Models: a Bayesian Perspective*, New York, pp. 57–72. Marcel Dekker.

Ferreira, M. A. R., D. Gamerman, and H. S. Migon (1997). Bayesian dynamic hierarchical models: Covariance matrices estimation and non-normality. *Brazilian Journal of Probability and Statistics 11*, 67–79.

Frühwirth-Schnatter, S. (1994). Data augmentation and dynamic linear models. *Journal of Time Series Analysis 15*(2), 183–202.

Frühwirth-Schnatter, S. (1995). Bayesian model discrimination and Bayes factors for linear Gaussian state-space models. *Journal of the Royal Statistical Society, Series B 57*, 237–246.

Gamerman, D. and H. F. Lopes (2006). *Markov Chain Monte Carlo: Stochastic Simulation for Bayesian Inference* (2nd ed.). Boca Raton, FL: Chapman and Hall/CRC.

Gelfand, A. E. and A. F. M. Smith (1990). Sampling-based approaches to calculating marginal densities. *Journal of the American Statistical Association 85*(410), 398–409.

Gelman, A. and D. B. Rubin (1992). Inference from iterative simulation using multiple sequences. *Statistical Science 7*, 457–511.

Ghosh, J. K., M. Delampady, and T. Samanta (2006). *An Introduction to Bayesian Analysis, Theory and Methods.* New York: Springer.

Knorr-Held, L. (2000). Bayesian modelling of inseparable space-time variation in disease risk. *Statistics in Medicine 19*, 2555–2567.

Knorr-Held, L. and S. Richardson (2003). A hierarchical model for space-time surveillance data on meningococcal disease incidence. *Applied Statistics 52*, 169–183.

Lafferty, K. D. (2009). The ecology of climate change and infectious diseases. *Ecology 90*(4), 888–900.

Lawson, A. B. (2018). *Bayesian Disease Mapping: Hierarchical Modeling in Spatial Epidemiology* (3rd ed.). Boca Raton: Chapman and Hall/CRC.

MacNab, Y. C. and C. B. Dean (2001). Autoregressive spatial smoothing and temporal spline smoothing for mapping rates. *Biometrics 57*, 949–956.

McCullagh, P. and J. A. Nelder (1989). *Generalized Linear Models* (2nd. ed.), Volume 37 of *Monographs on Statistics and Applied Probability*. London: Chapman and Hall.

Mugglin, A. S., N. Cressie, and I. Gemmell (2002). Hierarchical statistical modelling of influenza epidemic dynamics in space and time. *Statistics in Medicine 21*, 2703–2721.

Robert, C. P. and G. Casella (2004). *Monte Carlo Statistical Methods* (2nd ed.). Springer-Verlag.

Schmid, V. and L. Held (2004). Bayesian extrapolation of space-time trends in cancer registry data. *Biometrics 60*, 1034–1042.

Sun, D., R. K. Tsutakawa, H. Kim, and Z. He (2000). Spatio-temporal interaction with disease mapping. *Statistics in Medicine 19*, 2015–2035.

Tzala, E. and N. Best (2008). Bayesian latent variable modelling of multivariate spatio-temporal variation in cancer mortality. *Statistical Methods in Medical Research 17*, 97–118.

Vivar, J. C. and M. A. R. Ferreira (2009). Spatio-temporal models for Gaussian areal data. *Journal of Computational and Graphical Statistics 18*, 658–674.

Waller, L. A., B. P. Carlin, H. Xia, and A. E. Gelfand (1997). Hierarchical spatio-temporal mapping of disease rates. *Journal of the American Statistical Association 92*, 607–617.

West, M., P. J. Harrison, and H. S. Migon (1985). Dynamic generalized linear models and Bayesian forecasting (with discussion). *Journal of the American Statistical Association 80*, 73–96.

Wikle, C. K. and M. B. Hooten (2010). A general science-based framework for dynamical spatio-temporal models. *Test 19*, 417–451.

Williams, J., S. Xu, and M. A. R. Ferreira (2023). BGWAS: Bayesian variable selection in linear mixed models with nonlocal priors for genome-wide association studies. *BMC Bioinformatics 24*(1), 1–20.

Wu, H.-H., M. A. R. Ferreira, M. Elkhouly, and T. Ji (2020). Nonlocal priors for variable selection in generalized linear models. *Sankhya – Series A 82-A*, 147–185.

Wu, H.-H., M. A. R. Ferreira, and M. E. Gompper (2016). Consistency of hyper-g-prior-based Bayesian variable selection for generalized linear models. *Brazilian Journal of Probability and Statistics 30*, 691–709.

5

Dynamic Multiscale Spatio-Temporal Thresholding

Marco A. R. Ferreira

Virginia Tech

We propose thresholding methods in the context of dynamic multiscale spatio-temporal models (DMSTMs) for Gaussian univariate data (Ferreira et al., 2010, 2011). DMSTMs naturally accommodate data observed on irregular grids and heteroscedasticity. In addition, DMSTMs lead to computations that are parallelizable, scalable, and fast. These models assume a spatial hierarchical structure. Based on such structure, DMSTMs decompose spatial areal data with a multiscale decomposition. This decomposition yields empirical multiscale coefficients that are assumed to be noisy versions of latent multiscale coefficients. Typically, the DMSTM specification is completed by assuming that the latent multiscale coefficients evolve through time following state-space equations (e.g., see Prado et al., 2021). Conversely, in this chapter we also consider thresholding methods based on possible cases when some latent multiscale coefficients may be fixed through time. For example, some of the latent multiscale coefficients may be thresholded to zero vectors, allowing clustering of subregions. Another possible case estimates some latent multiscale coefficients to be different than zero but fixed through time, which may be used for data compression. We illustrate the application of our DMSTM thresholding methods with an analysis of agriculture production data.

DOI: 10.1201/9781032623443-5

5.1 Introduction

In dynamic multiscale spatio-temporal models, the latent multiscale coefficients may be thought of as representing the underlying differences among sibling subregions at each scale of resolution. As such, there may be three situations: there are no differences among the subregions at any time, there are differences among the subregions but those differences are constant through time, or there are temporally evolving differences among the subregions. Each of these three situations may have distinct practical implications.

For example, consider the case of a large corporation that owns a network of fast-food restaurants in the United States. Further, this network of restaurants is managed as a tree structure where each restaurant reports to a state manager, the state managers report to regional (e.g., Northeast, Southeast, Upper Midwest, Lower Midwest, Southwest, etc.) directors, and the regional directors report to the corporation headquarters. In this case, at each time there will be latent multiscale coefficients connecting each region to its descendant states, and there will be a latent multiscale coefficient connecting the headquarters to its descendant regions. Now consider the latent multiscale coefficient that connects the Southeast to its descendant states. If there are no underlying differences among these states at any time, then this latent multiscale coefficient will be fixed through time equal to a vector of zeros. In this situation, the Southeast regional director may manage the Southeast states as if they were homogenous, for example by setting the same marketing strategies for all the Southeast states. On another hand, if there are differences that are constant through time among these states then the latent multiscale coefficient will be fixed through time but not zero. In this situation, the Southeast regional director should probably set marketing strategies that consider the specificities of each state. Finally, if there are temporally evolving differences among the states then the latent multiscale coefficient will vary through time. In this situation, the Southeast regional director should use the estimated temporal evolution of the latent multiscale coefficient to guide temporally evolving dynamic marketing strategies that consider the specificities of each state.

Therefore, distinguishing among these three situations is of paramount importance. To that end, here we develop a Bayesian multiple hypothesis testing procedure which for each latent multiscale coefficient considers three hypotheses. The first hypothesis is that the latent multiscale coefficient varies through time. This first hypothesis corresponds to the case when the descendants differ, and their differences vary through time. The second hypothesis is that the latent multiscale coefficient is different from zero but constant through time. This second hypothesis corresponds to the case when the descendants differ from each other, but their differences are stable through time. Finally, the third hypothesis is that the latent multiscale coefficient is equal to zero at all time points. This third hypothesis corresponds to the case when the descendants do not differ from each other. The first hypothesis leads to unrestricted dynamic multiscale coefficients. The second hypothesis leads to thresholding of the dynamic multiscale coefficients in time. Finally, the third hypothesis leads to the thresholding of the dynamic multiscale coefficients in time and space. Further, we develop efficient Bayesian analysis based on a combination of analytic and fast numerical integration. We illustrate the usefulness and flexibility of our dynamic multiscale thresholding framework with an application to spatio-temporal modeling of agriculture production in the Brazilian state of Espírito Santo.

The remainder of this chapter is organized as follows. Section 5.2 describes multiscale factorizations for spatial processes. Section 5.3 presents an exploratory multiscale data analysis of agriculture production in the Brazilian state of Espírito Santo. The temporal evolution of the underlying latent multiscale coefficients is introduced in Section 5.4. Section 5.5 presents the Bayesian analysis based on multiscale spatio-temporal thresholding. Section 5.6 illustrates our methodology through a spatio-temporal analysis of agriculture production in the State of Espirito Santo, Brazil. Conclusions and possible extensions are presented in Section 5.7.

5.2 Multiscale Factorization

Here we present a multiscale factorization proposed by Ferreira et al. (2011) for spatial areal data. To define the factorization, the observations are aggregated at different levels of resolution using linear combinations. This multiscale factorization extends the multiscale factorization for Gaussian data proposed by Kolaczyk and Huang (2001) that is based on spatial aggregation by sums (see also Chapter 9 of Ferreira and Lee, 2007). Specifically, consider observations obtained at a given scale of resolution L on a partition of a domain $S \subset \mathbb{R}^k$, where typically $k = 2$ in spatial statistics. Denote this partition by $\{B_{L1}, \ldots, B_{Ln_L}\}$, with $B_{Lj} \in S$, $j = 1, \ldots, n_L$, $B_{Li} \cap B_{Lj} = \emptyset$, $i \neq j$, and $\cup_{j=1}^{n_L} B_{Lj} = S$. Let y_{Lj} be the observation associated with subregion B_{Lj} and denote its expected value by $\mu_{Lj} = E(y_{Lj})$. Here, we use as a motivating example the data on agriculture production per county in the Brazilian state of Espirito Santo. Ferreira et al. (2010) analyzed these data in the original scale whereas Ferreira et al. (2011) considered the logarithm scale. Here, we follow the latter and assume that y_{Lj} is the logarithm of the agriculture production in the jth county in Espirito Santo State.

Assume there exists a known multiscale spatial structure. This definition can be based on the geopolitical structure of a region of interest. For example, the multi-resolution organization of the Brazilian state of Espírito Santo is shown in Figure 5.1. Based on socio-economic variables, counties in each state

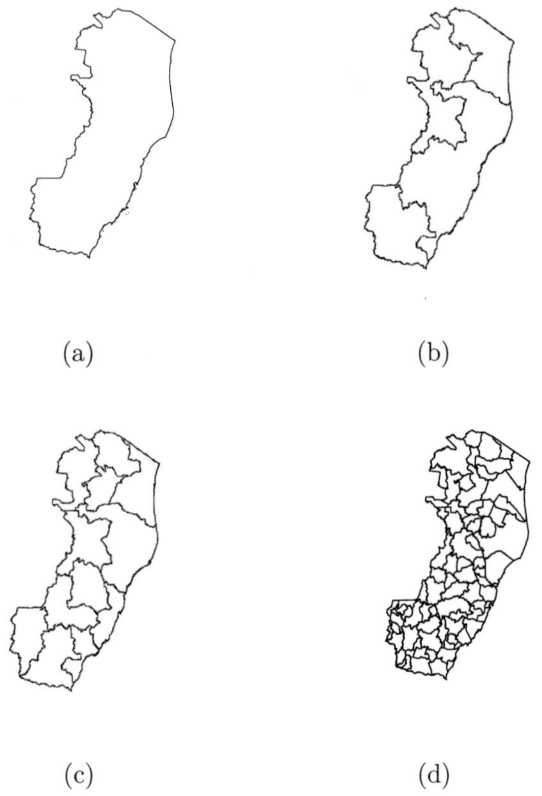

FIGURE 5.1
Geopolitical organization of the Brazilian state of Espírito Santo: (a) Entire Espírito Santo State, (b) macroregions, (c) microregions, and (d) counties.

in Brazil are aggregated by the Brazilian Institute of Geography and Statistics into microregions and macroregions. Specifically, counties are aggregated into microregions, and microregions are aggregated into macroregions, forming a multiscale spatial structure. We note that for cases when a multiscale spatial structure is not known a priori, Ferreira et al. (2011) have proposed a method for the estimation of the multiscale structure. For simplicity of exposition, in what follows we assume a known multiscale spatial structure.

In addition, assume that $y_{L1}, \ldots, y_{Ln_L}$ are conditionally independent given $\mu_{L1}, \ldots, \mu_{Ln_L}$, which is equivalent to assuming independent measurement errors. The construction presented below induces spatial correlation in the latent mean process μ. In turn, the dependence in μ carries over to the observations $y_{L1}, \ldots, y_{Ln_L}$. In addition to the observations $y_{L1}, \ldots, y_{Ln_L}$ and latent process

μ at the observed level of resolution L, we are interested in the process at aggregated coarser scales. At the lth scale of resolution, the domain S is partitioned in n_l subregions $B_{l1}, \ldots, B_{ln_l}$, $l = 1, \ldots, L$, where the partition at level l is a refinement of the partition at level $l+1$. Let D_{lj} be the set of descendants of subregion j at level l, with $D_{lj} \cap D_{li} = \emptyset, i \neq j$. Denote the number of descendants of subregion (l, j) by m_{lj}. Then, $B_{lj} = \cup_{(l+1,j') \in D_{lj}} B_{l+1,j'}$.

To define observations and the mean process at coarser resolutions, let $p_{lj}, l = 1, \ldots, L, j = 1, \ldots, n_l$, be known nonnegative weights. Then, the aggregated observation of subregion (l, j) at the lth level of resolution is recursively defined by $y_{lj} = \sum_{(l+1,j') \in D_{lj}} p_{l+1,j'} y_{l+1,j'}$ and, similarly, its expected value is $\mu_{lj} = \sum_{(l+1,j') \in D_{lj}} p_{l+1,j'} \mu_{l+1,j'}$. These equations include as special cases aggregation by totals, means and decimations. Aggregation by totals, considered by Kolaczyk and Huang (2001), is obtained when $p_{lj} = 1$, for all l and j. Aggregation by weighted averages is achieved when $\sum_{(l+1,j') \in D_{lj}} p_{l+1,j'} = 1$, for all l and j. Aggregation by decimation (which is common in the wavelets literature) is obtained when there is only one subregion $(l+1, j^*)$ in D_{lj} such that, for $(l+1, j') \in D_{lj}$, $p_{l+1,j'} = 1$ if $j' = j^*$, and $p_{l+1,j'} = 0$ if $j' \neq j^*$.

Let σ^2_{lj} denote the variance of y_{lj}. Additionally, define $\boldsymbol{y}_l = (y_{l1}, \ldots, y_{ln_l})'$, $\boldsymbol{\mu}_l = (\mu_{l1}, \ldots, \mu_{ln_l})'$, $\boldsymbol{\sigma}^2_l = (\sigma^2_{l1}, \ldots, \sigma^2_{ln_l})'$, $\boldsymbol{\rho}_l = (p_{l1}\sigma^2_{l1}, \ldots, p_{ln_l}\sigma^2_{ln_l})'$, and $\boldsymbol{\Sigma}_l = diag(\boldsymbol{\sigma}^2_l)$. Assuming that $y_{L1}, \ldots, y_{Ln_L}$ are conditionally independent given $\boldsymbol{\mu}_L$ and $\boldsymbol{\Sigma}^2_L$, the variance of y_{lj} can be recursively computed as $\sigma^2_{lj} = \sum_{(l+1,j') \in D_{lj}} p^2_{l+1,j'} \sigma^2_{l+1,j'}$. Further, let $\boldsymbol{\nu}_{lj} = \boldsymbol{\rho}_{D_{lj}} / \sigma^2_{lj}$, $\boldsymbol{\theta}_{lj} = \boldsymbol{\mu}_{D_{lj}} - \boldsymbol{\nu}_{lj} \mu_{lj}$, and $\boldsymbol{\Omega}_{lj} = \boldsymbol{\Sigma}_{D_{lj}} - \sigma^{-2}_{lj} \boldsymbol{\rho}_{D_{lj}} \boldsymbol{\rho}'_{D_{lj}}$. Conditional on the mean vector $\boldsymbol{\mu}_L$ and the vector of variances $\boldsymbol{\sigma}^2_L$, the joint distribution of the observations at the finest level L is multivariate Gaussian. As a consequence, the conditional distribution of $\boldsymbol{y}_{D_{lj}}$ given y_{lj} is $\boldsymbol{y}_{D_{lj}} \big| y_{lj}, \boldsymbol{\mu}_L, \boldsymbol{\Sigma}^2_L \sim N(\boldsymbol{\nu}_{lj} y_{lj} + \boldsymbol{\theta}_{lj}, \boldsymbol{\Omega}_{lj})$. The parameter vector $\boldsymbol{\theta}_{lj}$ is the difference between the mean level process at the descendants of subregion B_{lj} and the scaled mean level process at B_{lj}. Ferreira et al. (2011) call $\boldsymbol{\theta}_{lj}$ a multiscale coefficient. See Elkhouly and Ferreira (2021) for a connection between multiscale coefficients and spatial directional derivatives.

Let $\boldsymbol{\theta}^e_{lj} = \boldsymbol{y}_{D_{lj}} - \boldsymbol{\nu}_{lj} y_{lj}$ denote the empirical multiscale coefficient associated with $\boldsymbol{\theta}_{lj}$. Then, $\boldsymbol{\theta}^e_{lj} | y_{lj}, \boldsymbol{\mu}_L, \boldsymbol{\Sigma}^2_L \sim N(\boldsymbol{\theta}_{lj}, \boldsymbol{\Omega}_{lj})$, which is a singular Gaussian distribution (Anderson, 1984). Let $\boldsymbol{p}_l = (p_{l1}, \ldots, p_{ln_l})'$ be the vector of weights at level of resolution l. Here, $\boldsymbol{p}'_{D_{lj}} \boldsymbol{\theta}^e_{lj} = \boldsymbol{p}'_{D_{lj}} (\boldsymbol{y}_{D_{lj}} - \boldsymbol{\nu}_{lj} y_{lj}) = 0$ and, by a similar argument, $\boldsymbol{p}'_{D_{lj}} \boldsymbol{\theta}_{lj} = 0$. Moreover, the covariance matrix $\boldsymbol{\Omega}_{lj}$ has rank $m_{lj} - 1$, and $\boldsymbol{p}'_{D_{lj}} \boldsymbol{\Omega}_{lj} \boldsymbol{p}_{D_{lj}} = 0$ implying that $\boldsymbol{p}_{D_{lj}}$ is an (unnormalized) eigenvector of $\boldsymbol{\Omega}_{lj}$ corresponding to its zero eigenvalue. As a consequence, the singular Gaussian distribution above for $\boldsymbol{\theta}^e_{lj}$ implies that $\boldsymbol{p}'_{D_{lj}} \boldsymbol{\theta}^e_{lj} = 0$ with probability one and, therefore, naturally respects the aggregation constraint $\boldsymbol{p}'_{D_{lj}} \boldsymbol{\theta}^e_{lj} = 0$.

The approach we present here makes use of the fact that the analysis of observations from a singular Gaussian distribution may be performed by analyzing either subvectors of the observations or subvectors of an orthogonal transformation of the full vector of observations. Let $\boldsymbol{P}_{lj} \boldsymbol{D}_{lj} \boldsymbol{P}'_{lj}$ denote the spectral decomposition of the covariance matrix $\boldsymbol{\Omega}_{lj}$. Here, the columns of

$\boldsymbol{P}_{lj}$ are the eigenvectors of $\boldsymbol{\Omega}_{lj}$ and $\boldsymbol{D}_{lj} = \text{diag}(\lambda_{lj1}, \ldots, \lambda_{ljm_{lj}})$, where $\lambda_{lj1} \geq \cdots \geq \lambda_{lj,m_{lj}-1} > \lambda_{ljm_{lj}} = 0$ are the eigenvalues of $\boldsymbol{\Omega}_{lj}$. Note that the last column of $\boldsymbol{P}_{lj}$ is equal to the normalized vector of weights $\boldsymbol{p}_{D_{lj}}$, which is the eigenvector of $\boldsymbol{\Omega}_{lj}$ corresponding to its zero eigenvalue. Then, using standard results for the singular Gaussian distribution (e.g., see Ferreira et al., 2021), the density function of $\boldsymbol{\theta}^e_{lj}$ given $y_{lj}, \boldsymbol{\mu}_L, \boldsymbol{\Sigma}^2_L$ can be written as

$$\begin{aligned} p(\boldsymbol{\theta}^e_{lj}|y_{lj}, \boldsymbol{\theta}_{lj}, \boldsymbol{\Omega}_{lj}) &= (2\pi)^{m_{lj}-1} \left(\prod_{i=1}^{m_{lj}-1} \lambda_{lji} \right)^{-0.5} I(\boldsymbol{p}'_{D_{lj}} \boldsymbol{\theta}^e_{lj} = 0) \\ &\quad \exp\left\{ -0.5(\boldsymbol{\theta}^e_{lj} - \boldsymbol{\theta}_{lj})' \boldsymbol{\Omega}^+_{lj} (\boldsymbol{\theta}^e_{lj} - \boldsymbol{\theta}_{lj}) \right\} \\ &= (2\pi)^{m_{lj}-1} \left(\prod_{i=1}^{m_{lj}-1} \lambda_{lji} \right)^{-0.5} I(\boldsymbol{p}'_{D_{lj}} \boldsymbol{\theta}^e_{lj} = 0) \\ &\quad \exp\left\{ -0.5(\boldsymbol{\theta}^e_{lj} - \boldsymbol{\theta}_{lj})' \boldsymbol{P}_{lj} \boldsymbol{D}^+_{lj} \boldsymbol{P}'_{lj} (\boldsymbol{\theta}^e_{lj} - \boldsymbol{\theta}_{lj}) \right\}, \end{aligned}$$

where $I(\cdot)$ is the indicator function equal to 1 if the condition is true and equal to 0 otherwise, and $\boldsymbol{A}^+$ is the Monroe-Penrose inverse of matrix $\boldsymbol{A}$. As a consequence, this particular representation of the density function of $\boldsymbol{\theta}^e_{lj}$ given $y_{lj}, \boldsymbol{\mu}_L, \boldsymbol{\Sigma}^2_L$ considers in the exponent the first $m_{lj} - 1$ elements of the orthogonal transformation $\boldsymbol{P}'_{lj} \boldsymbol{\theta}^e_{lj}$. The last element, that corresponds to $\boldsymbol{p}'_{D_{lj}} \boldsymbol{\theta}^e_{lj} = 0$, appears in the indicator function. Thus, conditional on μ_{Lj} and σ^2_{Lj} all the distributions are also Gaussian and the multiscale factorization of the likelihood function may be expressed as

$$\begin{aligned} \prod_{j=1}^{n_L} p(y_{Lj}|\mu_{Lj}, \sigma^2_{Lj}) &\propto p(\boldsymbol{y}_1|\boldsymbol{\mu}_1, \boldsymbol{\Sigma}_1) \prod_{l=1}^{L-1} \prod_{j=1}^{n_l} p(\boldsymbol{y}_{D_{lj}}|y_{lj}, \boldsymbol{\theta}_{lj}, \boldsymbol{\Omega}_{lj}) \\ &= p(\boldsymbol{y}_1|\boldsymbol{\mu}_1, \boldsymbol{\Sigma}_1) \prod_{l=1}^{L-1} \prod_{j=1}^{n_l} p(\boldsymbol{\theta}^e_{lj}|y_{lj}, \boldsymbol{\theta}_{lj}, \boldsymbol{\Omega}_{lj}). \end{aligned}$$

The constant of proportionality in the first line of the equation above is the Jacobian of the transformation from the finest level $y_{L,1:n_L}$ to the aggregated data. This constant of proportionality is a function of the aggregation weights. Also, note that conditional on y_{lj}, the Jacobian of the transformation from $Y_{D_{lj}}$ to $\boldsymbol{\theta}^e_{lj}$ is equal to 1. Thus, the right hand side in the first line is exactly equal to the expression in the second line of the equation above. Finally, we note that the factorization above is equivalent to a reparametrization of the model, where $\mu_{L1}, \ldots, \mu_{Ln_L}$ are replaced by the mean level process at the coarsest level $\boldsymbol{\mu}_1$ and by the multiscale coefficients $\boldsymbol{\theta}_{lj}$s.

5.2.1 Relation to Haar Wavelets

The multiscale decomposition presented in Section 5.2 includes as a particular case the discrete version of the one-dimensional Haar wavelet transform (Haar, 1910). Moreover, the multiscale algorithm for the computation of the aggregated values at coarser levels and of the multiscale coefficients includes as a particular case a nondecimated version of the cascade algorithm (Mallat, 1989a,b) for Haar wavelets.

To see this, assume equally spaced observations $y_1, \ldots, y_n$, $n = 2^{(L-1)}$ on an interval which, without loss of generality, assume to be the interval $[0, 1]$. Define the scaling coefficients at the finest level to be $c_{Lj} = y_{Lj}$. Mallat's cascade algorithm applied to the discrete Haar wavelet transform recursively computes the scaling coefficients c_{lj} and the wavelet coefficients d_{lj}, $l = 1, \ldots, L-1$, $j = 1, \ldots, 2^{l-1}$, as $c_{lj} = 2^{-0.5}(c_{l+1,2j} + c_{l+1,2j+1})$ and $d_{lj} = 2^{-0.5}(c_{l+1,2j} - c_{l+1,2j+1})$.

The discrete Haar wavelet transform (DHWT) corresponds to the multiscale factorization of Section 5.2. Specifically, the DHWT corresponds to the use of a dyadic graph, that is, a graph where each node (l, j) has exactly two children, more precisely $D_{lj} = \{(l+1, 2j), (l+1, 2j+1)\}$. Moreover, for the DHWT the vector of weights is $\boldsymbol{p}_l = 2^{-0.5}\mathbf{1}'_{n_l}$ and the variance of the observations at the finest level is constant, that is, $\boldsymbol{\sigma}^2_L = \sigma^2\mathbf{1}'_{n_L}$. Then, the aggregation measurement at node (l, j) becomes $y_{lj} = 2^{-0.5}(y_{l+1,2j} + y_{l+1,2j+1}) = c_{lj}$, thus the scaling coefficient c_{lj} is equal to the aggregated value y_{lj}. In addition, straightforward algebra shows that the observational variance at node (l, j) is $\sigma^2_{lj} = \sigma^2$. As a consequence, $\boldsymbol{\rho}_l = 2^{-0.5}\sigma^2\mathbf{1}'_{n_l}$ and $\boldsymbol{\nu}_{lj} = 2^{-0.5}\mathbf{1}'_2$. Finally, the empirical multi-resolution coefficient at node (l, j) becomes $\boldsymbol{\theta}^e_{lj} = \boldsymbol{y}_{D_{lj}} - \boldsymbol{\nu}_{lj} y_{lj} = 2^{-0.5}(d_{lj}, -d_{lj})'$.

Therefore, the multiscale factorization presented in Section 5.2 may be thought of as a generalized Haar wavelet transform that is able to deal with distinct numbers of descendants across subregions, nonconstant weights, and spatial heterokedasticity.

5.3 Exploratory Multiscale Data Analysis

In Section 5.2, we considered multiscale factorizations for spatial data. However, we are interested in analyzing the spatio-temporal annual agriculture production data in the state of Espírito Santo, Brazil, from 1990 to 2005. In this section, we follow Ferreira et al. (2010) and perform an exploratory multiscale data analysis where we apply the multiscale factorization to the data at each time point. While Ferreira et al. (2010) performed this analysis for the original agriculture production dataset, here we perform an exploratory multiscale data analysis of the logarithm of agriculture production.

Figure 5.2 presents for each of the four macroregions the time series plots of the aggregate per macroregion (first column), the macroregion data disaggregated by microregion (second column), and the elements of the empirical multiscale coefficients (third column). Each row in Figure 5.2 represents one macroregion, with the top row representing macroregion 1, the second row representing macroregion 2, and so on. The aggregates at the microregion and macroregion levels are recursively computed from the logarithm of agriculture production at the county level. The first column of Figure 5.2 shows that each of the four macroregions presents increasing trends. The second column shows that all the twelve microregions also exhibit increasing trends.

Because the empirical multiscale coefficients have been computed for the logarithm of agriculture production, they represent the observed relative importance of the different microregions within each macroregion at each time point. To compute the empirical multiscale coefficients, we have used the estimated variances for each county based on the empirical Bayes methodology proposed by Ferreira et al. (2011). From the third column of Figure 5.2, for each macroregion, the empirical multiscale coefficient that relates the macroregion to its children microregions seems to be a noisy version of a latent process that either is fairly stable through time or varies slowly through time.

This motivates the use in this application of dynamic multiscale spatio-temporal models (Ferreira et al., 2010, 2011). These dynamic multiscale spatio-temporal models assume that each empirical multiscale coefficient is equal to the sum of a latent multiscale coefficient and an error vector, where the latent multiscale coefficients evolve through time according to state-space equations (West and Harrison, 1997; Prado et al., 2021). Further, examining again the third column of Figure 5.2, it seems that some of the latent multiscale coefficients may either be fixed through time or may evolve through time. Another possibility, that seems unlikely for this application based on the third column of Figure 5.2, would be that some of the latent multiscale coefficients may be fixed through time and equal to a zero vector.

To evaluate the support from the data for each of these three possible cases, in what follows we develop a Bayesian multiple hypothesis testing procedure which for each latent multiscale coefficient considers three hypotheses. The first hypothesis is that the latent multiscale coefficient evolves through time according to a state-space model. This first hypothesis corresponds to the case when the descendants differ, and their differences vary through time. The second hypothesis is that the latent multiscale coefficient is different from zero but constant through time. This second hypothesis corresponds to the case when the descendants differ from each other, but their differences are stable through time. Finally, the third hypothesis is that the latent multiscale coefficient is equal to zero at all time points. This third hypothesis corresponds to the case when the descendants do not differ from each other.

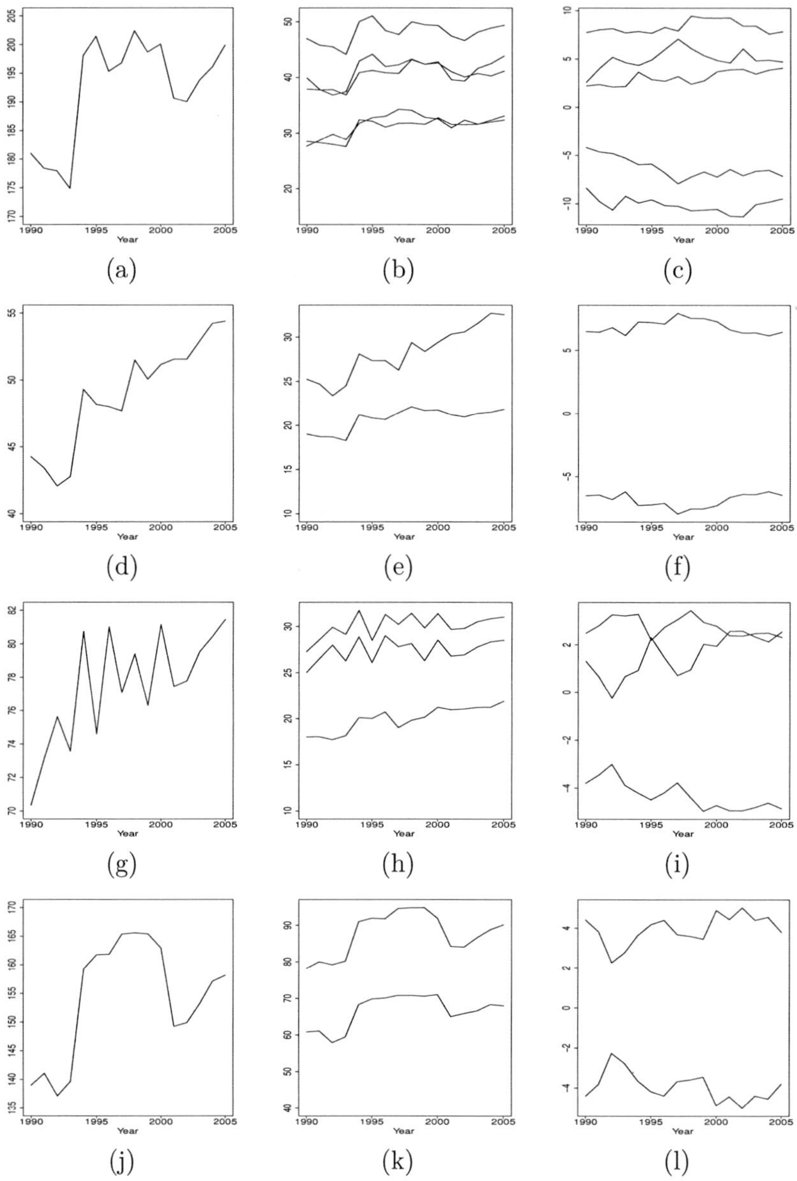

FIGURE 5.2
Time series plots per macroregion in the Brazilian state of Espírito Santo. Macroregions 1 (a-c), 2 (d-f), 3 (g-i), and 4 (j-l). Macroregion aggregate (a, d, g, j); macroregion disaggregated by microregion (b, e, h, k); elements of empirical multiscale coefficients (c, f, i, l). Time series at the microregion and macroregion resolutions are obtained by aggregating the time series of logarithm of agriculture production at the county level.

5.4 Dynamic Multiscale Spatio-Temporal Modeling

As we have seen in Section 5.3, the empirical multiscale coefficients for the agriculture production application seem be noisy versions of latent processes that vary slowly through time, motivating the use of dynamic multiscale spatio-temporal models (Ferreira et al., 2010, 2011). In addition, Section 5.2.1 shows that the empirical multiscale coefficients can be interpreted as generalized Haar wavelets that provide the detail information necessary to recover the mean process at resolution level $l+1$ from the mean process at resolution level l. Therefore, the dynamic multiscale spatio-temporal models of Ferreira et al. (2010, 2011) describe the temporal evolution of the detail information needed to recover finer resolutions from coarser resolutions.

Here we add a subscript t to indicate time and assume that the mean level process at the coarsest level μ_1 and the multiscale coefficient $\boldsymbol{\theta}_{lj}$ evolve through time according to state-space equations. This type of time evolution allows fairly flexible behavior such as local linear and quadratic trends, seasonality, and the inclusion of regressors. For example, one of the most widely used temporal evolutions is a random walk that can be expressed for the mean process at the coarsest level as $\mu_{t1} = \mu_{t-1,1} + \omega_{t1}, \omega_{t1} \sim N(0, \boldsymbol{W}_1)$. Similarly, a random walk for the latent multiscale coefficient $\boldsymbol{\theta}_{tlj}$ may be written as $\boldsymbol{\theta}_{tlj} = \boldsymbol{\theta}_{t-1,lj} + \boldsymbol{\Omega}_{tlj}, \boldsymbol{\Omega}_{tlj} \sim N(0, \boldsymbol{W}_{lj})$. Thus, fundamental to the construction of dynamic multiscale spatio-temporal models is the specification of the evolution covariance matrices $\boldsymbol{W}_1$ and $\boldsymbol{W}_{lj}$. These matrices have to be specified in a way that preserves the distinct correlation structure of the elements of the multiscale coefficient $\boldsymbol{\theta}_{tlj}$ and that respects the aggregation constraint $P[\boldsymbol{p}'_{D_{lj}}\boldsymbol{\theta}_{tlj} = 0] = 1$.

In terms of the covariance matrix at the coarsest level, we assume $\boldsymbol{W}_1 = \text{diag}(W_{11}, \ldots, W_{1n_1})$, with $W_{1k} = \xi_k \sigma^2_{1k}$. This representation is useful because ξ_k can be interpreted as the signal-to-noise ratio at the coarsest level. Thus, it is easier to assign a meaningful prior distribution for ξ_k than to assign a prior for W_{1k}. For the covariance matrix of the evolution innovations of the latent multiscale coefficients, we may also assume a signal-to-noise ratio parameterization with $\boldsymbol{W}_{lj} = \psi_{lj}\boldsymbol{\Omega}_{lj}$ as proposed by Ferreira et al. (2010, 2011). Note that this construction both preserves the special correlation structure of $\boldsymbol{\theta}_{tlj}$ and implies the aggregation constraint $P[\boldsymbol{p}'_{D_{lj}}\boldsymbol{\theta}_{tlj} = 0] = 1$. Another possible specification for $\boldsymbol{W}_{lj}$ that would also respect the aggregation constraint is based on the spectral decomposition of $\boldsymbol{\Omega}_{lj}$. Specifically, let $\boldsymbol{\Omega}_{lj} = \boldsymbol{P}_{lj}\boldsymbol{D}_{lj}\boldsymbol{P}'_{lj}$. Then, $\boldsymbol{W}_{lj}$ may be specified as $\boldsymbol{W}_{lj} = \boldsymbol{P}_{lj}\boldsymbol{D}^*_{lj}\boldsymbol{P}'_{lj}$ where, similarly to $\boldsymbol{D}_{lj}$, $\boldsymbol{D}^*_{lj}$ is a diagonal matrix with the last diagonal element equal to zero but with all other diagonal elements larger than zero. With this definition, $\boldsymbol{W}_{lj}$ has $\boldsymbol{p}_{D_{lj}}$ as an eigenvector with respective eigenvalue equal to zero. As a consequence, this specification of $\boldsymbol{W}_{lj}$ would also respect the aggregation constraint $P[\boldsymbol{p}'_{D_{lj}}\boldsymbol{\theta}_{tlj} = 0] = 1$.

Hence, the dynamic multiscale spatio-temporal model has an observation equation, a set of equations that decompose the observation equation, and a set of system equations. That is,

Observation equation:

$$\boldsymbol{y}_{tL} = \boldsymbol{\mu}_{tL} + \boldsymbol{v}_{tL}, \quad \boldsymbol{v}_{tL} \sim N(\boldsymbol{0}, \boldsymbol{V}_L), \tag{5.1}$$

where $\boldsymbol{V}_L = diag(\sigma^2_{L1}, \ldots, \sigma^2_{Ln_L})$.

Multiscale decomposition of the observation equation:

$$y_{t1k} \mid \mu_{t1k} \sim N(\mu_{t1k}, \sigma^2_{1k}). \tag{5.2}$$

$$\boldsymbol{\theta}^e_{tlj} \mid \boldsymbol{\theta}_{tlj} \sim N(\boldsymbol{\theta}_{tlj}, \boldsymbol{\Omega}_{lj}). \tag{5.3}$$

System equations:

$$\mu_{t1k} = \mu_{t-1,1k} + w_{t1k}, \quad w_{t1k} \sim N(0, \xi_k \sigma^2_{1k}). \tag{5.4}$$

$$\boldsymbol{\theta}_{tlj} = \boldsymbol{\theta}_{t-1,lj} + \boldsymbol{w}_{tlj}, \quad \boldsymbol{w}_{tlj} \sim N(\boldsymbol{0}, \psi_{lj}\boldsymbol{\Omega}_{lj}), \tag{5.5}$$

where w_{t1k} and w_{tlj} are independent noise.

Note that under the formulation above the aggregation constraints are satisfied. Nevertheless, we may be interested in other types of temporal evolution, such as time-evolving trend and seasonality. For example, we may assume that the trend component is composed of level and slope as proposed by Elkhouly and Ferreira (2021). As another example, to include level, slope, and seasonality, we may assume $\theta^e_{tlj} = [\boldsymbol{I0}\ldots\boldsymbol{0}]\theta_{tlj} + \boldsymbol{v}_{tlj}$, with $\boldsymbol{v}_{tlj} \sim N(\boldsymbol{0}, \Omega_{lj})$ and $\boldsymbol{\theta}_{tlj} = \boldsymbol{G}\boldsymbol{\theta}_{t-1,lj} + \boldsymbol{w}_{tlj}$, with $\boldsymbol{w}_{tlj} \sim N(\boldsymbol{0}, diag(\psi_{1lj}, \ldots, \psi_{clj}) \otimes \boldsymbol{\Omega}_{lj})$, where c is the number of components (level, slope, and seasonal factors) of the latent process. In this case, the evolution matrix $\boldsymbol{G}$ is a block-diagonal matrix with a block for the trend and another block for the seasonality as it is usually the case under the dynamic linear models framework (Prado et al., 2021). Because the covariance matrix of the system innovations is the Kronecker product of a diagonal matrix and the matrix $\boldsymbol{\Omega}_{lj}$, the aggregation constraints are satisfied for all the components of the latent process *a priori* and, as a result, also *a posteriori*.

5.4.1 Priors for Bayesian Spatio-Temporal Thresholding

Here we consider three types of phenomena that may lead to Bayesian spatio-temporal thresholding. First, we want to test if the relative importance of the descendants of node (l, j) changes through time or if it may be assumed constant. In terms of our multiscale spatio-temporal model, this situation translates into hypotheses about the multiscale coefficient $\boldsymbol{\theta}_{tlj}$ and its signal-to-noise parameter ψ_{lj}: either $\psi_{lj} = 0$ and thus $\boldsymbol{\theta}_{tlj} = \boldsymbol{\theta}_{1lj}$, $t = 1, \ldots, T$, is constant through time, or $\psi_{lj} > 0$ and θ_{tlj} evolves through time according to the System Equation (5.5).

In addition, in the case when $\boldsymbol{\theta}_{tlj}$ is constant through time, we want to test if the descendants of node (l, j) are different from each other, or if they behave similarly and we may without loss of relevant information merge them into node (l, j). In terms of our parameterization, this phenomenon translates into the case when $\psi_{lj} = 0$ and $\boldsymbol{\theta}_{1lj} = \mathbf{0}_{m_{lj}}$.

Thus, we consider the following three competing hypotheses:

$$H_{1lj} : \boldsymbol{\theta}_{1lj} \neq 0 \text{ and } \psi_{lj} > 0,$$
$$H_{2lj} : \boldsymbol{\theta}_{1lj} \neq 0 \text{ and } \psi_{lj} = 0,$$
$$H_{3lj} : \boldsymbol{\theta}_{1lj} = 0 \text{ and } \psi_{lj} = 0.$$

To allow for the possibility of constant $\boldsymbol{\theta}_{tlj}$ through time, we assume for the signal-to-noise parameter ψ_{lj} a prior that is a mixture of an inverse gamma distribution and a point mass at 0, that is,

$$\psi_{lj} \sim \omega_{lj}^{\psi} IG(0.5\tau_{lj}^{\psi}, 0.5\kappa_{lj}^{\psi}) + (1 - \omega_{lj}^{\psi})\delta_{\psi_{lj}}(0),$$

where $\delta_{\psi_{lj}}(0)$ is the Dirac delta function, ω_{lj}^{ψ} is the prior probability that ψ_{lj} is greater than zero, and τ_{lj}^{ψ} and κ_{lj}^{ψ} are fixed *a priori*.

In addition, to allow for the descendants of node (l, j) to behave similarly, we assume the following priors for $\boldsymbol{\theta}_{1lj}$ conditional on either $\psi_{lj} > 0$ or $\psi_{lj} = 0$:

$$\begin{aligned}
\boldsymbol{\theta}_{1lj}|\psi_{lj} > 0, D_1 &\sim N(\boldsymbol{\theta}_{1lj}^{e}, \boldsymbol{\Omega}_{lj}), \\
\boldsymbol{\theta}_{1lj}|\psi_{lj} = 0, D_1 &\sim \omega_{lj}^{\theta} N(\boldsymbol{\theta}_{1lj}^{e}, \boldsymbol{\Omega}_{lj}) + (1 - \omega_{lj}^{\theta})\delta_{\theta_{1lj}}(0),
\end{aligned}$$

where ω_{lj}^{θ} is the prior probability that $\boldsymbol{\theta}_{1lj}$ is different than zero conditional on ψ_{lj} being equal to zero. This prior is based on the application of the forward filter for the first time $t = 1$ for the limiting case $C_{0lj} \to \infty$. This prior eliminates the usual sensitivity of the posterior probabilities of competing hypotheses with respect to the priors. The behavior of the posterior probabilities when the variance of the prior goes to infinity is known in the Bayesian literature as Lindley's paradox (Lindley, 1957). For a detailed explanation of Lindley's paradox, see Chapter 5 of Robert (2001). Hence, the joint prior density for $(\psi_{lj}, \boldsymbol{\theta}_{1lj})$ becomes

$$\begin{aligned}
p(\psi_{lj}, \boldsymbol{\theta}_{1lj}|\boldsymbol{\theta}_{1lj}^{e}) &= \omega_{lj}^{\psi} IG(\psi_{lj}|0.5\tau_{lj}^{\psi}, 0.5\kappa_{lj}^{\psi}) N(\boldsymbol{\theta}_{1lj}|\boldsymbol{\theta}_{1lj}^{e}, \boldsymbol{\Omega}_{lj}) \\
&\quad + (1 - \omega_{lj}^{\psi})\delta_{\psi_{lj}}(0)\omega_{lj}^{\theta} N(\boldsymbol{\theta}_{1lj}|\boldsymbol{\theta}_{1lj}^{e}, \boldsymbol{\Omega}_{lj}) \\
&\quad + (1 - \omega_{lj}^{\psi})(1 - \omega_{lj}^{\theta})\delta_{\psi_{lj}}(0)\delta_{\theta_{1lj}}(0).
\end{aligned}$$

A discrete uniform prior on H_{1lj}, H_{2lj}, and H_{3lj} is obtained by assigning $\omega_{lj}^{\psi} = 1/3$ and $\omega_{lj}^{\theta} = 1/2$. For spatio-temporal models with few resolution levels such as in the application we present in Section 5.6, we recommend the use of this discrete uniform prior.

For models with many resolution levels, similar to the method proposed by Abramovich et al. (1998) for wavelet analysis, we may implement an empirical Bayes approach to estimate ω_{lj}^{ψ} and ω_{lj}^{θ} from the data. Specifically, this approach would write ω_{lj}^{ψ} and ω_{lj}^{θ} in terms of a few parameters and then maximize the likelihood function for those parameters. For example, we may assume the parameterization $\omega_{lj}^{\psi} = \min(1, 2^{-\alpha l} C_1)$ and $\omega_{lj}^{\theta} = \min(1, 2^{-\beta l} C_2)$. In this case, we would estimate α, β, C_1, and C_2 by maximizing the likelihood function. After that, this empirical Bayes approach would keep α, β, C_1, and C_2 fixed at their estimates when performing Bayesian analysis for the latent multiscale coefficients.

5.5 Posterior Analysis

As we discuss below, posterior analysis for dynamic multiscale spatio-temporal models may be performed using Markov Chain Monte Carlo or analytic numerical integration. One appealing advantage of the Bayesian paradigm is that it allows for a full account of uncertainty. In the discussion that follows, we assume that $\boldsymbol{\nu}_{lj}$ is known, which implies that $\boldsymbol{\Omega}_{lj}$ and $\boldsymbol{\theta}_{tlj}^{e}$ are well defined. In that case, $\boldsymbol{\theta}_{tlj}^{e}$ can be computed once, before the start of the posterior exploration algorithm. The weights $\boldsymbol{\nu}_{lj}$ are known in two important cases. The first case refers to homocedastic measurement errors, that is, when the variances at the fine level are all equal. The second case refers to heterocedastic measurement errors with $\sigma_{Lj}^2 = \kappa_{Lj}\sigma^2$ where $\kappa_{Lj}, j = 1, \ldots, n_L$, are known. In this latter case, σ^2 may be known or unknown.

Let

$$
\begin{aligned}
\boldsymbol{\theta}_{\bullet lj} &= (\boldsymbol{\theta}_{0lj}', \ldots, \boldsymbol{\theta}_{Tlj}')', \\
\boldsymbol{\theta}_{t\bullet j} &= (\boldsymbol{\theta}_{t1j}', \ldots, \boldsymbol{\theta}_{tLj}')', \\
\boldsymbol{\theta}_{\bullet\bullet\bullet} &= (\boldsymbol{\theta}_{\bullet 11}', \ldots, \boldsymbol{\theta}_{\bullet 1n_1}', \boldsymbol{\theta}_{\bullet 21}', \ldots, \boldsymbol{\theta}_{\bullet 2n_2}', \ldots, \boldsymbol{\theta}_{\bullet L1}', \ldots, \boldsymbol{\theta}_{\bullet Ln_L}')',
\end{aligned}
$$

with analogous definitions for $\boldsymbol{y}$ and $\boldsymbol{\mu}$. We may then use a Gibbs sampler to explore the posterior distribution (Ferreira et al., 2011). To increase the efficiency during posterior exploration, Ferreira et al. (2011) block the parameters of the model as follows: σ^2, $\xi_1, \ldots, \xi_{n_1}$, $\psi_{11}, \ldots, \psi_{1n_1}$, $\psi_{l1}, \ldots, \psi_{ln_l}$, $\boldsymbol{\mu}_{\bullet 11}, \ldots, \boldsymbol{\mu}_{\bullet 1n_1}$, $\boldsymbol{\theta}_{\bullet 11}, \ldots, \boldsymbol{\theta}_{\bullet 1n_1}, \ldots$, $\boldsymbol{\theta}_{\bullet L1}, \ldots$, $\boldsymbol{\theta}_{\bullet Ln_L}$. The following theorem from Ferreira et al. (2011) facilitates the exploration of the posterior distribution and leads to a highly parallelizable algorithm.

Theorem 5.1 (Ferreira et al., 2011) *Consider the dynamic multiscale spatio-temporal model defined by Equations (5.1), (5.2), (5.3), (5.4), and (5.5). Conditional on $\sigma^2, \xi_\bullet$, and $\psi_{\bullet\bullet}$, the vectors $\boldsymbol{\mu}_{\bullet 11}, \ldots, \boldsymbol{\mu}_{\bullet 1n_1}$, $\boldsymbol{\theta}_{\bullet 11}, \ldots$, $\boldsymbol{\theta}_{\bullet 1n_1}, \ldots$, $\boldsymbol{\theta}_{\bullet L1}, \ldots, \boldsymbol{\theta}_{\bullet Ln_L}$, are conditionally independent and normally distributed a posteriori.*

As a consequence of Theorem 5.1, simulation of $\boldsymbol{\mu}_{\bullet 11}, \ldots, \boldsymbol{\mu}_{\bullet 1n_1}$, $\boldsymbol{\theta}_{\bullet 11}, \ldots$, $\boldsymbol{\theta}_{\bullet 1n_1}, \ldots$, $\boldsymbol{\theta}_{\bullet L1}, \ldots, \boldsymbol{\theta}_{\bullet Ln_L}$, may be performed in parallel at each iteration of the MCMC algorithm. In particular, this algorithm may be implemented in a parallel computing environment, e.g. using the statistical language R (Rossini et al., 2007; R Core Team, 2023). In addition, each $\boldsymbol{\mu}_{\bullet 1k}, k = 1, \ldots, n_1$, can be simulated from its full conditional distribution using the Forward Filter Backward Sampler (FFBS) (Carter and Kohn, 1994; Frühwirth-Schnatter, 1994). Simulation of each $\boldsymbol{\theta}_{\bullet lj}, l = 1, \ldots, L; j = 1, \ldots, n_l$, is not trivial because the covariance matrix of the system equation is singular. Ferreira et al. (2011) have proposed a singular FFBS (SFFBS) for models with singular system equation covariance matrix that works well for the simulation of the latent multiscale coefficients. The SFFBS is presented in Theorem 5.2.

Theorem 5.2 (Singular forward filter backward sampler – SFFBS) (Ferreira et al., 2011) *Assume a dynamic multiscale spatio-temporal model with observation equation for the empirical multiscale coefficient* $\boldsymbol{\theta}^e_{tlj}$ *given by Equation (5.3) and system equation for the latent multiscale coefficient* $\boldsymbol{\theta}_{tlj}$ *given by Equation (5.5). Then,*

1. *Posterior at* $t-1$*:* $\boldsymbol{\theta}_{t-1,lj}|\mathcal{D}_{t-1} \sim N\left(\boldsymbol{m}_{t-1,lj}, C_{t-1,lj}\boldsymbol{\Omega}_{lj}\right)$;

2. *Prior at* t*:* $\boldsymbol{\theta}_{tlj}|\mathcal{D}_{t-1} \sim N\left(\boldsymbol{a}_{tlj}, R_{tlj}\boldsymbol{\Omega}_{lj}\right)$, *where* $\boldsymbol{a}_{tlj} = \boldsymbol{m}_{t-1,lj}$ *and* $R_{tlj} = C_{t-1,lj} + \psi_{lj}$;

3. *Posterior at* t*:* $\boldsymbol{\theta}_{tlj}|\mathcal{D}_{t} \sim N\left(\boldsymbol{m}_{tlj}, C_{tlj}\boldsymbol{\Omega}_{lj}\right)$, *where* $C_{tlj} = (1 + R^{-1}_{tlj})^{-1}$ *and* $\boldsymbol{m}_{tlj} = C_{tlj}\left(\boldsymbol{\theta}^e_{tlj} + R^{-1}_{tlj}\boldsymbol{a}_{tlj}\right)$.

4. *Backward sampling distribution:* $\boldsymbol{\theta}_{tlj}|\boldsymbol{\theta}_{(t+1):T,lj}, \mathcal{D}_T \equiv \boldsymbol{\theta}_{tlj}|\boldsymbol{\theta}_{t+1,lj}, \mathcal{D}_t \sim N(\boldsymbol{h}_{tlj}, H_{tlj}\boldsymbol{\Omega}_{lj})$, *where the multiplier in the covariance matrix is* $H_{tlj} = \left(C^{-1}_{tlj} + \psi^{-1}_{lj}\right)^{-1}$ *and* $\boldsymbol{h}_{tlj} = H_{tlj}\left(C^{-1}_{tlj}\boldsymbol{m}_{tlj} + \psi^{-1}_{lj}\boldsymbol{\theta}_{t+1,lj}\right)$.

We use the singular forward filter presented in Theorem 5.2 to perform analytic numerical integration, as we describe in the next section. We note that in this chapter we do not use the backward sampler that is part of Theorem 5.2. Instead, we use the backward smoother that we provide in the following theorem.

Theorem 5.3 (Backward smoother) *Assume a dynamic multiscale spatio-temporal model with observation equation for the empirical multiscale coefficient* $\boldsymbol{\theta}^e_{tlj}$ *given by Equation (5.3) and system equation for the latent multiscale coefficient* $\boldsymbol{\theta}_{tlj}$ *given by Equation (5.5). Then, the backward smoothing distribution is* $\boldsymbol{\theta}_{tlj}|\mathcal{D}_T \sim N(\boldsymbol{s}_{tlj}, S_{tlj}\boldsymbol{\Omega}_{lj})$, *where at the last time point* T *the mean vector is* $\boldsymbol{s}_{Tlj} = \boldsymbol{m}_{Tlj}$ *and the multiplier in the covariance matrix is* $S_{Tlj} = C_{Tlj}$, *and at time points* $T-1, \ldots, 1$ *these quantities are recursively computed as* $S_{tlj} = H_{tlj} + H^2_{tlj}\psi^{-2}_{lj}S_{t+1,lj}$ *and* $\boldsymbol{s}_{tlj} = H_{tlj}\left(C^{-1}_{tlj}\boldsymbol{m}_{tlj} + \psi^{-1}_{lj}\boldsymbol{s}_{t+1,lj}\right)$.

Proof *The proof of Theorem 5.3 is by induction. Assume that the smoothing distribution at time* $t+1$ *is* $\boldsymbol{\theta}_{t+1,lj}|\mathcal{D}_T \sim N(\boldsymbol{s}_{t+1,lj}, S_{t+1lj}\boldsymbol{\Omega}_{lj})$. *Note* *that after applying the forward filter, at the last time point T we have $\boldsymbol{\theta}_{Tlj}|\mathcal{D}_T \sim N(\boldsymbol{m}_{Tlj}, C_{Tlj}\boldsymbol{\Omega}_{lj})$. Thus, the backward smoother algorithm starts at time T with $\boldsymbol{s}_{Tlj} = \boldsymbol{m}_{Tlj}$ and $S_{Tlj} = C_{Tlj}$.*

The density of the smoothing distribution at time t is

$$p(\boldsymbol{\theta}_{tlj}|\mathcal{D}_T) = \int p(\boldsymbol{\theta}_{tlj}, \boldsymbol{\theta}_{t+1,lj}|\mathcal{D}_T)d\boldsymbol{\theta}_{t+1,lj}. \tag{5.6}$$

In addition, note that the integrand is

$$\begin{aligned} p(\boldsymbol{\theta}_{tlj}, \boldsymbol{\theta}_{t+1,lj}|\mathcal{D}_T) &= p(\boldsymbol{\theta}_{tlj}|\boldsymbol{\theta}_{t+1,lj}, \mathcal{D}_T)p(\boldsymbol{\theta}_{t+1,lj}|\mathcal{D}_T) && (5.7)\\ &= p(\boldsymbol{\theta}_{tlj}|\boldsymbol{\theta}_{t+1,lj}, \mathcal{D}_t)p(\boldsymbol{\theta}_{t+1,lj}|\mathcal{D}_T). && (5.8)\end{aligned}$$

Further, according to Theorem 5.2, the backward sampling distribution is $\boldsymbol{\theta}_{tlj}|\boldsymbol{\theta}_{t+1,lj}, \mathcal{D}_t \sim N(\boldsymbol{h}_{tlj}, H_{tlj}\boldsymbol{\Omega}_{lj})$ where $\boldsymbol{h}_{tlj} = H_{tlj}\left(C_{tlj}^{-1}\boldsymbol{m}_{tlj} + \psi_{lj}^{-1}\boldsymbol{\theta}_{t+1,lj}\right)$.

Therefore,

$$\begin{aligned} p(\boldsymbol{\theta}_{tlj}|\mathcal{D}_T) &= \int p(\boldsymbol{\theta}_{tlj}|\boldsymbol{\theta}_{t+1,lj}, \mathcal{D}_t)p(\boldsymbol{\theta}_{t+1,lj}|\mathcal{D}_T)d\boldsymbol{\theta}_{t+1,lj}\\ &= \int N(\boldsymbol{\theta}_{tlj}|H_{tlj}(C_{tlj}^{-1}\boldsymbol{m}_{tlj} + \psi_{lj}^{-1}\boldsymbol{\theta}_{t+1,lj}), H_{tlj}\boldsymbol{\Omega}_{lj})\\ &\qquad \times N(\boldsymbol{\theta}_{t+1,lj}|\boldsymbol{s}_{t+1,lj}, S_{t+1,lj}\boldsymbol{\Omega}_{lj})d\boldsymbol{\theta}_{t+1,lj}\\ &= N(\boldsymbol{\theta}_{tlj}|\boldsymbol{s}_{tlj}, S_{tlj}\boldsymbol{\Omega}_{lj}), && (5.9)\end{aligned}$$

where $S_{tlj} = H_{tlj} + H_{tlj}^2\psi_{lj}^{-2}S_{t+1,lj}$ and $\boldsymbol{s}_{tlj} = H_{tlj}(C_{tlj}^{-1}\boldsymbol{m}_{tlj} + \psi_{lj}^{-1}\boldsymbol{s}_{t+1,lj})$. □

5.5.1 Analytic Numerical Integration

Using the mixture prior for Bayesian thresholding, we obtain by Bayes' Theorem the following joint posterior distribution for $\boldsymbol{\theta}_{1lj}$ and ψ_{lj} :

$$\begin{aligned} p(\psi_{lj}, \boldsymbol{\theta}_{1lj}|\boldsymbol{\theta}_{1:T,lj}^e) &= \frac{p(\psi_{lj}, \boldsymbol{\theta}_{1lj}|\boldsymbol{\theta}_{1lj}^e)p(\boldsymbol{\theta}_{2:T,lj}^e|\psi_{lj}, \boldsymbol{\theta}_{1lj}, \boldsymbol{\theta}_{1lj}^e)}{p(\boldsymbol{\theta}_{2:T,lj}^e|\boldsymbol{\theta}_{1lj}^e)}\\ &= p_{1lj}p(\psi_{lj}, \boldsymbol{\theta}_{1lj}|\boldsymbol{\theta}_{1:T,lj}^e, M_{1lj})\\ &\quad + p_{2lj}p(\boldsymbol{\theta}_{1lj}|\boldsymbol{\theta}_{1:T,lj}^e, M_{2lj})\delta_{\psi_{lj}}(0)\\ &\quad + p_{3lj}\delta_{\psi_{lj}}(0)\delta_{\theta_{lj}}(0),\end{aligned}$$

where p_{1lj}, p_{2lj} and p_{3lj} are respectively the posterior probabilities of hypotheses H_{1lj}, H_{2lj} and H_{3lj}. These posterior probabilities are computed as

$$\begin{aligned} p_{1lj} &= k\omega_{lj}^{\psi}\int_{\Psi_{lj}} IG(\psi_{lj}|0.5\tau_{lj}^{\psi}, 0.5\kappa_{lj}^{\psi})p(\boldsymbol{\theta}_{2:T,lj}^e|\boldsymbol{\theta}_{1lj}^e, \psi_{lj} > 0)d\psi_{lj},\\ p_{2lj} &= k(1-\omega_{lj}^{\psi})\omega_{lj}^{\theta}p(\boldsymbol{\theta}_{2:T,lj}^e|\boldsymbol{\theta}_{1lj}^e, \psi_{lj} = 0, \boldsymbol{\theta}_{1lj} \neq \boldsymbol{0}),\\ p_{3lj} &= k(1-\omega_{lj}^{\psi})(1-\omega_{lj}^{\theta})N(\boldsymbol{\theta}_{2:T,lj}^e|\boldsymbol{1}_{T-1}\otimes\boldsymbol{0}_{m_{lj}}, \boldsymbol{I}_{T-1}\otimes\boldsymbol{\Omega}_{lj}),\end{aligned}$$

where k is a normalizing constant such that $p_{1lj} + p_{2lj} + p_{3lj} = 1$. Let $\boldsymbol{M} = min(i,j)$, $i = 1,\ldots,T$, $j = 1,\ldots,T$. For the evaluation of p_{1lj} and p_{2lj}, we efficiently compute the predictive density $p(\boldsymbol{\theta}^e_{2:T,lj}|\boldsymbol{\theta}^e_{1lj}, \psi_{lj})$ with the singular forward filter, that is

$$\begin{aligned} p(\boldsymbol{\theta}^e_{2:T,lj}|\boldsymbol{\theta}^e_{1lj}, \psi_{lj}) &= \int_{\Theta_{1lj}} N\left(\boldsymbol{\theta}_{1lj}|\boldsymbol{\theta}^e_{1lj}, \boldsymbol{\Omega}_{lj}\right) N\left(\boldsymbol{\theta}^e_{2:T,lj}|\mathbf{1}_{T-1} \otimes \boldsymbol{\theta}_{1lj},\right. \\ &\qquad \left.(\boldsymbol{I}_{T-1} + \psi_{lj}\boldsymbol{M}) \otimes \boldsymbol{\Omega}_{lj}\right) d\boldsymbol{\theta}_{1lj} \\ &= \prod_{t=2}^{T} p(\boldsymbol{\theta}^e_{tlj}|\boldsymbol{\theta}^e_{1:(t-1),lj}, \psi_{lj}) \\ &= \prod_{t=2}^{T} N(\boldsymbol{\theta}^e_{tlj}|\boldsymbol{f}_{tlj}, Q_{tlj}\boldsymbol{\Omega}_{lj}), \end{aligned} \tag{5.10}$$

where $\boldsymbol{f}_{tlj} = \boldsymbol{a}_{tlj}$ and $Q_{tlj} = R_{tlj} + 1$ are recursively computed with the singular forward filter. Moreover,

$$\begin{aligned} p(\boldsymbol{\theta}_{1lj}, \psi_{lj}|\boldsymbol{\theta}^e_{1:T,lj}, H_{1lj}) \propto\ & IG(\psi_{lj}|0.5\tau^{\psi}_{lj}, 0.5\kappa^{\psi}_{lj}) N(\boldsymbol{\theta}_{1lj}|\boldsymbol{\theta}^e_{1lj}, \boldsymbol{\Omega}_{lj}) \\ & N\left(\boldsymbol{\theta}^e_{2:T,lj}|\mathbf{1}_{T-1} \otimes \boldsymbol{\theta}_{1lj},\right. \\ & \qquad \left.(\boldsymbol{I}_{T-1} + \psi_{lj}\boldsymbol{M}) \otimes \boldsymbol{\Omega}_{lj}\right). \end{aligned}$$

Thus, conditional on $\psi_{lj} > 0$, the marginal posterior density of ψ_{lj} is

$$\begin{aligned} p(\psi_{lj}|\boldsymbol{\theta}^e_{1:T,lj}, H_{1lj}) &= \int_{\Theta_{1lj}} p(\boldsymbol{\theta}_{1lj}, \psi_{lj}|\boldsymbol{\theta}^e_{1:T,lj}, M_{1lj}) d\boldsymbol{\theta}_{1lj} \\ &\propto IG(\psi_{lj}|0.5\tau^{\psi}_{lj}, 0.5\kappa^{\psi}_{lj}) p(\boldsymbol{\theta}^e_{2:T,lj}|\boldsymbol{\theta}^e_{1lj}, \psi_{lj}). \end{aligned}$$

As explained in Equation (5.10), for each value of ψ_{lj} we can efficiently compute $p(\boldsymbol{\theta}^e_{2:T,lj}|\boldsymbol{\theta}^e_{1lj}, \psi_{lj})$ with the singular forward filter. Therefore, posterior summaries for ψ_{lj} such as posterior mean and variance can be efficiently computed with the composite trapezoid Newton-Cotes rule (Press et al., 1992).

In addition, under hypothesis H_{1lj}, the conditional posterior density of $\boldsymbol{\theta}_{1lj}$ given ψ_{lj} is

$$\begin{aligned} p(\boldsymbol{\theta}_{1lj}|\boldsymbol{\theta}^e_{1:T,lj}, \psi_{lj}, H_{1lj}) &\propto N(\boldsymbol{\theta}_{1lj}|\boldsymbol{\theta}^e_{1lj}, \boldsymbol{\Omega}_{lj}) N\left(\boldsymbol{\theta}^e_{2:T,lj}|\mathbf{1}_{T-1} \otimes \boldsymbol{\theta}_{1lj},\right. \\ &\qquad \left.(\boldsymbol{I}_{T-1} + \psi_{lj}\boldsymbol{M}) \otimes \boldsymbol{\Omega}_{lj}\right) \\ &\propto N(\boldsymbol{\theta}_{1lj}|\boldsymbol{s}_{1lj}, S_{1lj}\boldsymbol{\Omega}_{lj}), \end{aligned}$$

where $\boldsymbol{s}_{1lj}$ and S_{1lj} (which are functions of ψ_{lj}) are computed with the singular forward filter and the backward smoother as explained in Theorems 5.2 and 5.3.

Finally, under model H_{2lj} the posterior density of $\boldsymbol{\theta}_{1lj}$ is

$$p(\boldsymbol{\theta}_{1lj}|\boldsymbol{\theta}^e_{1:T,lj}, H_{2lj}) \propto N(\boldsymbol{\theta}_{1lj}|\boldsymbol{\theta}^e_{1lj}, \boldsymbol{\Omega}_{lj}) \prod_{t=2}^{T} N(\boldsymbol{\theta}^e_{tlj}|\boldsymbol{\theta}_{1lj}, \boldsymbol{\Omega}_{lj}).$$

Therefore, $\boldsymbol{\theta}_{1lj}|\boldsymbol{\theta}^e_{1:T,lj}, H_{2lj} \sim N(T^{-1}\sum_{t=1}^{T} \boldsymbol{\theta}^e_{tlj}, T^{-1}\boldsymbol{\Omega}_{lj})$.

5.6 Application: Agriculture Production in Espírito Santo State

This section demonstrates the use of our multiscale thresholding method with an application to the data on agriculture production in Espírito Santo State explored in Section 5.3.

Table 5.1 presents results for the multiscale coefficients that connect each macroregion to its children microregions. Specifically, Table 5.1 presents posterior probabilities of the hypotheses of each multiscale coefficient being fixed zero, fixed not zero, or temporally dynamic. As explained in Section 5.5, these probabilities take into account the temporal variability in the empirical multiscale coefficients through time. According to Table 5.1, the data strongly supports the hypotheses that the multiscale coefficients for Macroregions 1 and 3 are time-varying dynamic. Meanwhile, evidence from the data reflects very low posterior probability that the multiscale coefficients for Macroregions 2 and 4 are equal to zero vectors. Instead, the data provides evidence that the multiscale coefficients for macroregions 2 and 4 are either fixed nonzero or time-varying dynamic, but without strongly supporting one of these hypotheses over the other.

Table 5.2 presents results for the multiscale coefficients that connect each microregion to its children counties. Specifically, Table 5.2 presents posterior probabilities of the hypotheses of each multiscale coefficient being fixed zero, fixed not zero, or temporally dynamic. From this table, there is strong evidence that none of the multiscale coefficients that connect each microregion to its descendants is fixed and zero. In addition, there is strong evidence that the multiscale coefficient for Microregion 3 is temporally dynamic. Further, there is some evidence that the multiscale coefficient for Microregion 2 may be fixed but not zero.

TABLE 5.1
Agriculture production application. Multiscale coefficients from macroregions to microregions. Posterior probabilities of the hypotheses of each multiscale coefficient being fixed zero, fixed not zero, or temporally dynamic.

Macroregion	Hypothesis		
	Temporally dynamic	Fixed not zero	Fixed zero
1	0.99	0.01	0.00
2	0.67	0.33	0.00
3	1.00	0.00	0.00
4	0.49	0.51	0.00

TABLE 5.2
Agriculture production application. Multiscale coefficients from microregions to counties. Posterior probabilities of the hypotheses of each multiscale coefficient being fixed zero, fixed not zero, or temporally dynamic.

Microregion	Hypothesis		
	Temporally dynamic	Fixed not zero	Fixed zero
1	0.87	0.13	0.00
2	0.78	0.22	0.00
3	1.0	0.00	0.00
4	0.35	0.65	0.00
5	0.84	0.16	0.00
6	0.86	0.14	0.00
7	0.61	0.39	0.00
8	0.36	0.64	0.00
9	0.37	0.63	0.00
10	0.52	0.48	0.00
11	0.33	0.67	0.00
12	0.09	0.91	0.00

To further examine the results for Microregion 3, each panel in Figure 5.3 presents the time series plots of one of the elements of the multiscale coefficient for Microregion 3. Specifically, Figure 5.3 presents the empirical multiscale coefficient (circles), the posterior mean (solid line), and 95% credible intervals (dashed lines). As we can see in the figure, the elements of the multiscale coefficient for Microregion 3 seem to be changing through time. Indeed, Table 5.2 informs that the probability of the latent multiscale coefficient for Microregion 3 being a time-evolving dynamic multiscale coefficient is about one. Hence, the data provides very definite evidence that the latent multiscale coefficient for Microregion 3 is time-evolving dynamic. Therefore, there is definite evidence that, during the study period, the differences among the counties within Microregion 3 changed over time.

Similarly, each panel in Figure 5.4 presents the time series plots of one of the elements of the multiscale coefficient for Microregion 12. As we can see in the figure, the elements of the multiscale coefficient for Microregion 12 seem to be either fixed or changing slightly through time. In fact, Table 5.2 informs that the probability of the latent multiscale coefficient for Microregion 12 being fixed through time but not zero is 0.91. Hence, the data provides some evidence that the latent multiscale coefficient for Microregion 12 is different than the zero vector but fixed through time. Therefore, there is some evidence that, during the study period, there were differences among the counties within Microregion 3 but those differences remained fixed over time.

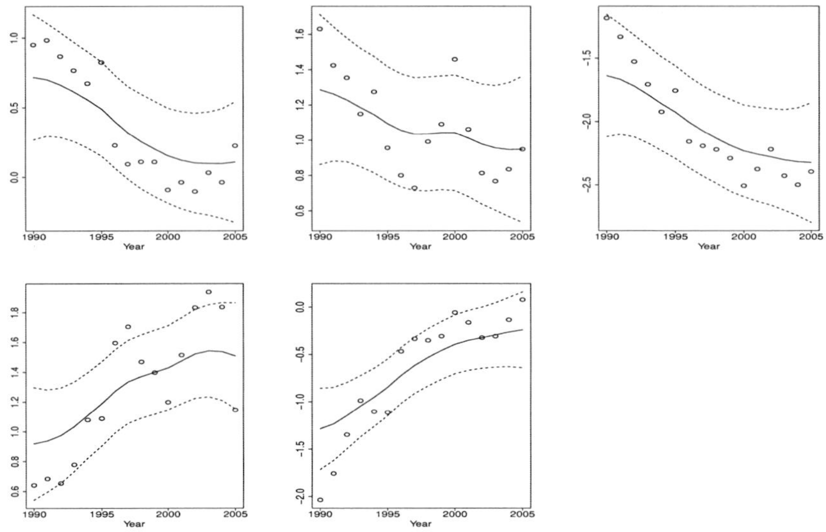

FIGURE 5.3
Time series plots of the five elements of the multiscale coefficient for Microregion 3. Empirical multiscale coefficient (circles), posterior mean (solid line), and 95% credible interval (dashed lines). According to our method, the probability of the latent multiscale coefficient being a dynamic multiscale coefficient is about one (see Table 5.2).

5.7 Conclusions

We have introduced here a novel threshold method for spatio-temporal Gaussian data based on dynamic multiscale spatio-temporal models. We note that while we have presented this method in the context of spatio-temporal analysis, the method may be applied to any situation where the data are on a tree structure. For example, we may approximate a network with a tree. In that context, one possible application of this threshold method would be for streaming data on a network where the analyst needs to decide on what parts of the network to focus because of temporally dynamic changes.

There are many possible avenues for future research. A promising research direction is the extension of the threshold method proposed here to spatio-temporal nonGaussian data. In particular, extensions to Poisson data may be achieved by considering the dynamic multiscale spatio-temporal models

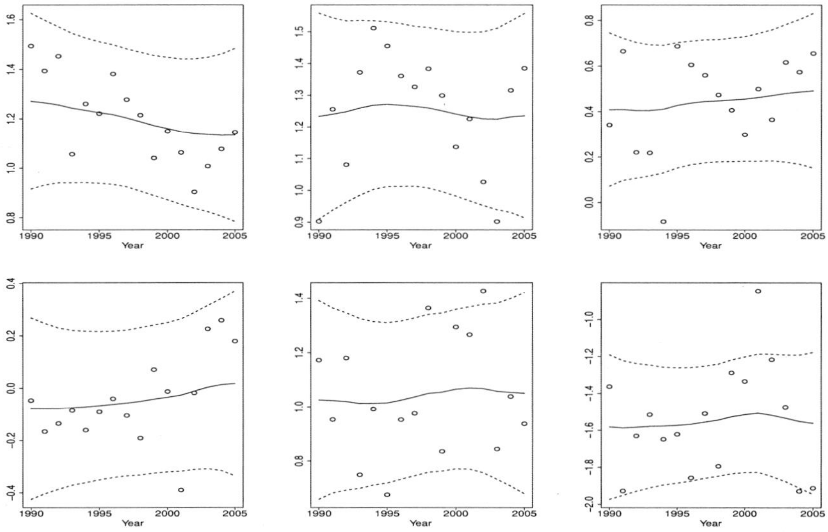

FIGURE 5.4
Time series plots of six of the elements of the multiscale coefficient for Microregion 12. Empirical multiscale coefficient (circles), posterior mean (solid line), and 95% credible interval (dashed lines). According to our method, the probability of the latent multiscale coefficient being fixed through time but not zero is 0.91 (see Table 5.2).

for Poisson data proposed by Hoegh et al. (2016) and Fonseca and Ferreira (2017). While the latter proposes a dynamic Dirichlet evolution for the multiscale coefficient vectors, the former evolves the multiscale coefficient vectors using generalized dynamic linear models (West et al., 1985; Ferreira and Gamerman, 2000; Alves et al., 2010). Another possible research direction is the development of threshold methods for dynamic multiscale multivariate spatio-temporal models (Elkhouly and Ferreira, 2021). Other possible research direction is to develop methods for the design of adaptive multiscale data collection, possibly using ideas and results from optimal Bayesian design of monitoring networks (Ruiz-Cárdenas et al., 2010, 2012; Ferreira and Sanyal, 2014; Ferreira, 2015). Finally, another possible direction is the development of multiscale spatio-temporal models and thresholding for lognormal processes, which may be facilitated by approximations to aggregation of lognormal processes (Salazar and Ferreira, 2011).

Bibliography

Abramovich, F., T. Sapatinas, and B. W. Silverman (1998). Wavelet thresholding via a Bayesian approach. *Journal of the Royal Statistical Society 60*, 725–749.

Alves, M. B., D. Gamerman, and M. A. R. Ferreira (2010). Transfer functions in dynamic generalized linear models. *Statistical Modelling 10*, 03–40.

Anderson, T. W. (1984). *An Introduction to Multivariate Statistical Analysis* (2nd ed.). New York: John Wiley & Sons.

Carter, C. K. and R. Kohn (1994). On Gibbs sampling for state space models. *Biometrika 81*, 541–553.

Elkhouly, M. and M. A. R. Ferreira (2021). Dynamic multiscale spatiotemporal models for multivariate Gaussian data. *Spatial Statistics 41*, 100475.

Ferreira, M. A. R. (2015). Inhomogeneous evolutionary MCMC for Bayesian optimal sequential environmental monitoring. *Environmental and Ecological Statistics 22*(4), 705–724.

Ferreira, M. A. R., A. I. Bertolde, and S. Holan (2010). Analysis of economic data with multi-scale spatio-temporal models. In A. O'Hagan and M. West (Eds.), *Handbook of Applied Bayesian Analysis*, pp. 295–318. Oxford University Press.

Ferreira, M. A. R. and D. Gamerman (2000). Dynamic generalized linear models. In G. Dey and Mallick (Eds.), *Generalized Linear Models: a Bayesian Perspective*, New York, pp. 57–72. Marcel Dekker.

Ferreira, M. A. R., S. H. Holan, and A. I. Bertolde (2011). Dynamic multiscale spatio-temporal models for Gaussian areal data. *Journal of the Royal Statistical Society (Series B: Methodological) 73*, 663–688.

Ferreira, M. A. R. and H. K. H. Lee (2007). *Multiscale Modeling: A Bayesian Perspective.* Springer Series in Statistics. New York: Springer.

Ferreira, M. A. R., E. M. Porter, and C. T. Franck (2021). Fast and scalable computations for Gaussian hierarchical models with intrinsic conditional autoregressive spatial random effects. *Computational Statistics and Data Analysis 162*, 107264.

Ferreira, M. A. R. and N. Sanyal (2014). Bayesian optimal sequential design for nonparametric regression via inhomogeneous evolutionary MCMC. *Statistical Methodology 18*, 131–141.

Fonseca, T. C. O. and M. A. R. Ferreira (2017). Dynamic multiscale spatiotemporal models for Poisson data. *Journal of the American Statistical Association 112*, 215–234.

Frühwirth-Schnatter, S. (1994). Data augmentation and dynamic linear models. *Journal of Time Series Analysis 15*(2), 183–202.

Haar, A. (1910). Zur theorie der orthogonalen funktionensysteme. *Mathematische Annalen 69*, 331–371.

Hoegh, A., M. A. R. Ferreira, and S. Leman (2016). Spatiotemporal model fusion: multiscale modelling of civil unrest. *Journal of the Royal Statistical Society: Series C (Applied Statistics) 65*, 529–545.

Kolaczyk, E. D. and H. Huang (2001). Multiscale statistical models for hierarchical spatial aggregation. *Geographical Analysis 33*, 95–118.

Lindley, D. V. (1957). A statistical paradox. *Biometrika 44*, 187–192.

Mallat, S. G. (1989a). Multiresolution approximations and the wavelet orthonormal bases of $l^2(r)$. *Transactions of the American Mathematical Society 315*, 69–87.

Mallat, S. G. (1989b). A theory for multiresolution signal decomposition: the wavelet representation. *IEEE Transactions on Pattern Analysis and Machine Intelligence 11*, 674–693.

Prado, R., M. A. R. Ferreira, and M. West (2021). *Time Series: Modeling, Computation, and Inference* (2nd ed.). Boca Raton: CRC Press.

Press, W. H., S. A. Teukolsky, W. T. Vetterling, and B. P. Flannery (1992). *Numerical Recipes in C* (2nd ed.). Cambridge: Cambridge University Press.

R Core Team (2023). *R: A Language and Environment for Statistical Computing.* Vienna, Austria: R Foundation for Statistical Computing.

Robert, C. P. (2001). *The Bayesian Choice* (2nd ed.). New York: Springer.

Rossini, A. J., L. Tierney, and N. Li (2007). Simple parallel statistical computing in R. *Journal of Computational and Graphical Statistics 16*, 399–420.

Ruiz-Cárdenas, R., M. A. R. Ferreira, and A. M. Schmidt (2010). Stochastic search algorithms for optimal monitoring network designs. *Environmetrics 21*, 102–112.

Ruiz-Cárdenas, R., M. A. R. Ferreira, and A. M. Schmidt (2012). Evolutionary Markov chain Monte Carlo algorithms for optimal monitoring network designs. *Statistical Methodology (Special Issue on Astrostatistics and Spatial Statistics) 9*, 185–194.

Salazar, E. and M. A. R. Ferreira (2011). Temporal aggregation of lognormal autoregressive processes. *Journal of Time Series Analysis 32*, 661–671.

West, M. and P. J. Harrison (1997). *Bayesian Forecasting and Dynamic Models* (2nd. ed.). New York: Springer.

West, M., P. J. Harrison, and H. S. Migon (1985). Dynamic generalized linear models and Bayesian forecasting (with discussion). *Journal of the American Statistical Association 80*, 73–96.

6

Multiscale Spatio-Temporal Data Assimilation

Ana G. Rappold

U.S. Environmental Protection Agency

Marco A. R. Ferreira

Virginia Tech

Across scientific disciplines, there is a strong interest in developing methodology for assimilation of computer model predictions with observed data. In particular, in the assessment of air quality we are interested in assimilating data from models such as the Community Multiscale Air Quality model (CMAQ) from the United States Environment Protection Agency (EPA) with data from monitoring stations. Here, we present a Bayesian hierarchical model for such data assimilation in both time and space, where the joint distribution of the two data types is modeled through a common spatio-temporal latent process. The model accommodates the change of support problem and handles spatially and temporally distributed additive bias associated with CMAQ data through a multiscale spatio-temporal model. The multiscale approach to estimating bias gives insight into the similarities and discrepancies of two data types at different scales of resolution through time. We illustrate the value of data assimilation in the imputation of daily time series across all locations, and in prediction over areas with sparse data collection. Our framework places strong emphasis on prediction, interpretability, and computational feasibility.

DOI: 10.1201/9781032623443-6

6.1 Introduction

A substantial focus of air pollution research is concerned with the nature of the adverse health effects associated with human exposure to particulate matter (PM). The US Environmental Protection Agency (EPA) is interested in developing detailed maps of daily fluctuations of PM. Such maps are used in epidemiological research of health effects, in the formulation of regulatory policies, and in the evaluation of regional attainment of regulatory polices. Assessments of exposure to particulate matter at the population level are largely limited to the ambient monitoring networks. These are networks of stations sparsely located across the US and they collect data at distinct temporal intervals. In addition to the monitoring network data, current advances in mathematical models such as the Community Multiscale Air Quality model are producing potentially valuable air quality predictions at dense spatial and temporal scales. Thus, there is an opportunity to combine the monitoring network data with the CMAQ data to obtain detailed maps of daily fluctuations of PM. In addition, a proper assimilation of the monitoring network data and the CMAQ data can enhance our understanding of fluctuations in ambient air quality. However, one important point to note is that the CMAQ data are based on deterministic mathematical models and may be biased. Therefore, the goal of this chapter is to present a framework for monitoring and CMAQ data assimilation on both temporal and spatial domains that corrects for possible CMAQ biases, is computationally feasible, and produces detailed daily maps of pollution.

EPA maintains air monitoring networks across the United States independently and in collaboration with local, tribal, and state governments. This monitoring program encompasses various fractions and components of particulate matter, ozone, and other pollutants. In this chapter we focus on the mass concentration of solid and liquid airborn particles with a diameter less than or equal to $2.5\,\mu m$, also called the fine fraction portion $PM_{2.5}$ of particulate matter. The size of these particles allows for deep absorption into the airways, causing health concerns such as increased respiratory symptoms, bronchitis, aggravation of asthma, and mortality (Pope III et al., 2002). The main components of $PM_{2.5}$ are sulfates, nitrates, carbon, soil, and dust, and ambient concentration depends on the relative distribution of these components. Therefore, the ambient concentration depends on the regional, local and seasonal emissions, and on the ability of particle components to form secondary pollutants in the atmosphere. Sulfates, for example, are regional and seasonal pollutants in the American Southeast, primarily formed by the conversion of power plant generated sulfate dioxide emissions. Their size allows for long-scale transport and their chemical reactivity under regular atmospheric conditions accounts for increased levels during the summer months even at a constant emission rate. In the American Southwest, on the other hand, dominant components of $PM_{2.5}$ are elemental and organic carbon emitted by on-road and stationary sources. Forest fires can also emit large amounts of carbon causing spikes in $PM_{2.5}$ concentration. Thus, the spatio-temporal dynamics of $PM_{2.5}$ depends on non-trivial relationships in the emissions, composition of particles, meteorological conditions, and the geographic background of the region. As a consequence, learning about the spatio-temporal dynamics is difficult on the basis of sparsely distributed observed monitoring station data alone.

In partnership with the National Oceanic and Atmospheric Administration (NOAA), EPA has developed an Air Quality Forecasting model by incorporating NOAA's meteorological models into EPA's Community Multiscale Air Quality (CMAQ) model (`www.epa.gov/cmaq`). CMAQ is a numerical model which simulates atmospheric processes guiding the conversion, transport, and deposition of air pollutants (Byun and Ching, 1999). CMAQ predictions are biased because the model depends on uncertain inputs and deterministic mechanisms. However, diagnostic analysis of CMAQ data suggests reliable performance and promising forecasting utility in a number of airborne pollutants. Previous analyses by Jun and Stein (2004), Eder and Yu (2006), and others reveal that modeling of emissions' fate and transport in the atmosphere, together with meteorological conditions and geographical classifications can capture temporal and spatial dependencies. In the analysis of sulfate concentrations, for example, Jun and Stein (2004) show that CMAQ captures the space-time correlation of the monitored concentrations. Thus, the results obtained by Jun and Stein (2004), Eder and Yu (2006), and others point to the potential usefulness of CMAQ data for the estimation of detailed daily maps of pollution. With this latter goal in mind, the combination of monitoring station data and CMAQ data requires sophisticated statistical modeling. Two primary reasons for the need of sophisticated models are the sparseness of the monitoring data and the different spatial support between the two data types. The monitoring data are sparse in space and time but are points in continuous space, while CMAQ data represent pollutant volume averages over a given time period within a grid. We handle the difference in support between the two data types by representing the monitoring station data as noisy realizations from a latent pollution process, and the CMAQ data as biased realizations from the grid cell integrated latent pollution process.

In this chapter we present a fully Bayesian hierarchical model for data assimilation in both time and space that combines monitoring station $PM_{2.5}$ data and CMAQ $PM_{2.5}$ data. Following the Bayesian melding framework of Fuentes and Raftery (2005), we model the two data types jointly by assuming that they are noisy observations of a latent *true* $PM_{2.5}$ process. The part of this latent pollution process that is shared by both data types is a common latent process constructed as a convolution of a white noise process whose parameters evolve dynamically in time (West and Harrison, 1997; Higdon, 1998; Stroud et al., 2001; Ferreira, 2015; Prado et al., 2021). The model accommodates the change of support problem and handles spatially and temporally distributed additive bias associated with CMAQ data through a dynamic multiscale spatio-temporal model. The dynamic multiscale approach to estimating bias gives insight into the similarities and discrepancies of the two data types at different scales of resolution through time. This is an important aspect of data validation, as the posterior summaries of the spatio-temporal bias can be used explicitly for feedback to CMAQ modelers. Our model-based framework provides estimates of missing daily values of particulate matter, complete bias-corrected spatial surfaces, and the associated uncertainty estimates, within a fully probabilistic framework.

Dynamic multiscale spatio-temporal models have been successfully developed and applied in several research areas. For example, Berliner et al. (1999) analyzed turbulent flow by decomposing the turbulence data using wavelets and then modeling the underlying wavelet coefficients with state-space models. Johannesson et al. (2007) developed a multiscale modeling framework with temporal dynamics only at a level of resolution coarser than the level of resolution of the observations, and applied their method to daily total column ozone satellite data. Ferreira et al. (2010, 2011) developed dynamic multiscale spatio-temporal models for univariate Gaussian data allowing heteroscedasticity and temporal dynamics at all scales of resolution, and applied their method to agriculture and epidemiological data. Building on the models of Ferreira et al. (2010, 2011), Wang et al. (2021) developed Gaussian dynamic multiscale spatio-temporal models applied to consumer purchases in marketing. Hoegh et al. (2016) and Fonseca and Ferreira (2017) developed dynamic multiscale spatio-temporal models for Poisson data. Hoegh et al. (2016) presented an application to spatio-temporal data on civil unrest whereas Fonseca and Ferreira (2017) presented applications to epidemiology and tornado occurrence. Elkhouly and Ferreira (2021) developed dynamic multiscale spatio-temporal models for multivariate Gaussian data, and applied their framework to stratospheric temperatures. Ferreira (2021) provides a review of the literature on dynamic multiscale spatio-temporal models and other models based on multiscale factorizations.

The framework we propose here can be thought of as spatio-temporal Bayesian melding. In the context of pollution, Bayesian melding was first proposed by Fuentes and Raftery (2005) for combining SO_2 data from monitoring stations and a numerical model output. While Fuentes and Raftery (2005) proposed a method for spatial data, here we propose a method for spatio-temporal data. Another distinction is that while the approach of Fuentes and Raftery (2005) models the bias as a polynomial function of the spatial location at a single time point, our approach models the bias through time using a dynamic multiscale spatio-temporal model (Ferreira et al., 2010, 2011; Ferreira, 2021) which provides more flexibility and allows for fast computations. Finally, the proposed spatio-temporal assimilation model gives us the ability to evaluate CMAQ performance and make predictions in a unified manner.

An alternative to joint modeling of disparate data is a conditional approach such as found in Brown et al. (2001), Wikle et al. (2001), and Berrocal et al.

(2010). In such an approach, a conditional relationship is specified between the observed data and the deterministic counterpart, that is, the output of the computer model is used as a predictor variable. Conditional relationships are often more intuitive to specify and are consequently interpretable, especially in the context of model validation. In applications such as ours, a drawback of the approaches proposed by Brown et al. (2001) and Wikle et al. (2001) is that predictions are limited to the support or dimension of the conditioning variables (Royle and Berliner, 1999). However, the spatio-temporal downscaler proposed by Berrocal et al. (2010) is able to provide predictions at the spatial point level. Specifically, Berrocal et al. (2010) relate the monitoring station data to the computer model output through a linear regression model with spatially varying coefficients. In addition, the spatially varying coefficients are modeled with a correlated spatial Gaussian process through coregionalization (Schmidt and Gelfand, 2003; Gelfand et al., 2004). This approach may be faster than the Bayesian melding approach that we propose here. However, as the data from each monitoring station is regressed on the computer model output for the grid cell that contains the monitoring station, the predicted surface in the spatio-temporal downscaler may not be spatially continuous. In contrast, the approach we propose here yields a spatially continuous predicted surface.

In addition to the literature cited above, there is substantial literature on space-time modeling of environmental exposure. Hierarchical Bayesian approaches for spatial prediction of air pollution have been developed in Brown et al. (1994), Sun et al. (2000), and Huerta et al. (2004). Specific spatial and temporal modeling and prediction of $PM_{2.5}$ can be found in Kibria et al. (1997), Cressie et al. (1999) , Zidek et al. (2002), Shaddick and Wakefield (2002), Smith et al. (2003), and Sahu and Mardia (2005). Wikle (2003) provides an overview of hierarchical modeling in environmental science. Gotway and Young (2002) give a historical perspective of the inferential problems associated with combining data from different spatial resolutions and an overview of geostatistical solutions. In the same context, Swall and Foley (2009) give a specific illustration of difficulties in CMAQ model validation. Chapter 7 of the book by Banerjee et al. (2014) gives examples of spatial change of support problems, related spatial and spatio-temporal models, and numerous references.

Finally, we note that multiscale models and methods have been developed and applied successfully in many other scientific contexts. For example, multiscale approximations may provide considerable computational speed-ups in the analysis of spatial data (Higdon, 2002; Johannesson and Cressie, 2004; Cressie and Johannesson, 2008; Katzfuss, 2013; Nychka et al., 2015; Katzfuss, 2017; Katzfuss and Gong, 2020; Zammit-Mangion and Rougier, 2020; Ferreira, 2021). Another example are multiscale random field models that have been successfully applied in subsurface hydrology (Ferreira et al., 2005; Ferreira and Lee, 2007). In addition, Ferreira and Jaramillo (2014) have proposed entropy-based multiscale phylogenetic models. Another important area is the development of multiscale models based on Pólya tree and its extensions for nonparametric estimation of density functions (Ferguson, 1973; Lavine, 1992; Müller et al., 1996; Hanson and Johnson, 2002; Hanson, 2006; Wong and Ma, 2010; Jara and Hanson, 2011; Ma, 2017a,b; Ma and Soriano, 2018)

The remainder of this chapter is organized as follows. Section 6.2 describes the data used in the study and presents an exploratory data analysis which motivates the model we propose. Section 6.3 details our joint model for CMAQ and monitoring network data, which includes a spatio-temporal multiscale model component for the additive bias. Section 6.4 presents results of the spatio-temporal analysis of the CMAQ and monitoring network data using our proposed methodology. Finally, Section 6.5 offers a brief summary and a discussion of future work to be done in this setting. Details about the multiscale decomposition are presented in the Appendix.

6.2 Data and Exploratory Analysis

In this study, we analyze $PM_{2.5}$ data collected at 200 stations within the region of the United States between latitudes $30°N$ and $40°N$ and longitudes $80°W$ and $90°W$ in year 2001. This region includes the states of Alabama, Georgia, Kentucky, and Tennessee, as well as parts of the states of Florida, Illinois, Indiana, Mississippi, Missouri, North Carolina, Ohio, South Carolina, Virginia, and West Virginia. These 200 stations are part of the combined network program observing the Federal Reference Method for data collection. Collection of 24-hour integrated concentrations follows a site-specific protocol for collection every one, three, or six days. The $PM_{2.5}$ values are skewed to the right; we normalize and stabilize the variance in the data by using the logarithm transformation. The observed levels of ambient $PM_{2.5}$ vary by location and at daily and seasonal time scales. Short and long range trends in $PM_{2.5}$ can be seen from the plot of daily log concentration at the monitoring stations in the region, through time, in Figure 6.1. The figure reveals increased concentration values during the summer months and a cyclical behavior with a period of approximately one week. The within-day variability in Figure 6.1 is attributed to the spatial dimension.

The geographical region of interest is covered by 777 CMAQ grid cells all of size 36km $\times$ 36km. Geographic positions of the monitoring sites and the centroids of the CMAQ prediction grid are shown in Figure 6.2. Black triangles, circles, and squares represent the locations of monitoring stations. Monitors with daily collection ($\sim$ 12%) correspond to black triangles, collection every third day to black circles ($\sim$ 70%), and every sixth day to black squares ($\sim$ 18%). Thus, not only the CMAQ and monitoring station data have different spatial support, but the monitoring station data have a substantial amount of missing data. Therefore, as a consequence of the data collection

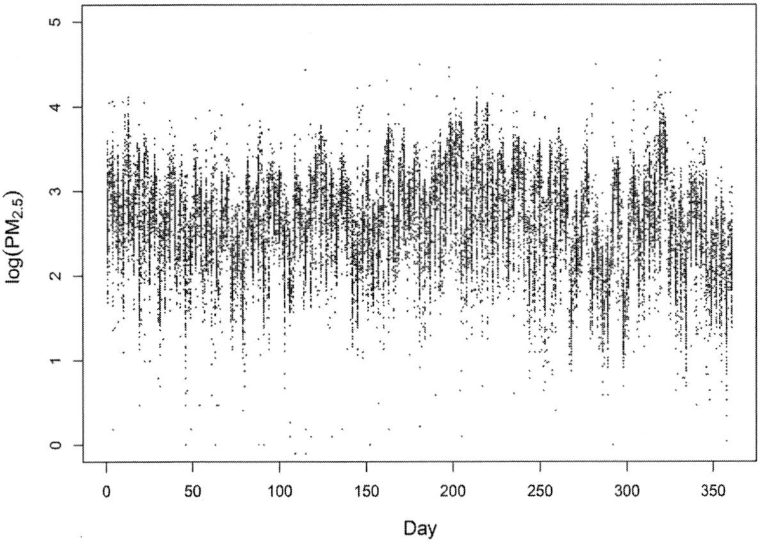

FIGURE 6.1
Time series of log-transformed $PM_{2.5}$ concentrations ($\mu g/m^3$) observed at monitoring stations located between $30°N - 40°N$ and $80°W - 90°W$ in year 2001.

schedule, on two out of three days 62% to 88% percent of the monitoring station data are missing.

We propose a spatio-temporal model for data assimilation that takes into account the amount of missing data in the observed series, the change of spatial support, and the spatio-temporal bias of CMAQ predictions. Focusing on the data from the first two weeks of 2001, Figure 6.3 illustrates the main points that provide the motivation for our proposed spatio-temporal model. Panel (a) presents box-plots of the daily values at stations collecting samples on a daily basis along with those of the corresponding CMAQ predictions. Panel (b) presents box-plots of values from stations collecting data every third day and the corresponding daily CMAQ predictions. CMAQ predictions are summarized only for the grid cells that contain monitors in the same panel. In both panels, we observe that CMAQ output underestimates but tracks daily patterns of $PM_{2.5}$ well. The observed series in Panel (a) are on average higher than those in Panel (b) because the stations with daily sampling are located in urban areas whereas most stations in Panel (b) are positioned in remote locations. Distribution of $PM_{2.5}$ in remote areas is closely approximated by CMAQ predictions since the homogeneous distribution of $PM_{2.5}$ in remote areas allows better approximation by areal averages such as those given by CMAQ data. Higher concentrations of $PM_{2.5}$ in urban areas coupled with a

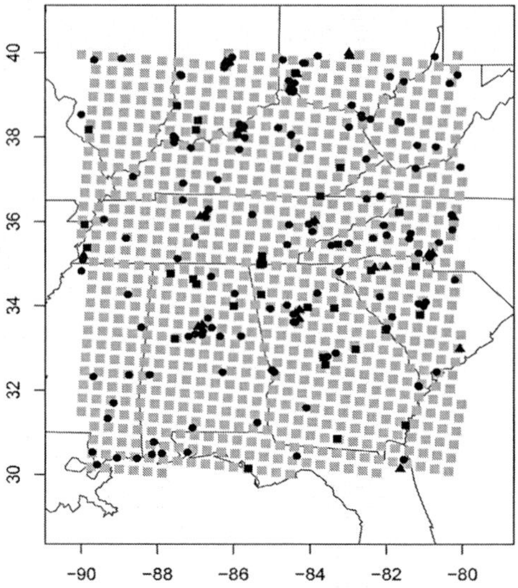

FIGURE 6.2
Locations of monitoring stations and CMAQ grid cells. Monitoring stations collecting data every one, three, and six days are indicated by black triangles, black circles, and black squares, respectively. Gray boxes are drawn around the centroids of CMAQ grid cells.

relatively large cell size of CMAQ grid partly explain the larger bias reflected in Panel (a). Therefore, the network design and the data collection protocol explain some of the space-time bias associated with CMAQ values.

Evaluation of the CMAQ bias by direct comparison of CMAQ data and monitoring station data is nontrivial because at most few monitoring stations observations are available to approximate any grid cell areal values. Specifically, a handful of grid cells contain more than one monitoring station, and the vast majority of the cells contain none. Aggregation of the data to coarser temporal resolutions also poses difficulties because on two out of any three consecutive days 62% to 88% percent of the data are missing. For example, due to weekly patterns of variation in $PM_{2.5}$ levels, aggregated weekly values can vary by 30% depending on the number of 3rd day monitoring data used.

In any case, the exploratory analysis in Figure 6.3 provides insight about the temporal dependence in the data and the mutual agreement between CMAQ and monitoring data types. Further, Figure 6.3 illustrates the difficulties of empirical validation of CMAQ predictions, and consequently difficulties in data assimilation, due to the amount of missing data from monitoring

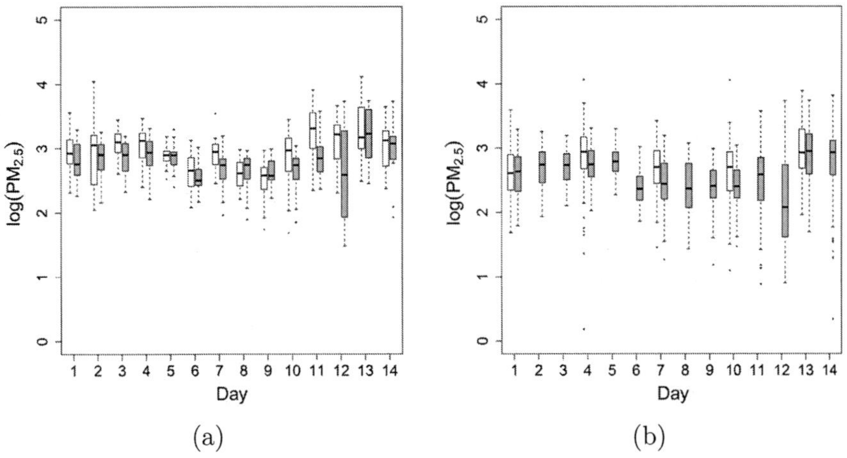

FIGURE 6.3
Box-plots of $PM_{2.5}$ values for the first two weeks of 2001 for monitoring stations (white) and CMAQ data for cells containing the corresponding monitoring stations (gray). Two groups of monitoring stations are considered according to collection frequency: daily (Panel (a)), and every third day (Panel (b)).

stations, and the difference in spatial support among the two data types. Therefore, the goal of our dynamic multiscale spatio-temporal framework is to account for the change of support, the missing data, the CMAQ bias, and the spatio-temporal dependence in the data, to produce detailed daily maps of $PM_{2.5}$.

6.3 Data Assimilation Model

Henceforth, we denote log transformed $PM_{2.5}$ values measured by a monitoring station at location $s \in \mathbb{R}^2$ and time step t by $\tilde{Y}_t(s)$, for $s = s_1, \ldots, s_n$, $t = 0, \ldots, T$. Log transformed $PM_{2.5}$ CMAQ values for grid cell B for $B \in S = \{B_1, \ldots, B_m\}$ at time t are denoted by $Y_t(B)$. In this application, there are $T = 363$ days, $n = 200$ monitoring stations, and $m = 777$ CMAQ gridblocks.

The response variables are first modeled with a measurement error model, where the two data types are assumed to be noisy observations of a latent spatio-temporal pollution process. We note from Figure 6.1 that the monitoring station observations seem to have a common mean level that changes through time. Here we denote this common mean level at time t by α_t. In addition, at each time point, spatial variations around α_t of the pollution process follow a spatial process that we denote by $\mu_t(s)$. Thus, we model the pollution process at location s as $\alpha_t + \mu_t(s)$ and we assume that the monitoring station observations are unbiased measurements of the pollution process. Further, we assume that the CMAQ data are biased measurements of the pollution process, with the bias process at time t and grid cell B equal to $\beta_t(B) - \alpha_t$. Furthermore, we assume that, given α_t, $\mu_t(s)$, and $\beta_t(B)$, the monitoring station data and the CMAQ data are conditionally independent, and thus are observed with mutually independent and temporally and spatially uncorrelated zero mean Gaussian errors with variances $\tilde{\sigma}^2$ and σ^2, respectively. Therefore, the measurement error model is as follows:

$$\begin{aligned} \tilde{Y}_t(s) &= \alpha_t + \mu_t(s) + \tilde{\epsilon}_t, & \tilde{\epsilon}_t &\sim N(0, \tilde{\sigma}^2), \\ Y_t(B) &= \beta_t(B) + \mu_t(B) + \epsilon_t, & \epsilon_t &\sim N(0, \sigma^2), \end{aligned}$$

for any location s and grid cell B. Here, the common latent spatio-temporal process $\mu_t(s)$ is defined as a convolution of a white noise process (Higdon, 1998)

$$\mu_t(s) = \int_{\mathbb{R}^2} k(s - \omega) x_t(\omega) d\omega. \tag{6.1}$$

The mean of the common latent spatio-temporal process $\mu_t(s)$ for grid cell B is then defined as the integral over the cell region

$$\mu_t(B) = \frac{1}{|B|} \int_B \mu_t(s) ds. \tag{6.2}$$

Combining Equations (6.1) and (6.2), we can write $\mu_t(B)$ as

$$\mu_t(B) = \frac{1}{|B|} \int_B \int_{\mathbb{R}^2} k(s - \omega) x_t(\omega) d\omega ds. \tag{6.3}$$

We handle the difference in support between the two data types by representing the monitoring station data as noisy realizations from the pollution process, and the CMAQ data as biased realizations from the grid cell integrated process. In practice, we discretize the white noise process $x(\omega)$ to an equally spaced grid at approximately one degree latitude and longitude, i.e. $x(\omega_1), \cdots, x(\omega_M)$ with $M = 100$. Further, based on exploratory data analysis of the monitoring station data, in this application we take the kernel function $k(.)$ to be a Gaussian density with mean zero and standard deviation of 150km. Furthermore, the kernel smoothing weights are centered and scaled so that the following cross-product approximates the spatial correlation between sites s and s':

$$\sum_{l=1}^{M} k(\omega_l - s) k(\omega_l - s') \approx Cor(\mu(s), \mu(s')).$$

Because the kernel function $k(.)$ is fixed, computations can be further simplified by performing the integration only once prior to the start of the Gibbs sampler. Specifically, we create a grid of $N = 10,000$ equally spaced locations $u_1, \ldots, u_N$, and approximate the integrals in Equation (6.3) with averages yielding

$$\begin{aligned}
\mu_t(B) &\approx n_B^{-1} \sum_{l=1}^{N} \mu_t(u_l) 1_{\{u_l \in B\}} && (6.4)\\
&\approx n_B^{-1} \sum_{l=1}^{N} \sum_{i=1}^{M} k(u_l - \omega_i) x_t(\omega_i) 1_{\{u_l \in B\}} && (6.5)\\
&= n_B^{-1} \sum_{i=1}^{M} x_t(\omega_i) \sum_{l=1}^{N} k(u_l - \omega_i) 1_{\{u_l \in B\}} && (6.6)\\
&= \sum_{i=1}^{M} x_t(\omega_i) k^*(B, \omega_i), && (6.7)
\end{aligned}$$

where $n_B = \sum_{l=1}^{N} 1_{\{u_l \in B\}}$ is the number of u_l's in B and $k^*(B, \omega_i) = n_B^{-1} \sum_{l=1}^{N} k(u_l - \omega_i) 1_{\{u_l \in B\}}$. Therefore, after changing the order of summation, the resulting inner summation needs to be performed only once before the start of the Gibbs sampler.

6.3.1 Multiscale Spatial Model for CMAQ Bias

In order to accommodate the spatio-temporal variation in bias associated with CMAQ data, we model the bias process $\beta_t(B) - \alpha_t$ with a multiscale spatio-temporal model (Ferreira et al., 2010, 2011; Ferreira, 2021). For each time point, we use a multiscale decomposition first proposed by Kolaczyk and Huang (2001) (see also Ferreira and Lee, 2007; Ferreira et al., 2010, 2011). For notation simplicity, in this section we focus on a particular time point and omit the time subscript t.

First, define multiscale partitions of the CMAQ domain S at L levels of resolution such that the partition at level $l+1$ is a refinement of the partition at level l:

$$B_{lj} = \cup_{(l+1,j') \in D_{lj}} B_{l+1,j'}, \text{ with } B_{lj} \cap B_{lk} = \text{‰}, j \neq k,$$

where D_{lj} is the set of descendants of subregion j at level l. CMAQ data are observed at the finest level L on m gridblocks and are assumed to be conditionally independent given the mean process:

$$Y(B_{Lj}) | \, \mu, \beta, \sigma_L^2 \sim N(\beta(B_{Lj}) + \mu(B_{Lj}), \sigma_L^2),$$

for $j = 1, \ldots, m$. The variance at the finest level $\sigma_L^2 = \sigma^2$ is assumed to be constant, and $\mu(B) = \frac{1}{|B|} \int_B \mu(s) ds$ as above. Then we aggregate recursively the data minus the mean process, the CMAQ bias, and variances to coarser resolution levels by

$$Y'(B_{lj}) = \frac{1}{n_{lj}} \sum_{(l+1,j') \in D_{lj}} Y(B_{l+1,j'}) - \mu(B_{l+1,j'}), \tag{6.8}$$

$$\beta(B_{lj}) = \frac{1}{n_{lj}} \sum_{(l+1,j') \in D_{lj}} \beta(B_{l+1,j'}), \tag{6.9}$$

$$\sigma_{lj}^2 = \frac{1}{n_{lj}^2} \sum_{(l+1,j') \in D_{lj}} \sigma_{l+1,j'}^2 \tag{6.10}$$

where j indexes subregions at each resolution l. Summations are over all descendants of subregion j at resolution l, that is, summations are over the set $\{(l+1, j') : (l+1, j') \in D_{lj}\}$. For each level of resolution $l > 1$, empirical multiscale coefficients are defined to link the scaled responses at different resolutions. These are differences between the bias process at the descendants $B_{D_{lj}}$ of subregion B_{lj} and the scaled bias process at B_{lj}. More details can be found in the Appendix, Kolaczyk and Huang (2001), Ferreira and Lee (2007), and Ferreira et al. (2010, 2011). For each level l and subregion j, a vector of empirical multiscale coefficients is defined as

$$\theta_{lj}^e = \{Y'(B_{D_{lj}}) - \nu_{lj} Y'(B_{lj})\},$$

where $\nu_{lj} = \sigma_{D_{lj}}^2 / \sigma_{lj}^2 = n_{D_{lj}}^{-1} \mathbf{1}_{n_{D_{lj}}}$, $n_{D_{lj}}$ is the number of descendants of subregion j at resolution level l, and $\mathbf{1}_n$ is the n-dimensional vector of ones. Finally, empirical multiscale coefficients are distributed as

$$\theta_{lj}^e | Y_{lj}', \beta_L, \sigma_L^2 \sim N(\theta_{lj}, \Omega_{lj}),$$

for $j = 1, \ldots, n_l$ and $l = 1, \ldots, L-1$. Here, $\Omega_{lj} = \Sigma_{D_{lj}} - \sigma_{D_{lj}}^2 \sigma_{lj}^{-2} \sigma_{D_{lj}}^2$ is the covariance matrix of the conditional distribution of $Y_{D_{lj}}' | Y_{lj}', \beta_L, \sigma_L^2$. The covariance matrix Ω_{lj} is singular, which requires appropriate modifications in the specialized algorithms (Ferreira et al., 2010, 2011).

6.3.2 Spatio-Temporal Model

Now we extend the model to account for the temporal dependence in the observed data. We cast the above model within a dynamic linear model framework (West and Harrison, 1997; Prado et al., 2021) which provides flexibility in modeling dependencies in parameters at each time step and through time.

Specifically, we start with the observation equations

$$\tilde{Y}_t(s) = \alpha_t + \mu_t(s) + \tilde{\epsilon}_t, \qquad \tilde{\epsilon}_t \sim N(0, \tilde{\sigma}^2),$$
$$Y_t(B) = \beta_t(B) + \mu_t(B) + \epsilon_t, \qquad \epsilon_t \sim N(0, \sigma^2).$$

The common latent spatio-temporal process $\mu_t(s)$ is modeled with a spatio-temporal convolution approach (Higdon, 2002) similar to the spatial convolution approach explained in Section 6.3. Specifically, we approximate $\mu_t(s)$ with

$$\mu_t(s) \approx \sum_{i=1}^{M} k(s - \omega_i) x_t(\omega_i),$$

where, as explained below, the coefficients at time t are collected in a vector $x_t = (x_t(\omega_1), \ldots, x_t(\omega_M))$ that evolves through time with a state-space model.

To model the CMAQ data, analogously to Section 6.3.1, we define in an obvious way for each time t the quantities $Y_t'(B)$, $\theta_t(B_{lj})$, and $\theta_t^e(B_{lj})$. This implies the following multiscale CMAQ observation equations

$$Y_t(B_1) = \beta_t(B_1) + \mu_t(B_1) + \epsilon_{t1}, \qquad \epsilon_{t1} \sim N(0, \sigma_1^2), \tag{6.11}$$
$$\theta_t^e(B_{lj}) = \theta_t(B_{lj}) + \epsilon_{tl}, \qquad \epsilon_{tl} \sim N(0, \Omega_{tlj}), \tag{6.12}$$

where (6.11) is the CMAQ observation equation at the coarsest resolution level, and (6.12) represents observation equations for the empirical multiscale coefficients that connect the observations at the several resolution levels. After that, we incorporate time evolution for the several parameters through the system evolution equations

$$\alpha_t = \alpha_{t-1} + w_\alpha, \qquad w_\alpha \sim N(0, \tau^2),$$
$$x_t = x_{t-1} + w_x, \qquad w_x \sim N(\mathbf{0}_M, \tau_x^2 \boldsymbol{I}_M),$$
$$\beta_t(B_1) = \beta_{t-1}(B_1) + w_\beta, \qquad w_\beta \sim N(0, \xi\sigma_1^2),$$
$$\theta_t(B_{lj}) = \theta_{t-1}(B_{lj}) + w_\theta, \qquad w_\theta \sim N(0, \psi_{lj}\Omega_{lj}),$$

for $l = 1, \ldots, L$ and $j = 1, \ldots, n_{lj}$. The covariance matrix $\psi_{lj}\Omega_{lj}$ of the evolution innovations for the multiscale coefficient $\theta_t(B_{lj})$ are specified in a signal-to-noise parameterization. The assumption of constant variance across all grid cells at the observed resolution level L allows for factorization of the covariance matrices Ω_{lj} into the product $\sigma_l^2 \times \Omega_{lj}^*$ where Ω_{lj}^* needs to be computed only once, before the start of the Gibbs sampler.

6.3.3 Prior Distributions

When using dynamic linear models as we do here, the choice of prior distributions is very important (Prado et al., 2021). In particular, an appropriate choice of priors allows the inference procedure to correctly separate the signal from the noise. Of particular interest is the correct estimation of the variance

$\tilde{\sigma}^2$ of the measurement error for the monitoring station data. We note that $\tilde{\sigma}^2$ may be estimated in an empirical Bayes manner prior to the application of our spatio-temporal multiscale assimilation method. Specifically, we may regard the data from the monitoring stations observations on each day as realizations from Gaussians processes. In that case, $\tilde{\sigma}^2$ will be what is known in the geostatistical literature as the nugget variance. With that in mind, we have fitted a Gaussian process to the monitoring station data on each day. That has yielded an estimate of $\tilde{\sigma}^2$ on each day. We then set $\tilde{\sigma}^2$ to be the median of these nugget variance estimates over the observed days, resulting in $\widehat{\tilde{\sigma}}^2 = 0.023$. For the variance σ^2 of the CMAQ measurement error, we assign a vague conditionally conjugate prior $IG(0.001, 0.001)$. For the signal-to-noise ratio parameters ξ and ψ_{lj}, we assign vague conditionally conjugate priors $IG(1, 0.1)$. For the state-space equations variances τ^2 and τ_x^2, we also assign vague conditionally conjugate priors $IG(1, 0.1)$. We note that priors for the variances of the state space equations cannot be too vague, and that improper priors for those parameters may lead to improper posterior distributions. We further note that, although proper, the $IG(1, 0.1)$ priors do not have any positive integer moments and, thus, are fairly vague. Finally, the prior distributions for α_t and for $x_t(\omega_1), \ldots, x_t(\omega_M)$ at time $t = 0$ are vague independent Gaussian distributions with mean zero and variance 10^{20}.

6.3.4 Computation

Posterior simulations of the parameters in the model are performed using a Gibbs sampler. Parameters x_t of the common latent spatio-temporal process are sampled from their joint full conditional distribution using the forward information filter backward sampler algorithm (FIFBS) (Vivar and Ferreira, 2009), which combines the information filter (Anderson and Moore, 1979) and the forward filter backward sampler algorithm (FFBS) (Frühwirth-Schnatter, 1994; Carter and Kohn, 1994). Samples from the full conditional distribution of the multiscale coefficients $\theta_t(B_{lj})$ are obtained via the Singular Forward Filter Backward Sampler algorithm (SFFBS), which was developed for parameters with singular system equation covariance matrix (Ferreira et al., 2010, 2011). At each iteration, the coarsest level parameter $\beta_t(B_{1j})$ is sampled using the FFBS. The simulated values for $\theta_t(B_{lj})$ and $\beta_t(B_{1j})$ are then used to recursively compute the corresponding simulated values of the process $\beta_t(B_{lj})$ at resolution level l, $l = 1, \ldots, L$. The values of $\beta_t(B_{Lj})$ sampled at the finest level are then used for the simulation of the other model parameters.

Due to the data collection scheme, on any given day the monitoring data are characterized by the large amount of missing data. In order to deal with this issue, we include in the Gibbs sampler the simulation of the missing observations from their full conditional distribution, which is easily obtained from the observation equation for $\tilde{Y}_t(s)$.

6.4 Application to Fine Particulate Matter Data

The distribution of particles in the ambient air is influenced by local and regional scale atmospheric processes and by the type of land cover of the area. Broadly speaking, there are few monitors positioned in remote areas often creating difficulty in modeling spatial non-stationarity. Accurate geographic information, meteorological conditions, and other factors used by the CMAQ model help capture the dynamic distribution of $PM_{2.5}$ and provide full spatial and temporal coverage. As such, data assimilation methods represent a promising tool for prediction of ambient air quality. Here, we illustrate predictions from the data assimilation model introduced in Section 6.3.

The exploratory analysis of the monitoring and CMAQ data shows that CMAQ does not predict equally well across time and space. To account for this CMAQ bias, the spatio-temporal model introduced in Section 6.3.2 contains a bias term modeled with a multiscale spatio-temporal process. The approach allows for parameter estimation at different regional levels. Here we use three levels of resolution with approximately equally sized subregions at each resolution. The finest resolution is the observed level with 777 cells, the mid-level resolution is composed of 9 subregions, and the coarsest resolution is the aggregation of the entire region. Other examples of multiscale partitions may include states, counties or other spatial regions of interest. Increasing the number of resolution levels will improve computational speed as the dimension of the multiscale coefficient vector at each subregion decreases. This is an important aspect as the CMAQ predictions become available on a smaller grid. Here we consider data obtained from runs of CMAQ that predict ambient levels at a grid size of 36km $\times$ 36km.

Posterior means of the bias process at each level of resolution provide information about CMAQ performance at regional and local levels through time. Figure 6.4 shows posterior means for the bias $\beta_t(B_{lj}) - \alpha_t$ through time at all levels of resolution. Each of the panels corresponds to one of the nine subregions at the intermediate resolution level. Posterior means at the coarsest, intermediate, and finest resolution levels are given in solid black, dashed black, and solid gray lines, respectively. The bias is similar at the coarsest and intermediate resolutions. However, at the finest resolution, we observe grid cells for which CMAQ systematically under or over-predicts the monitoring station data levels. Further examination shows that the most biased CMAQ grid cells are found in the south-eastern portion of the region (bottom right panel of Figure 6.4), with most CMAQ grid cells systematically underestimating $PM_{2.5}$ levels. In the whole study region, the average estimated bias over all CMAC grid cells in the logarithm scale is 0.25. Possible reasons for systematic under/over prediction through time may be incomplete emission inventories or specific geographic characteristics.

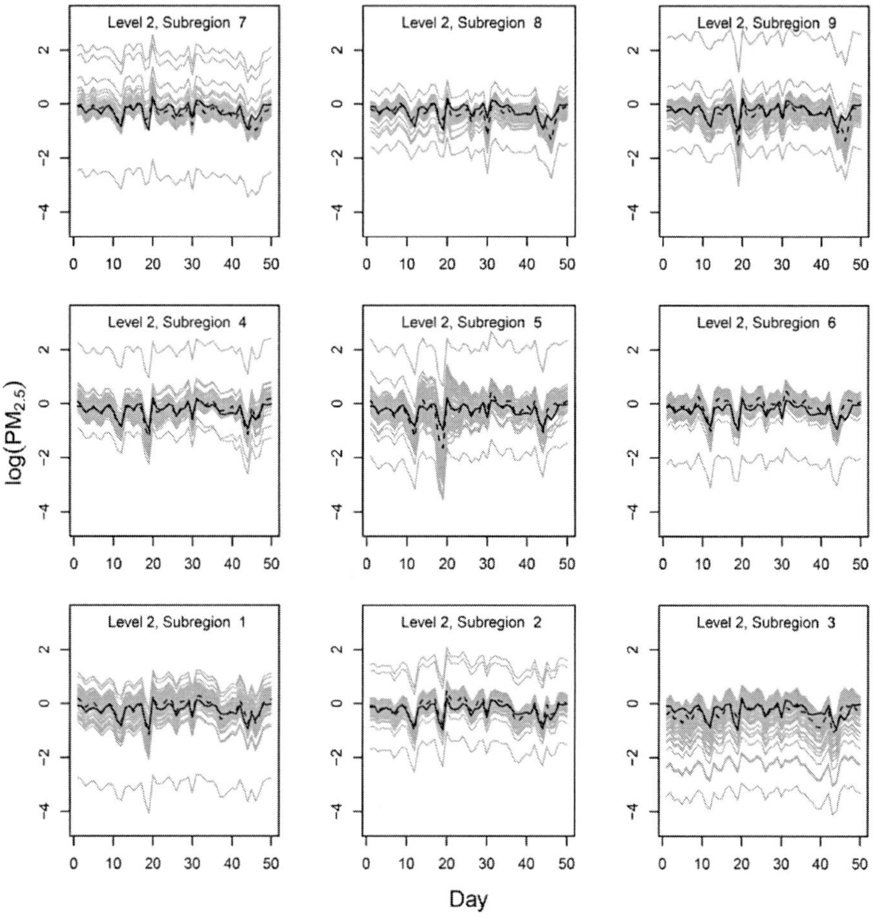

FIGURE 6.4
Posterior means for bias $\beta_t(B_{lj}) - \alpha_t$ through time at all three levels of resolution. Coarsest, intermediate, and finest level resolutions are given in solid black, dashed black, and solid gray lines, respectively.

Figures 6.5 and 6.6 illustrate model-based prediction in the spatial domain on two consecutive days, January 19 and 20, 2001. In both Figures 6.5 and 6.6, the CMAQ data and observed monitoring data for that day are given in panels (a) and (b) respectively, whereas the posterior mean of the latent $PM_{2.5}$ process is given in panel (c). January 19 had data from 88% of the monitoring stations, whereas January 20 had data from only 14% of the monitoring stations. On the first day, a storm moved from the southwest corner

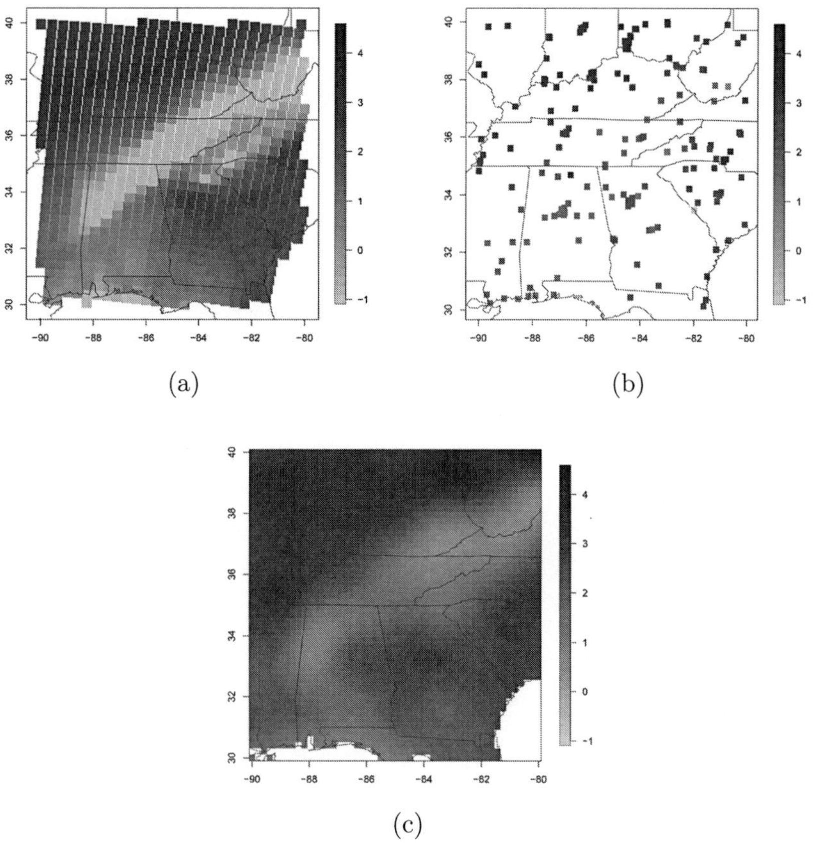

FIGURE 6.5
Spatial prediction on January 19, 2001. CMAQ data (Panel (a)), monitoring station data (Panel (b)), and posterior mean for $\log(PM_{2.5})$ based on our multiscale spatio-temporal assimilation model (Panel (c)). The multiscale spatio-temporal assimilation approach corrects the CMAQ bias in panel (a) by using the monitoring data in panel (b) to obtain reduced-bias, continuous predictions of PM2.5 across the region of interest, as shown in panel (c).

toward the north-east corner of the region lowering the amount of $PM_{2.5}$ in the air. On the next day, the effects of the storm dimished as the storm moved further to the north-east corner of the region.

From panels (a) and (b) of Figure 6.5, it is clear that the CMAQ data underpredicts the values observed at the monitoring stations. As shown in

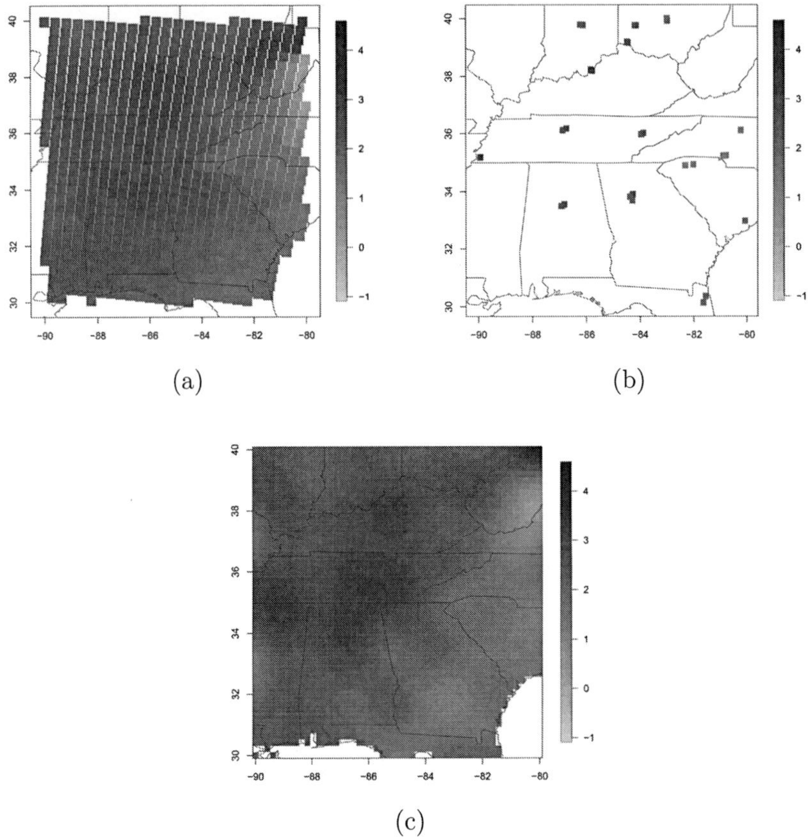

FIGURE 6.6
Spatial prediction on January 20, 2001. CMAQ data (Panel (a)), monitoring station data (Panel (b)), and posterior mean for $\log(PM_{2.5})$ based on our multiscale spatio-temporal assimilation model (Panel (c)). This figure shows the great importance of assimilating CMAQ data for days with a high percentage of missing data.

panel (c), our multiscale spatio-temporal assimilation framework takes into account both the CMAQ and monitoring station data to produce an estimated map of $PM_{2.5}$ that corrects the CMAQ bias, has finer resolution than the CMAQ spatial resolution, and has much better spatial coverage than the monitoring station data.

Figure 6.6 shows the great importance of assimilating CMAQ data for days with a high percentage of missing data. We note that methods based only on the information from the monitoring stations would not be able to correctly predict the spatial pattern of $PM_{2.5}$ and the effect of the storm. In contrast, our multiscale spatio-temporal assimilation method incorporates the information provided by CMAQ and clearly shows the effect of the storm moving in one direction on the resulting levels and spatial distribution of $PM_{2.5}$ on the second day.

To validate the predictive performance of our dynamic multiscale spatio-temporal assimilation model, we selected and removed from the estimation six monitoring sites, two at each temporal sampling protocol. Specifically, we selected the following monitoring stations: 40 and 137 (daily collection), 37 and 133 (collection every third day), and 95 and 171 (collection every sixth day). For each of the six sites, we removed the entire time series, ran the MCMC algorithm, and computed prediction intervals. The predictive posterior distributions at these validation sites are summarized in Figure 6.7. Given the amount of missing data in this application, the model does a reasonable job of predicting the hold-out data. The top row of Figure 6.7 summarizes the predictive distribution for the two sites with daily collection of samples. Combined for both sites, the predictive intervals did not include the observed value in 7 out of 80 days. In the second row, the same summaries are given for two sites collecting samples every three days. For monitoring station 37, in the two days when the observed values were not within the predictive interval, we note that the observed values were uncharacteristically low. Lastly, the model-based predictions are less reliable at the remote sites where data are typically collected every 6th day (last row of Figure 6.7). Overall, of the 120 monitoring station observations at these 6 sites, 89% were contained in the corresponding 95% predictive intervals. Thus, even though our spatio-temporal multiscale assimilation framework does a reasonable job in terms of predictions, clearly there is room for improvement. For example, including information about land cover may improve the predictive ability of the dynamic multiscale spatio-temporal assimilation model.

6.5 Discussion

We have introduced here a novel multiscale spatio-temporal assimilation model. This model specifies a joint distribution for two sources of data as arising from a latent spatio-temporal pollution process. The part of this latent spatio-temporal pollution process that is shared by the two data types is, in turn, modeled as a convolution of Gaussian white noise processes which evolve through time. The specification of the model allows for the input data to be observed at different spatial resolutions. In particular, in this application the monitoring station data are observed daily averages continuously distributed in space, and CMAQ data are the computer model output representing areal

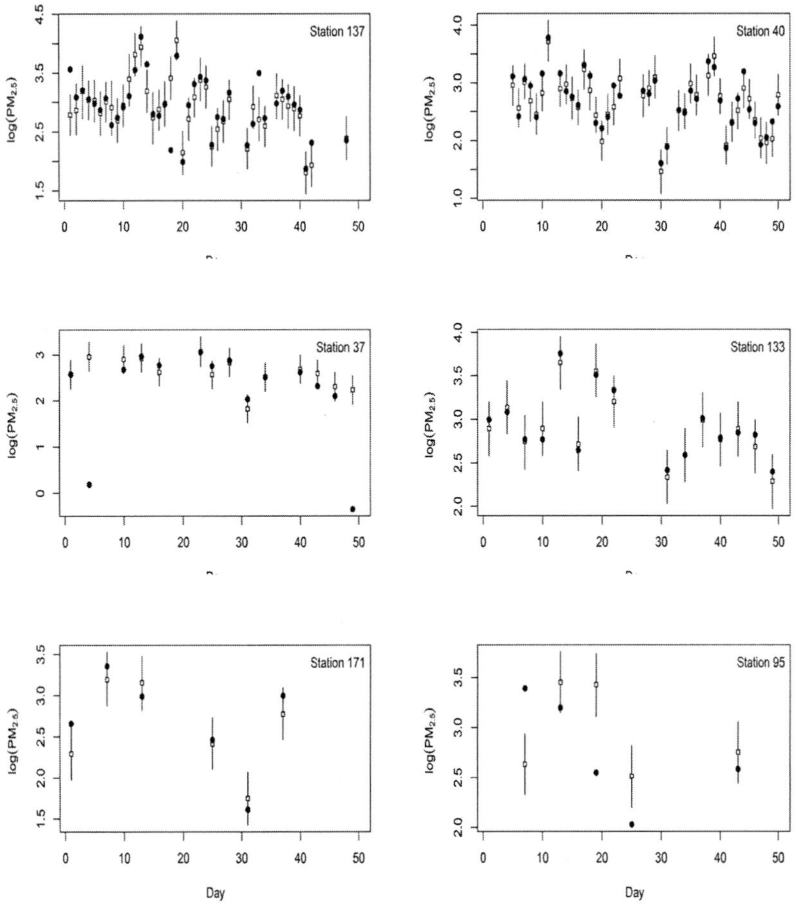

FIGURE 6.7
Predictive means and 95% predictive intervals at six monitoring station sites used in model validation. The observed values are marked with black circles.

averages over equally sized grid cells. Our framework can be adapted to other spatio-temporal data assimilation problems in any scientific area. Extensions to assimilate more than two data sources are conceptually straightforward but may be computationally challenging. Extensions such as those developed by Higdon et al. (1999) and Fuentes and Raftery (2005) to account for spatial nonstationarity are possible.

We have demonstrated through validation the predictive performance of our multiscale spatio-temporal assimilation model for time series at monitoring stations. Our assimilation model for $PM_{2.5}$ produces in a straightforward manner the imputation of daily time series across all locations. Posterior summaries of daily levels of particulate matter at the finest resolution level can provide information for epidemiological researchers to link pollutant exposure to health effects. Specifically, the information can be readily incorporated with human health data in studies of temporal lag effects of pollution on cardiopulmonary responses of populations at risk. Daily prediction surfaces at the finest resolution level can also be used in attainment studies by regulatory bodies. Scaling to coarser temporal resolutions, along with the associated uncertainty estimates, follow immediately from Bayesian posterior analysis. Finally, posterior summaries of the bias process may be used for evaluation of the CMAQ predictions. Of particular interest may be the identification of conditions under which bias exceeds pre-specified limits.

Over the last several decades, due to advances in cross-disciplinary science and technology, the research community has experienced an increase in computational/mathematical modeling approaches for prediction of natural phenomena. In the instances when data collection procedures are manually and fiscally intense, society can benefit significantly from the use of the output from these computer models. In addition to air pollution, other fields which can benefit from data generated by computer models include ocean temperatures, hurricanes, forest fires, volcano eruption, and earthquakes. However, in all these examples computer models have limitations that we desire to overcome. These computer models may be, in part or as a whole, deterministic, they take imprecise inputs, and sometimes are incomplete. Nevertheless, they can be re-run, improved, and can provide unlimited amounts of data. Moreover, computer model output may be assimilated with experimental data using data assimilation methods such as our multiscale spatio-temporal assimilation approach. Our approach provides insights into the computer model shortfalls, provides quantification of the associated uncertainty, and provides accurate spatial predictions with complete spatial coverage at a pre-specified finest resolution level.

6.6 Appendix: Multiscale Decomposition

Consider the aggregation as in Equations (6.8), (6.9), and (6.10). Let D_{lj} denote the index of descendants of subregion $j, j = 1, \ldots, n_l$ at resolution $l, l < L$. Then, the joint distribution of the Y'_{lj} and descendants $Y'_{D_{lj}}$ is

$$\begin{pmatrix} Y'_{lj} \\ Y'_{D_{lj}} \end{pmatrix} \Big| \beta_L, \sigma_L^2 \sim N\left[\begin{pmatrix} \beta(B_{lj}) \\ \beta(B_{D_{lj}}) \end{pmatrix}, \begin{pmatrix} \sigma_{lj}^2 & (\sigma^2_{D_{lj}})' \\ \sigma^2_{D_{lj}} & \Sigma_{D_{lj}} \end{pmatrix}\right],$$

and the conditional distribution of $Y'_{D_{lj}}$ given Y'_{lj} is given by

$$
\begin{aligned}
Y'_{D_{lj}} | Y'_{lj}, \beta_L, \sigma_L^2 &\sim N(\nu_{lj} Y'_{lj} + \theta_{lj}, \Omega_{lj}) \text{ with} \\
\nu_{lj} &= \sigma^2_{D_{lj}} / \sigma^2_{lj}, \\
\theta_{lj} &= \beta_{D_{lj}} - \nu_{lj} \beta_{lj}, \\
\Omega_{lj} &= \Sigma_{D_{lj}} - \sigma^2_{D_{lj}} \sigma^{-2}_{lj} \sigma^2_{D_{lj}}.
\end{aligned}
$$

Now the empirical estimates are defined as $\theta^e_{lj} = Y'_{D_{lj}} - \nu_{lj} Y'_{lj}$ of θ_{lj} with

$$\theta^e_{lj} | Y_{lj}, \beta_L, \sigma^2_L \sim N(\theta_{lj}, \Omega_{lj}).$$

The joint distribution of the observed data is then factorized into the product of the marginal density at the coarsest level and the conditional densities of differences between the parents and their descendants at each level of resolution (Kolaczyk and Huang, 2001)

$$\prod_{j=1}^{m} p(Y'_{Lj} | \beta_{lj}, \sigma^2_{Lj}) \propto p(Y'_1 | \beta_1, \sigma^2_1) \times \prod_{l=1}^{L-1} \prod_{j=1}^{n_l} p(\theta^e_{lj} | Y'_{lj}, \theta_{lj}, \Omega_{lj}).$$

Bibliography

Anderson, B. D. and J. B. Moore (1979). *Optimal Filtering.* New Jersey: Prentice Hall.

Banerjee, S., B. P. Carlin, and A. E. Gelfand (2014). *Hierarchical Modeling and Analysis for Spatial Data* (2nd ed.). Boca Raton, FL: CRC Press.

Berliner, L. M., C. K. Wikle, and R. F. Milliff (1999). Multiresolution wavelet analyses in hierarchical Bayesian turbulence models. In P. Müller and B. Vidakovic (Eds.), *Bayesian Inference in Wavelet Based Models*, pp. 341–359. New York: Springer-Verlag.

Berrocal, V. J., A. E. Gelfand, and D. M. Holland (2010). A spatio-temporal downscaler for output from numerical models. *Journal of Agricultural, Biological, and Environmental Statistics 15*(2), 176–197.

Brown, P. E., P. J. Diggle, M. E. Lord, and P. C. Young (2001). Space-time calibration of radar-rainfall data. *Applied Statistics 50*, 221–241.

Brown, P. J., N. D. Le, and J. V. Zidek (1994). Multivariate spatial interpolation and exposure to air pollutants. *The Canadian Journal of Statistics 22*, 489–509.

Byun, D. W. and J. K. S. Ching (1999). *Science Algorithms of the EPA Models-3 Community Multiscale Air Quality (CMAQ) Modeling System.* RTP, NC: EPA/600/R-99/030.

Carter, C. K. and R. Kohn (1994). On Gibbs sampling for state space models. *Biometrika 81*, 541–553.

Cressie, N. and G. Johannesson (2008). Fixed rank Kriging for very large spatial data sets. *Journal of the Royal Statistical Society: Series B (Statistical Methodology) 70*(1), 209–226.

Cressie, N., M. S. Kaiser, M. Daniels, J. Aldworth, J. Lee, S. Lahiri, and L. H. Cox (1999). *Spatial Analysis of Particulate Matter in an Urban Environment.* J. Gomez-Hernandez, A. Soares, and R. Froidevaux, Dordrecht: Kluwer: *GeoEnvII:* Geostatistics for Environmental Applications.

Eder, B. and S. Yu (2006). A performance evaluation of the 2004 release of Models-3 CMAQ. *Atmospheric Environment 40*, 4811–4824.

Elkhouly, M. and M. A. R. Ferreira (2021). Dynamic multiscale spatiotemporal models for multivariate Gaussian data. *Spatial Statistics 41*, 100475.

Ferguson, T. S. (1973). A Bayesian analysis of some nonparametric problems. *The Annals of Statistics 1*(2), 209–230.

Ferreira, M. A. R. (2015). Inhomogeneous evolutionary MCMC for Bayesian optimal sequential environmental monitoring. *Environmental and Ecological Statistics 22*(4), 705–724.

Ferreira, M. A. R. (2021). Bayesian spatial and spatiotemporal models based on multiscale factorizations. *Wiley Interdisciplinary Reviews: Computational Statistics 13*(2), e1509.

Ferreira, M. A. R., A. I. Bertolde, and S. Holan (2010). Analysis of economic data with multi-scale spatio-temporal models. In A. O'Hagan and M. West (Eds.), *Handbook of Applied Bayesian Analysis*, pp. 295–318. Oxford University Press.

Ferreira, M. A. R., D. Higdon, H. K. H. Lee, and M. West (2005). Multiscale random field models. Technical report, UFRJ-DME.

Ferreira, M. A. R., S. H. Holan, and A. I. Bertolde (2011). Dynamic multiscale spatio-temporal models for Gaussian areal data. *Journal of the Royal Statistical Society (Series B: Methodological) 73*, 663–688.

Ferreira, M. A. R. and M. A. Jaramillo (2014). Bayesian multiscale phylogenetics. *Journal of the Indian Society of Agricultural Statistics 68*, 285–292.

Ferreira, M. A. R. and H. K. H. Lee (2007). *Multiscale Modeling: A Bayesian Perspective.* Springer Series in Statistics. New York: Springer.

Fonseca, T. C. O. and M. A. R. Ferreira (2017). Dynamic multiscale spatiotemporal models for Poisson data. *Journal of the American Statistical Association 112*, 215–234.

Frühwirth-Schnatter, S. (1994). Data augmentation and dynamic linear models. *Journal of Time Series Analysis 15*(2), 183–202.

Fuentes, M. and A. Raftery (2005). Model evaluation and spatial interpolation by Bayesian combination of observations with outputs from numerical models. *Biometrics 61*, 36–45.

Gelfand, A. E., A. M. Schmidt, S. Banerjee, and C. Sirmans (2004). Nonstationary multivariate process modeling through spatially varying coregionalization. *Test 13*(2), 263–312.

Gotway, C. and L. Young (2002). Combining incompatible spatial data. *Journal of the American Statistical Association 97*, 632–648.

Hanson, T. E. (2006). Inference for mixtures of finite Pólya tree models. *Journal of the American Statistical Association 101*(476), 1548–1565.

Hanson, T. E. and W. O. Johnson (2002). Modeling regression error with a mixture of Pólya trees. *Journal of the American Statistical Association 97*(460), 1020–1033.

Higdon, D. (1998). A process-convolution approach to modeling temperatures in the North Atlantic Ocean. *Journal of Environmental and Ecological Statistics 5*, 173–190.

Higdon, D., J. Swall, and J. Kern (1999). Non-stationary spatial modeling. In J. Bernardo, J. Berger, A. Dawid, and A. Smith (Eds.), *Bayesian Statistics 6*. London: Oxford University Press.

Higdon, D. M. (2002). Space and space-time modeling using process convolutions. In C. W. Anderson, V. Barnett, P. C. Chatwin, and A. H. El-Shaarawi (Eds.), *Quantitative Methods for Current Environmental Issues*, pp. 37–54. New York: Springer-Verlag.

Hoegh, A., M. A. R. Ferreira, and S. Leman (2016). Spatiotemporal model fusion: multiscale modelling of civil unrest. *Journal of the Royal Statistical Society: Series C (Applied Statistics) 65*, 529–545.

Huerta, G., B. Sansó, and J. R. Stroud (2004). A spatiotemporal model for Mexico City ozone levels. *Journal of the Royal Statistical Society, Series C 53*, 231–248.

Jara, A. and T. E. Hanson (2011). A class of mixtures of dependent tail-free processes. *Biometrika 98*(3), 553–566.

Johannesson, G. and N. Cressie (2004). Variance-covariance modeling and estimation for multi-resolution spatial models. In *geoENV IV – Geostatistics for Environmental Applications*, pp. 319–330. Springer.

Johannesson, G., N. Cressie, and H. Huang (2007). Dynamic multi-resolution spatial models. *Environmental and Ecological Statistics 14*, 5–25.

Jun, M. and M. L. Stein (2004). Statistical comparison of observed and CMAQ modeled daily sulfate levels. *Atmospheric Environment 38*, 4427–4436.

Katzfuss, M. (2013). Bayesian nonstationary spatial modeling for very large datasets. *Environmetrics 24*(3), 189–200.

Katzfuss, M. (2017). A multi-resolution approximation for massive spatial datasets. *Journal of the American Statistical Association 112*(517), 201–214.

Katzfuss, M. and W. Gong (2020). A class of multi-resolution approximations for large spatial datasets. *Statistica Sinica 30*(4), 2203–2226.

Kibria, B., M. Golam, L. Sun, J. Zidek, and N. Le (1997). Bayesian spatial prediction of random space-time fields with application to mapping $PM_{2.5}$ exposure. *Journal of American Statistica Association 97*, 112–124.

Kolaczyk, E. D. and H. Huang (2001). Multiscale statistical models for hierarchical spatial aggregation. *Geographical Analysis 33*, 95–118.

Lavine, M. (1992). Some aspects of Pólya tree distributions for statistical modelling. *The Annals of Statistics 20*(3), 1222–1235.

Ma, L. (2017a). Adaptive shrinkage in Pólya tree type models. *Bayesian Analysis 12*(3), 779–805.

Ma, L. (2017b). Recursive partitioning and multi-scale modeling on conditional densities. *Electronic Journal of Statistics 11*(1), 1297–1325.

Ma, L. and J. Soriano (2018). Analysis of distributional variation through multi-scale beta-binomial modeling. *Journal of Computational and Graphical Statistics 27*, 529–541.

Müller, P., A. Erkanli, and M. West (1996). Bayesian curve fitting using multivariate normal mixtures. *Biometrika 83*(1), 67–79.

Nychka, D., S. Bandyopadhyay, D. Hammerling, F. Lindgren, and S. Sain (2015). A multiresolution Gaussian process model for the analysis of large spatial datasets. *Journal of Computational and Graphical Statistics 24*(2), 579–599.

Pope III, A. C., R. T. Burnett, M. J. Thun, E. E. Calle, D. Krewski, K. Ito, and G. D. Thurston (2002). Lung cancer, cardiopulmonary mortality and long-term exposure to fine particulate air pollution. *Journal of American Medical Association 292*, 2372–2378.

Prado, R., M. A. R. Ferreira, and M. West (2021). *Time Series: Modeling, Computation, and Inference* (2nd ed.). Boca Raton: CRC Press.

Royle, A. and M. Berliner (1999). A hierarchical approach to multivariate spatial modeling and prediction. *Journal of Agricultural, Biological and Environmental Statistics 4*, 29–56.

Sahu, S. K. and K. V. Mardia (2005). A Bayesian Kriged-Kalman model for short-term forecasting of air pollution levels. *Journal of the Royal Statistical Society, Series C 54*, 223–244.

Schmidt, A. M. and A. E. Gelfand (2003). A Bayesian coregionalization approach for multivariate pollutant data. *Journal of Geophysical Research: Atmospheres 108*, D24.

Shaddick, G. and J. Wakefield (2002). Modelling daily multivariate pollutant data at multiple sites. *Journal of the Royal Statistica Society, Series C 51*, 351–372.

Smith, R. L., S. Kolenikov, and L. H. Cox (2003). Spatio-temporal modeling of $PM_{2.5}$ data with missing values. *Journal of Geophysical Research-Atmospheres 108 D24 9004*, doi:https://doi.org/10.1029/2002JD002914.

Stroud, J. R., P. Müller, and B. Sansó (2001). Dynamic models for spatiotemporal data. *Journal of the Royal Statistical Society, Series B 63*, 673–689.

Sun, L., J. V. Zidek, N. D. Le, and H. Ozkaynak (2000). Interpolating Vancouver's daily ambient PM_{10} field. *Environmetrics 11*, 651–663.

Swall, J. L. and K. M. Foley (2009). The impact of spatial correlation and incommensurability on model evaluation. *Atmospheric Environment 43*, 1204–1217.

Vivar, J. C. and M. A. R. Ferreira (2009). Spatio-temporal models for Gaussian areal data. *Journal of Computational and Graphical Statistics 18*, 658–674.

Wang, X., J. Pancras, and D. K. Dey (2021). Investigating emergent nested geographic structure in consumer purchases: a Bayesian dynamic multiscale spatiotemporal modeling approach. *Journal of Applied Statistics 48*, 410–433.

West, M. and P. J. Harrison (1997). *Bayesian Forecasting and Dynamic Models* (2nd. ed.). New York: Springer.

Wikle, C. K. (2003). Hierarchical models in environmental science. *International Statistical Review 71*, 181–199.

Wikle, C. K., R. F. Milliff, D. Nychka, and L. M. Berliner (2001). Spatio-temporal hierarchical Bayesian modeling: Tropical ocean surface winds. *Journal of the American Statistical Association 96*, 382–397.

Wong, W. H. and L. Ma (2010). Optional Pólya tree and Bayesian inference. *The Annals of Statistics 38*(3), 1433–1459.

Zammit-Mangion, A. and J. Rougier (2020). Multi-scale process modelling and distributed computation for spatial data. *Statistics and Computing 30*(6), 1609–1627.

Zidek, J. V., L. Sun, N. D. Le, and H. Ozkaynak (2002). Contending with space-time interaction in the spatial prediction of pollution: Vancouver's hourly ambient $pm_{2.5}$ field. *Environmetrics 13*, 595–613.

7

Multiscale Heteroscedastic Multivariate Spatio-Temporal Models

Mohamed Elkhouly

Cleveland State University

Marco A. R. Ferreira

Virginia Tech

Technological advances are generating massive multivariate spatio-temporal datasets. However, modeling of such highly structured multivariate datasets

DOI: 10.1201/9781032623443-7

is not trivial because computations may scale cubicly with spatio-temporal dataset size. Here, we propose a fast and scalable Bayesian dynamic multiscale spatio-temporal framework that accounts for the spatio-temporal and inter-variables correlation structures. Our framework builds on the Bayesian dynamic multiscale spatio-temporal models for multivariate data that we have recently proposed in Elkhouly and Ferreira (2021). Note that Elkhouly and Ferreira (2021) assume that the covariance matrices at the finest resolution level are constant spatially. While this assumption of homoskedasticity is reasonable for many spatio-temporal datasets such as thc dataset on *stratospheric temperatures* considered in Elkhouly and Ferreira (2021), this assumption is not reasonable for other datasets such as the dataset on *tropospheric temperatures* considered in this chapter. Hence, here we extend the Bayesian dynamic multivariate multiscale spatio-temporal models proposed in Elkhouly and Ferreira (2021) to the heteroskedastic case when the covariance matrices at the finest resolution level vary spatially. Finally, we employ our proposed framework to model tropospheric temperatures at different altitudes. Our results indicate persistent long-term trends for temperatures at the upper and middle tropospheric altitudes.

7.1 Introduction

Advances in technology continue to create massive multivariate spatio-temporal datasets in diverse fields of application such as economy, human health, and climate science. As an example, here we consider environmental data on temperature at different altitudes in the troposphere. Consequently, a realistic analysis of such datasets requires a framework that considers the spatio-temporal dependence in addition to the inter-dependence among the considered variables. Meanwhile, modeling multivariate spatio-temporal data is computationally expensive and thus, depending on the computation resources, may not be feasible (Stein, 2008). This difficulty is due to the high dimensionality and the fact that computations for models based on dense covariance matrices scale cubicly with dataset size (Sun et al., 2012; Stein, 2014). One way to deal with such high dimensionality is by using factor models that perform dimension reduction such as those proposed by Lopes et al. (2008, 2011) and Shin and Ferreira (2023) for univariate spatio-temporal data, and by Calder (2007) for multivariate spatio-temporal data.

Another way to deal with the high dimensionality is by using dynamic multiscale spatio-temporal models such as those proposed by Ferreira et al. (2010, 2011) for univariate Gaussian data, by Fonseca and Ferreira (2017) and Hoegh et al. (2016) for univariate Poisson data, and by Elkhouly and Ferreira (2021) for multivariate Gaussian data. For a review of literature on Bayesian spatial and spatio-temporal models based on multiscale factorizations, see Ferreira

(2021). In particular, Elkhouly and Ferreira (2021) assume homoskedasticity with covariance matrices at the finest resolution level being constant spatially and over time. Elkhouly and Ferreira (2021) present an analysis of stratospheric temperatures for which the assumption of homoskedasticity is reasonable. However, such an assumption is not reasonable for many spatio-temporal datasets. In particular, the assumption of homoskedasticity does not hold for the dataset of tropospheric temperatures considered in this chapter. Thus, here we extend the Bayesian dynamic multiscale spatio-temporal models proposed in Elkhouly and Ferreira (2021) to the heteroskedastic case when the covariance matrices at the finest resolution level vary spatially. This chapter is based on results first presented in a Ph.D. dissertation by Elkhouly (2019). We call our proposed model the dynamic multivariate multiscale spatio-temporal model (DMMSTM).

To develop the DMMSTM framework, first we propose a novel multiscale decomposition for heteroskedastic multivariate spatial data. This novel multiscale decomposition yields aggregated observation vectors at the coarsest scale of resolution as well as empirical multiscale coefficients at multiple scales of resolution. These empirical multiscale coefficients connect each pair of subsequent resolutions. In addition, we assume that the empirical multiscale coefficients are noisy versions of latent multiscale coefficients. Further, we assume that the latent multiscale coefficients evolve through time according to dynamic linear models (West and Harrison, 1997; Prado et al., 2021). Finally, to explore the posterior distribution of the latent multiscale coefficients and other unknown parameters at the multiple scales of resolution, we have developed a Markov chain Monte Carlo (MCMC) algorithm that is scalable, parallelizable, and fast.

Our multivariate multiscale spatio-temporal model framework for Gaussian data extends the univariate Gaussian framework by Ferreira et al. (2010, 2011) in three ways. First, we consider the multivariate case and analyze the available variables jointly. In doing so, DMMSTM takes into account the spatial, temporal, and inter-variable dependence structure when estimating the mean vectors at different resolutions. Second, we account for possible rough discrepancies among the covariance matrices of adjacent regions. Finally, we consider more general temporal dynamics to model the temporal evolution of the latent process at different scales of resolution. This allows our framework to take into account possible nonstationary long-term trends as well as stationary short-term fluctuations. In particular, our framework allows the estimation of mean level fields and temporal gradient fields at multiple scales of resolution. Finally, the mean level fields and temporal gradient fields are coherently estimated across resolutions.

We have applied the proposed DMMSTM framework to analyze the annual summer temperature anomalies of upper and middle tropospheric altitudes from the NCEP/NCAR Reanalysis-I dataset (Kalnay et al., 1996) over North America from 1951 to 2016. Previously published studies have focused on one or two dimensions of the spatio-temporal data from multiple altitudes.

For example, several previous studies have analyzed the temporal evolution of the column average of tropospheric temperatures (Wentz and Schabel, 1998; Christy and Norris, 2004; Perlwitz et al., 2015). Other studies have considered multiple tropospheric altitudes either to explore the univariate temporal changes (Douglass et al., 2004) or to explore the spatial discrepancies of temporal averages (Tian et al., 2013).

To the best of our knowledge, this is the first work that jointly analyzes tropospheric air temperatures at different altitudes in a spatio-temporal framework. While climatc change is expected to involve changes not only in temperature but also in precipitation (Tebaldi et al., 2011; Ferreira, 2011), strength of tropical cyclones (Walsh et al., 2023), and risk of tornados (Karpman et al., 2013; Elkhouly et al., 2023), here we focus on temperature. For the altitudes we consider, the corresponding temperatures have correlated temporal changes, and therefore a joint analysis with our DMMSTM framework increases the precision in the estimates of these trends. Case in point, as we show in this chapter, our proposed DMMSTM framework identifies statistically and practically significant long-term trends of temperatures at the upper and middle tropospheric altitudes over North America.

The remainder of this chapter is organized as follows. Section 7.2 presents the multiscale factorization for heteroskedastic multivariate spatial data that we propose. Section 7.3 presents our dynamic multivariate multiscale model. In Section 7.4, we present the MCMC algorithm for the exploration of the posterior distribution. Section 7.5 presents a conditional Bayes factor for model selection. Section 7.6 presents an application of the DMMSTM framework to tropospheric temperatures over North America. Section 7.7 presents a brief discussion. The Appendix 7.8 presents proofs of two theorems and the definition of a singular Gaussian matrix-variate distribution.

7.2 Multivariate Multiscale Factorization

We present here a novel multiscale factorization for heteroskedastic multivariate spatial data. We are interested in a study region S. We assume the existence of a multiscale structure for S such as the one presented in Figure 7.1. We have built this multiscale structure based on the data available in the NCEP/NCAR Reanalysis-I dataset. Specifically, the multiscale structure we consider in Figure 7.1 has four levels of resolution. The finest resolution – which is the resolution considered in the NCEP/NCAR Reanalysis-I dataset – is the fourth resolution and subdivides the study region into 12 latitude slices and 24 longitude slices in a total of 288 subregions. Then, the third resolution aggregates each 4 subregions at the fourth resolution level into 1 subregion, resulting in a total of 72 subregions. After that, the second resolution aggregates each 4 subregions of the third resolution level into 1 subregion, resulting in a total of 18 subregions. Finally, the first resolution – which is the coarsest resolution – aggregates each 2 subregions of the second resolution into 1 subregion, resulting in a total of 9 subregions at the first resolution level. There are two important features of this multiscale structure: each level of resolution is a partition of the study region, and the multiscale partitions are nested.

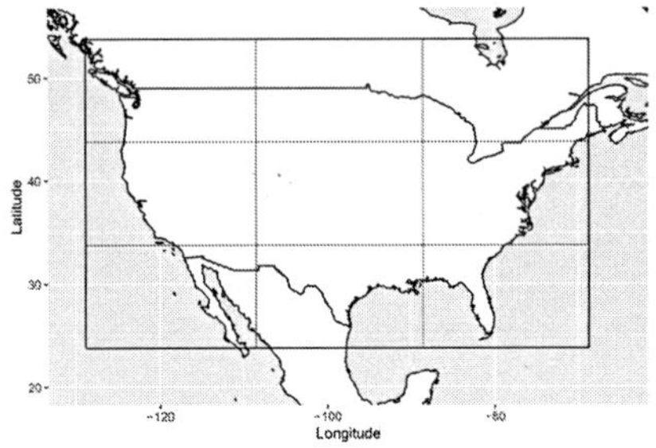

(a) First resolution

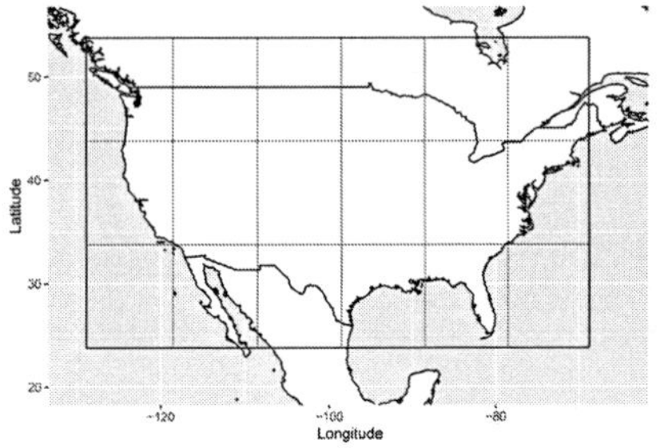

(b) Second resolution

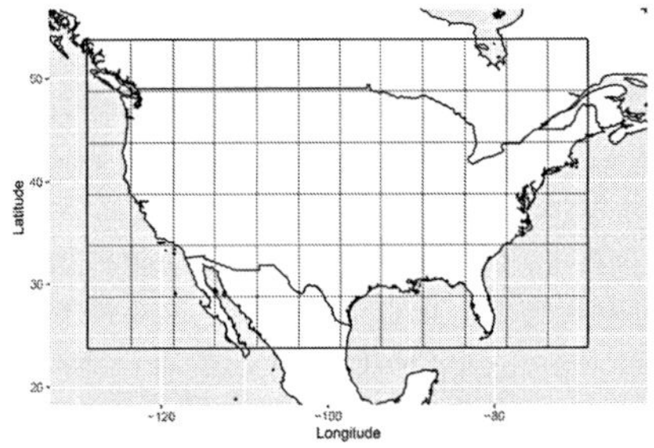

(c) Third resolution

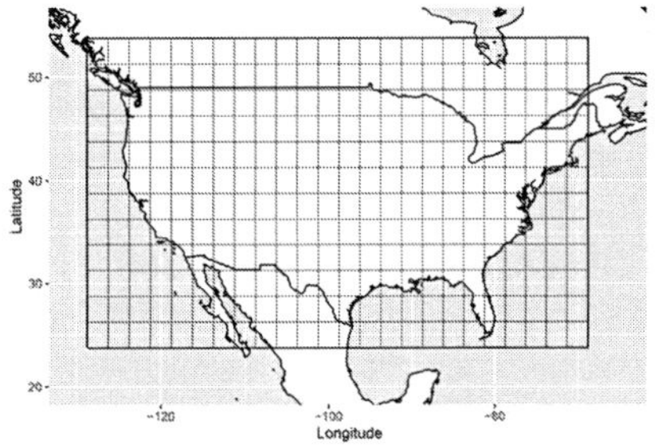

(d) Fourth resolution

FIGURE 7.1
Maps of the study region overlaid with the four spatial resolutions of the multiscale structure. (a) The first – coarsest – level of resolution has 9 subregions. (b) The second level of resolution has 18 subregions. (c) The third level of resolution has 72 subregions. (d) The fourth – finest – level of resolution has 288 subregions.

We assume that we observe q-dimensional data vectors only at the finest level of resolution. We denote the finest level of resolution by L, and denote progressively coarser levels by $L-1, L-2, \ldots, 1$. The l-th level of resolution has n_l subregions. Let $\boldsymbol{y}_{Lj}$ be the q-dimensional data vector observed in subregion j at the finest resolution level L. We assume that $\boldsymbol{y}_{Lj}$ follows a multivariate normal distribution with mean vector $\boldsymbol{\mu}_{Lj}$ and positive definite covariance matrix $\boldsymbol{V}_{Lj}$, that is, $\boldsymbol{y}_{Lj} \sim N_q(\boldsymbol{\mu}_{Lj}, \boldsymbol{V}_{Lj})$. Further, we assume that $\boldsymbol{y}_{L1}, \boldsymbol{y}_{L2}, \ldots, \boldsymbol{y}_{Ln_L}$ are conditionally independent given their corresponding mean vectors $\boldsymbol{\mu}_{L1}, \boldsymbol{\mu}_{L2}, \ldots, \boldsymbol{\mu}_{Ln_L}$ and covariance matrices

$\boldsymbol{V}_{L1}, \boldsymbol{V}_{L2}, \ldots, \boldsymbol{V}_{Ln_L}$. However, we note that in what follows the mean vectors $\boldsymbol{\mu}_{L1}, \boldsymbol{\mu}_{L2}, \ldots, \boldsymbol{\mu}_{Ln_L}$ are spatially correlated and, therefore, after integrating out $\boldsymbol{\mu}_{L1}, \boldsymbol{\mu}_{L2}, \ldots, \boldsymbol{\mu}_{Ln_L}$ the observed data vectors become $\boldsymbol{y}_{L1}, \boldsymbol{y}_{L2}, \ldots, \boldsymbol{y}_{Ln_L}$ spatially correlated. Finally, our framework allows for spatially varying covariances matrices, which is particularly important for the application we consider to tropospheric temperatures.

To define the multiscale structure, at each level of resolution l we partition the spatial domain S into n_l subregions $B_{l1}, B_{l2}, \ldots, B_{ln_l}$ such that $B_{li} \cap B_{lj} = \emptyset$ for $i \neq j$ and $\cup_{i=1}^{n_l} B_{li} = S$. In addition, each subregion B_{lj} at level l has a set $\mathcal{D}_{lj}$ of descendant subregions at level $l+1$. Let d_{lj} be the number of descendants of subregion B_{lj}. Further, the partitions at levels l and $l+1$ are nested such that $\cup_{j=1}^{n_l} \mathcal{D}_{lj} = S$ and $D_{li} \cap D_{lj} = \emptyset$ for $i \neq j$. From the observed data vectors at the finest resolution level L, we recursively compute the data vectors at progressively coarser resolutions. Specifically, the data vector of subregion B_{lj} is a weighted average of the data vectors of the descendant subregions $\mathcal{D}_{lj}$. As a consequence of the assumptions of normality and conditional independence, the conditional distribution of the data vector $\boldsymbol{y}_{lj}$ of subregion B_{lj} is multivariate normal with mean vector $\boldsymbol{\mu}_{lj}$ and covariance matrix $\boldsymbol{V}_{lj}$, thus, $\boldsymbol{y}_{lj} | \boldsymbol{\mu}_{lj}, \boldsymbol{V}_{lj} \sim N_q(\boldsymbol{\mu}_{lj}, \boldsymbol{V}_{lj})$.

To obtain the formulas for the computation of $\boldsymbol{\mu}_{lj}$ and $\boldsymbol{V}_{lj}$, we first stack the data vectors of the set of descendants $\mathcal{D}_{lj}$ into one vector $\boldsymbol{y}_{D_{lj}} = (\boldsymbol{y}'_{\mathcal{D}_{lj1}}, \ldots, \boldsymbol{y}'_{\mathcal{D}_{ljd_{lj}}})'$ of length qd_{lj}. Denote the distribution of the data vector $\boldsymbol{y}_{\mathcal{D}_{ljr}}$ for the r-th descendant by $\boldsymbol{y}_{\mathcal{D}_{ljr}} | \boldsymbol{\mu}_{\mathcal{D}_{ljr}}, \boldsymbol{V}_{\mathcal{D}_{ljr}} \sim N_q(\boldsymbol{\mu}_{\mathcal{D}_{ljr}}, \boldsymbol{V}_{\mathcal{D}_{ljr}})$, $r = 1, \ldots, d_{lj}$. Accordingly, the vector of stacked data vectors of descendant subregions has distribution $\boldsymbol{y}_{\mathcal{D}_{lj}} | \boldsymbol{\mu}_{\mathcal{D}_{lj}}, \boldsymbol{V}_{\mathcal{D}_{lj}} \sim N_{qd_{lj}}(\boldsymbol{\mu}_{\mathcal{D}_{lj}}, \boldsymbol{V}_{\mathcal{D}_{lj}})$, where $\boldsymbol{\mu}_{\mathcal{D}_{lj}} = (\boldsymbol{\mu}'_{\mathcal{D}_{lj1}}, \ldots, \boldsymbol{\mu}'_{\mathcal{D}_{ljd_{lj}}})'$ and $\boldsymbol{V}_{\mathcal{D}_{lj}}$ is a block diagonal matrix with blocks $\boldsymbol{V}_{\mathcal{D}_{lj1}}, \ldots, \boldsymbol{V}_{\mathcal{D}_{ljd_{lj}}}$. Now, let $\boldsymbol{P}_{lj}$ be a weight matrix of dimension $q \times (qd_{lj})$ for the descendants of subregion B_{lj}. Here, each row of $\boldsymbol{P}_{lj}$ adds up to 1 and aggregates specific variables across the different descendant subregions. For example, as proposed by Ferreira et al. (2011) we may aggregate using weights proportional to the precisions of the considered variables at the different descendant subregions. Let $\boldsymbol{I}_q$ denote the q-dimensional identity matrix. Then, as another example, we may use $\boldsymbol{P}_{lj} = (1/d_{lj})[\boldsymbol{I}_q, \ldots, \boldsymbol{I}_q]$ for equal weights among the different d_{lj} descendants. To aggregate the data vectors of the descendants in $\mathcal{D}_{lj}$ to the data vector $\boldsymbol{y}_{lj}$ of subregion B_{lj}, we employ the linear aggregation equation $\boldsymbol{y}_{lj} = \boldsymbol{P}_{lj} \boldsymbol{y}_{\mathcal{D}_{lj}}$. Consequently, the mean vector and covariance matrix of $\boldsymbol{y}_{lj}$ are obtained, respectively, using the aggregation equations

$$\boldsymbol{\mu}_{lj} = \boldsymbol{P}_{lj} \boldsymbol{\mu}_{\mathcal{D}_{lj}} \tag{7.1}$$

and

$$\boldsymbol{V}_{lj} = \boldsymbol{P}_{lj} \boldsymbol{V}_{\mathcal{D}_{lj}} \boldsymbol{P}'_{lj}. \tag{7.2}$$

While the previous paragraph provided information on how to aggregate from finer to coarser levels, our framework focuses on modeling the details that allow the disaggregation from coarser to finer levels. To obtain these details, which in the wavelets literature are known as wavelet coefficients (Vidakovic, 1999), we consider the conditional distribution of the data vectors of the descendant subregions given the data vector of the parent subregion. Using standard results for the multivariate normal distribution, the conditional distribution of interest is

$$\boldsymbol{y}_{\mathcal{D}_{lj}} | \boldsymbol{y}_{lj}, \boldsymbol{\mu}_{\mathcal{D}_{lj}}, \boldsymbol{V}_{D_{lj}} \sim N_{qd_{lj}}(\boldsymbol{\nu}_{lj}\boldsymbol{y}_{lj} + \boldsymbol{\theta}_{lj}, \boldsymbol{\Omega}_{lj}),$$

where $\boldsymbol{\nu}_{lj} = \boldsymbol{V}_{\mathcal{D}_{lj}} \boldsymbol{P}'_{lj} \boldsymbol{V}_{lj}^{-1}$, $\boldsymbol{\theta}_{lj} = \boldsymbol{\mu}_{\mathcal{D}_{lj}} - \boldsymbol{\nu}_{lj}\boldsymbol{\mu}_{lj}$, and $\boldsymbol{\Omega}_{lj} = \boldsymbol{V}_{\mathcal{D}_{lj}} - \boldsymbol{V}_{\mathcal{D}_{lj}} \boldsymbol{P}'_{lj} \boldsymbol{V}_{lj}^{-1} \boldsymbol{P}_{lj} \boldsymbol{V}_{\mathcal{D}_{lj}}$. Analogously to the univariate case (Ferreira et al., 2010, 2011), we call $\boldsymbol{\theta}_{lj}$ the vector of multiscale coefficients. See Elkhouly and Ferreira (2021) for a connection between multiscale coefficients and spatial directional derivatives. Writing the stacked mean vector of descendants in terms of the disaggregation equation $\boldsymbol{\mu}_{\mathcal{D}_{lj}} = \boldsymbol{\nu}_{lj}\boldsymbol{\mu}_{lj} + \boldsymbol{\theta}_{lj}$ helps us see that the matrix $\boldsymbol{\nu}_{lj}$ distributes the mean vector $\boldsymbol{\mu}_{lj}$ of parent subregion B_{lj} among its descendants while $\boldsymbol{\theta}_{lj}$ provides details specific to each descendant that are needed for the disaggregation of the mean process $\boldsymbol{\mu}_{lj}$ at level l onto the vectorized mean process $\boldsymbol{\mu}_{\mathcal{D}_{lj}}$ at level $l+1$.

A naïve way to estimate the multiscale coefficient vector is to substitute $\boldsymbol{\mu}_{\mathcal{D}_{lj}}$ by $\boldsymbol{y}_{\mathcal{D}_{lj}}$ in the formula for $\boldsymbol{\theta}_{lj}$, thus obtaining an empirical multiscale coefficient vector $\boldsymbol{\theta}^e_{lj} = \boldsymbol{y}_{\mathcal{D}_{lj}} - \boldsymbol{\nu}_{lj}\boldsymbol{y}_{lj}$. As a consequence of the conditional distribution of $\boldsymbol{y}_{\mathcal{D}_{lj}}$ given $\boldsymbol{y}_{lj}$, it follows that the conditional distribution of $\boldsymbol{\theta}^e_{lj}$ is $\boldsymbol{\theta}^e_{lj} | \boldsymbol{y}_{lj}, \boldsymbol{\mu}_{\mathcal{D}_{lj}}, \boldsymbol{V}_{D_{lj}} \equiv \boldsymbol{\theta}^e_{lj} | \boldsymbol{\theta}_{lj}, \boldsymbol{\Omega}_{lj} \sim N_{qd_{lj}}(\boldsymbol{\theta}_{lj}, \boldsymbol{\Omega}_{lj})$. Note that the covariance matrix $\boldsymbol{\Omega}_{lj}$ is such that

$$\begin{aligned}
\boldsymbol{\Omega}_{lj}\boldsymbol{P}'_{lj} &= \boldsymbol{V}_{\mathcal{D}_{lj}}\boldsymbol{P}'_{lj} - \boldsymbol{V}_{\mathcal{D}_{lj}}\boldsymbol{P}'_{lj}\boldsymbol{V}_{lj}^{-1}\boldsymbol{P}_{lj}\boldsymbol{V}_{\mathcal{D}_{lj}}\boldsymbol{P}'_{lj} \\
&= \boldsymbol{V}_{\mathcal{D}_{lj}}\boldsymbol{P}'_{lj} - \boldsymbol{V}_{\mathcal{D}_{lj}}\boldsymbol{P}'_{lj}\boldsymbol{V}_{lj}^{-1}\boldsymbol{V}_{lj} \\
&= \boldsymbol{V}_{\mathcal{D}_{lj}}\boldsymbol{P}'_{lj} - \boldsymbol{V}_{\mathcal{D}_{lj}}\boldsymbol{P}'_{lj} \;=\; \mathbf{0}.
\end{aligned}$$

Thus, each of the q rows of $\boldsymbol{P}_{lj}$ is an eigenvector of $\boldsymbol{\Omega}_{lj}$ with a corresponding zero eigenvalue. Hence, the covariance matrix $\boldsymbol{\Omega}_{lj}$ is a singular matrix of rank $q(d_{lj}-1)$. Therefore, the conditional distribution of $\boldsymbol{\theta}^e_{lj}$ given $\boldsymbol{\theta}_{lj}$ and $\boldsymbol{\Omega}_{lj}$ is a singular normal distribution (Muirhead, 1982; Anderson, 1984; Siotani et al., 1985) that automatically encodes q aggregation constraints. See also Section 7.8.1 in this chapter for more information about the singular normal distribution. To see what these aggregation constraints are, let us consider the conditional distribution of $\boldsymbol{P}_{lj}\boldsymbol{\theta}^e_{lj}$. The conditional expected value of $\boldsymbol{P}_{lj}\boldsymbol{\theta}^e_{lj}$ is

$$\begin{aligned}
E[\boldsymbol{P}_{lj}\boldsymbol{\theta}^e_{lj} | \boldsymbol{\theta}_{lj}, \boldsymbol{\Omega}_{lj}] &= \boldsymbol{P}_{lj}\boldsymbol{\theta}_{lj} \\
&= \boldsymbol{P}_{lj}(\boldsymbol{\mu}_{\mathcal{D}_{lj}} - \boldsymbol{\nu}_{lj}\boldsymbol{\mu}_{lj}) \\
&= \boldsymbol{P}_{lj}\boldsymbol{\mu}_{\mathcal{D}_{lj}} - \boldsymbol{P}_{lj}\boldsymbol{V}_{\mathcal{D}_{lj}}\boldsymbol{P}'_{lj}\boldsymbol{V}_{lj}^{-1}\boldsymbol{\mu}_{lj} \\
&= \boldsymbol{\mu}_{lj} - \boldsymbol{V}_{lj}\boldsymbol{V}_{lj}^{-1}\boldsymbol{\mu}_{lj} \;=\; \mathbf{0}
\end{aligned}$$

and, because the $\boldsymbol{P}_{lj}$ are eigenvectors of $\boldsymbol{\Omega}_{lj}$ corresponding zero eigenvalues, the conditional covariance matrix of $\boldsymbol{P}_{lj}\boldsymbol{\theta}^e_{lj}$ is

$$Var[\boldsymbol{P}_{lj}\boldsymbol{\theta}^e_{lj}|\boldsymbol{\theta}_{lj},\boldsymbol{\Omega}_{lj}] = \boldsymbol{P}_{lj}\boldsymbol{\Omega}_{lj}\boldsymbol{P}'_{lj} = \boldsymbol{P}_{lj}\boldsymbol{0} = \boldsymbol{0}.$$

Thus, the conditional distribution of $\boldsymbol{P}_{lj}\boldsymbol{\theta}^e_{lj}$ is a degenerate distribution with $P[\boldsymbol{P}_{lj}\boldsymbol{\theta}^e_{lj} = \boldsymbol{0}] = 1$. In addition, as we have shown above, by construction $\boldsymbol{P}_{lj}\boldsymbol{\theta}_{lj} = \boldsymbol{0}$. Therefore, there are q aggregation constraints that are encoded in the conditional distribution $\boldsymbol{\theta}^e_{lj}|\boldsymbol{\theta}_{lj},\boldsymbol{\Omega}_{lj}$. These constraints are related to the fact that the mean vector at subregion B_{lj} is a weighted average of the mean vectors at the descendants $\mathcal{D}_{lj}$. The framework we propose here respects these aggregation constraints and, thus, estimates the mean process at each scale of resolution in a coherent manner.

The development above provides a multiscale factorization of the data $\boldsymbol{y}_{L1},\ldots,\boldsymbol{y}_{Ln_L}$ observed at the finest resolution level in terms of the data $\boldsymbol{y}_{11},\ldots,\boldsymbol{y}_{1n_1}$ at the coarsest resolution level and the empirical multiscale coefficients $\boldsymbol{\theta}^e_{lj}$, $l = 1,\ldots,L-1$, $j = 1,\ldots,n_l$. In addition, the same linear operations provide a multiscale factorization of the mean process $\boldsymbol{\mu}_{L1},\ldots,\boldsymbol{\mu}_{Ln_L}$ at the finest level of resolution in terms of the mean process $\boldsymbol{\mu}_{11},\ldots,\boldsymbol{\mu}_{1n_1}$ at the coarsest level and the multiscale coefficients $\boldsymbol{\theta}_{lj}$, $l = 1,\ldots,L-1$, $j = 1,\ldots,n_l$. This latter factorization reparameterizes the mean process at the finest level in terms of the mean process at the coarsest level and the multiscale coefficients. Taken together, these two factorizations allow the likelihood function at the finest level of resolution to be factorized as

$$\begin{aligned}\prod_{j=1}^{n_L} p(\boldsymbol{y}_{Lj}|\boldsymbol{\mu}_{Lj},\boldsymbol{V}_{Lj}) &\propto p(\boldsymbol{y}_1|\boldsymbol{\mu}_1,\boldsymbol{V}_1)\prod_{l=1}^{L-1}\prod_{j=1}^{n_l} p(\boldsymbol{y}_{D_{lj}}|\boldsymbol{y}_{lj},\boldsymbol{\theta}_{lj},\boldsymbol{\Omega}_{lj}) \\ &= p(\boldsymbol{y}_1|\boldsymbol{\mu}_1,\boldsymbol{V}_1)\prod_{l=1}^{L-1}\prod_{j=1}^{n_l} p(\boldsymbol{\theta}^e_{lj}|\boldsymbol{\theta}_{lj},\boldsymbol{\Omega}_{lj}). \qquad (7.3)\end{aligned}$$

The constant of proportionality in the first line of Equation (7.3) is the Jacobian of the transformation from the finest level $y_{L,1:n_L}$ to the aggregated data. Thus, the constant of proportionality depends only on the aggregation weights. In addition, conditional on y_{lj}, the Jacobian of the transformation from $Y_{D_{lj}}$ to $\boldsymbol{\theta}^e_{lj}$ is equal to 1. Thus, the right hand side in the first line is exactly equal to the expression in the second line of Equation (7.3). The factorization of the likelihood function given in Equation (7.3) allows for the development of computational algorithms that are parallelizable, scalable, and fast. In what follows, we use this factorization in the context of spatio-temporal modeling.

7.3 Dynamic Multivariate Multiscale Spatio-Temporal Models

In the previous section, we propose a multiscale factorization for multivariate spatial data. In that context, the model for the data observed at subregion j at the finest resolution L can be written as $\boldsymbol{y}_{Lj} = \boldsymbol{\mu}_{Lj} + \boldsymbol{v}_{Lj}, \boldsymbol{v}_{Lj} \sim N_q(\boldsymbol{0}, \boldsymbol{V}_{Lj})$. In this section, we add a subscript t to include a time component and model spatio-temporal data. Thus, $\boldsymbol{y}_{tLj}$ is the q-dimensional vector of observations of the jth subregion at the finest resolution L at time t. Then, we can write the model for the data vector $\boldsymbol{y}_{tLj}$ as the observation equation

$$\boldsymbol{y}_{tLj} = \boldsymbol{\mu}_{tLj} + \boldsymbol{v}_{tLj}, \qquad \boldsymbol{v}_{tLj} \sim N_q(\boldsymbol{0}, \boldsymbol{V}_{Lj}), \tag{7.4}$$

where $\boldsymbol{\mu}_{tLj} = E(\boldsymbol{y}_{tLj}|\boldsymbol{\mu}_{tLj}, \boldsymbol{V}_{Lj})$ is a q-dimensional vector and $\boldsymbol{V}_{Lj} = Var(\boldsymbol{y}_{tLj}|\boldsymbol{\mu}_{tLj}, \boldsymbol{V}_{Lj})$ is a $q \times q$ covariance matrix. In addition, we assume that the error vectors $\boldsymbol{v}_{tLj}$, $t = 1, \ldots, T$, $j = 1, \ldots, n_L$, are mutually independent.

Figure 7.2 presents the multiple time series of troposphere air temperature anomalies at the finest level of resolution, with each panel presenting the data for one of the five altitude pressure levels. Two features are evident from the figure: the existence of a possibly stochastic long term trend, and the existence of transitory fluctuations around the long term trend. We account for these two features in the data by expressing the mean process as a sum of two components. Specifically, we decompose the mean at the finest resolution $\boldsymbol{\mu}_{tLj}$ into a sum of a nonstationary component $\boldsymbol{\alpha}^{\mu}_{tLj}$ and a mean-zero stationary component $\boldsymbol{\gamma}^{\mu}_{tLj}$, that is, $\boldsymbol{\mu}_{tLj} = \boldsymbol{\alpha}^{\mu}_{tLj} + \boldsymbol{\gamma}^{\mu}_{tLj}$. The q-dimensional nonstationary component $\boldsymbol{\alpha}^{\mu}_{tLj}$ represents the long-term trend, and the q-dimensional stationary component $\boldsymbol{\gamma}^{\mu}_{tLj}$ accounts for the transitory fluctuations around the long-term trend. As a result, we can rewrite the observation equation at the finest resolution as

$$\boldsymbol{y}_{tLj} = \boldsymbol{\alpha}^{\mu}_{tLj} + \boldsymbol{\gamma}^{\mu}_{tLj} + \boldsymbol{v}_{tLj}, \qquad \boldsymbol{v}_{tLj} \sim N_q(\boldsymbol{0}, \boldsymbol{V}_{Lj}), \tag{7.5}$$

$t = 1, \ldots, T$, $j = 1, 2, \ldots, n_L$. Because of the stationarity, long-term predictions for $\boldsymbol{\gamma}^{\mu}_{tLj}$ will converge to zero. Thus, long-term predictions for $\boldsymbol{y}_{tLj}$ are going to depend mostly on long-term predictions for $\boldsymbol{\alpha}^{\mu}_{tLj}$. Meanwhile, short-term predictions for $\boldsymbol{y}_{tLj}$ are going to depend on both $\boldsymbol{\alpha}^{\mu}_{tLj}$ and $\boldsymbol{\gamma}^{\mu}_{tLj}$.

Next, we apply the multivariate multiscale factorization from Section 7.2 to both sides of Equation (7.5). This results in multiscale factorizations for the observations, the long-term non-stationary component, and the mean-zero stationary component. In particular, the mean $\boldsymbol{\mu}_{t1k}$ at the coarsest level of resolution can be written as a sum of an aggregated non-stationary long-term component and an aggregated stationary short-term component, that is, $\boldsymbol{\mu}_{t1k} = \boldsymbol{\alpha}^{\mu}_{t1k} + \boldsymbol{\gamma}^{\mu}_{t1k}$. Similarly, the (qd_{lj})-dimensional latent multiscale coefficient $\boldsymbol{\theta}_{tlj}$ can be written as $\boldsymbol{\theta}_{tlj} = \boldsymbol{\alpha}^{\boldsymbol{\theta}}_{tlj} + \boldsymbol{\gamma}^{\boldsymbol{\theta}}_{tlj}$, where $\boldsymbol{\alpha}^{\boldsymbol{\theta}}_{tlj}$ is a latent multiscale

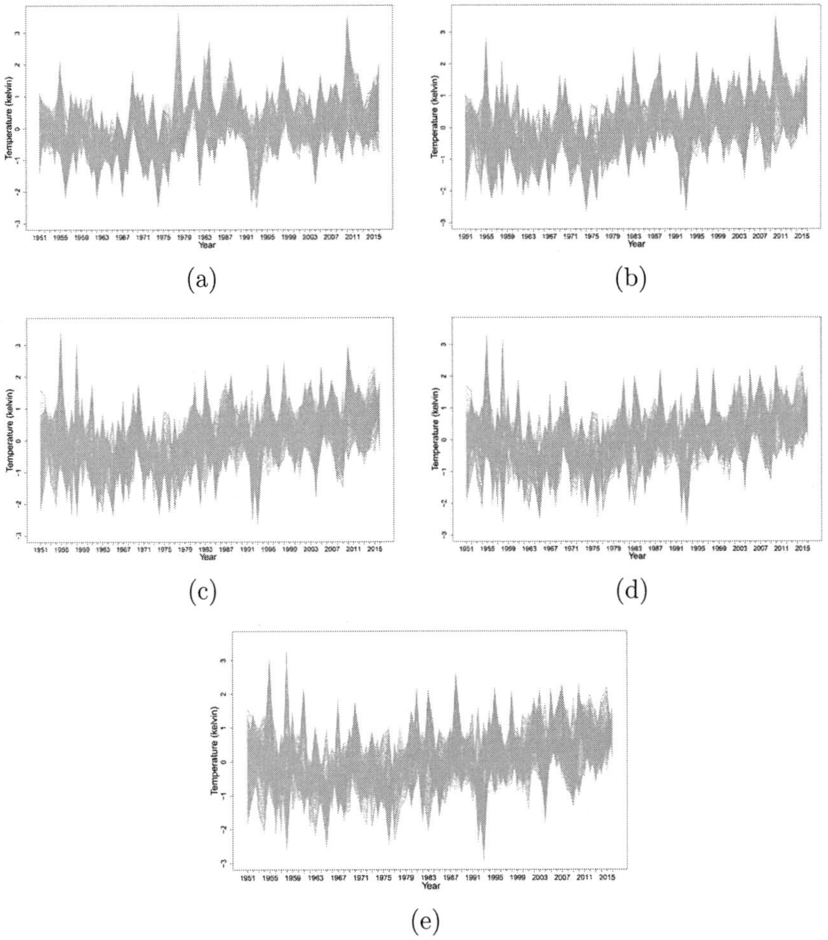

FIGURE 7.2
Time series plots of troposphere air temperature anomalies in Kelvin over North America at 5 different altitude pressure levels: (a) 250hPa; (b) 300hPa; (c) 400hPa; (d) 500hPa; (e) 600hPa.

coefficient of the non-stationary long-term component and $\gamma^{\boldsymbol{\theta}}_{tlj}$ is a latent multiscale coefficient of the stationary short-term component. Thus, we obtain the following observation equations for the coarsest resolution aggregated observations and for the empirical multiscale coefficients through time

$$\boldsymbol{y}_{t1k} = \boldsymbol{\alpha}^{\mu}_{t1k} + \boldsymbol{\gamma}^{\mu}_{t1k} + \boldsymbol{v}_{t1k}, \qquad \boldsymbol{v}_{t1k} \sim N_q(\mathbf{0}, \boldsymbol{V}_{1k}), \tag{7.6}$$

$$\boldsymbol{\theta}^{e}_{tlj} = \boldsymbol{\alpha}^{\boldsymbol{\theta}}_{tlj} + \boldsymbol{\gamma}^{\boldsymbol{\theta}}_{tlj} + \boldsymbol{e}_{tlj}, \qquad \boldsymbol{e}_{tlj} \sim N_{qd_{lj}}(\mathbf{0}, \boldsymbol{\Omega}_{lj}), \tag{7.7}$$

$k = 1, \ldots, n_1$, $l = 1, \ldots, L-1$, $j = 1, \ldots, n_l$, $t = 1, \ldots, T$.

Our framework allows the covariance matrices $\boldsymbol{V}_{Lj}$, $j = 1, \ldots, n_L$, at the finest level of resolution L to vary spatially. We present in Section 7.6.1 an empirical Bayes approach to estimate these covariance matrices. This empirical Bayes approach then assumes that these covariance matrices are fixed equal to their estimates. After that, this empirical Bayes approach uses these estimates to compute $\boldsymbol{V}_{1k}$, $k = 1, \ldots, n_1$, and $\boldsymbol{\Omega}_{lj}$, $l = 1, \ldots, L-1$, $j = 1, \ldots, n_l$. This approach allows our framework to account for the interactions and the co-movements of the multivariate time series.

7.3.1 Evolution through Time

To model the temporal evolution of the mean process at the coarsest resolution and the temporal evolution of the various multiscale coefficients, we use dynamic linear models (DLM). We assume that each of the multiple time series shares the same type of latent evolution structure over time (Prado et al., 2021). Based on the exhibited observational temporal patterns of tropospheric temperatures, we adapt the DLM to model the long-term nonstationary and short-term stationary components of the mean process at the coarsest level and of the latent multiscale coefficients. In particular, we use the linear growth (or local linear trend) state-space model to model the nonstationary component. The linear growth model adds a time varying gradient (or slope $\boldsymbol{\beta}_t$) to the dynamics of the trend (or level) $\boldsymbol{\alpha}_t$. Note that if we were to remove the gradient from the model, we would obtain a random walk (local level) model. For the short-term fluctuations $\boldsymbol{\gamma}_t$, we use a stationary first-order autoregressive (AR1) process.

To model the temporal evolution of the latent processes $\boldsymbol{\alpha}^{\mu}_{t1k}$, $\boldsymbol{\beta}^{\mu}_{t1k}$, and $\boldsymbol{\gamma}^{\mu}_{t1k}$ at the coarsest level, we stack them as rows of a $3 \times q$ matrix $\boldsymbol{H}_{t1k}$ defined as

$$\boldsymbol{H}_{t1k} = \begin{pmatrix} (\boldsymbol{\alpha}^{\mu}_{t1k})' \\ (\boldsymbol{\beta}^{\mu}_{t1k})' \\ (\boldsymbol{\gamma}^{\mu}_{t1k})' \end{pmatrix}. \tag{7.8}$$

We can then describe the temporal evolution of $\boldsymbol{H}_{t1k}$ with the matrix-variate system equation (Quintana and West, 1987; Landim and Gamerman, 2000; Carvalho and West, 2007)

$$\boldsymbol{H}_{t1k} = \mathbf{G}^{\mu}_{1k}\boldsymbol{H}_{t-1,1k} + \boldsymbol{w}_{t1k}, \tag{7.9}$$

where the error matrix $\boldsymbol{w}_{t1k}$ follows a matrix-normal (MN) distribution. To specify the dependence structure among the elements of the innovation matrix $\boldsymbol{w}_{t1k}$, we use a signal-to-noise parameterization. First, we assume that the dependence among the columns of $\boldsymbol{w}_{t1k}$ is proportional to the dependence among the elements of the corresponding observation equation error vector $\boldsymbol{v}_{t1k}$. This implies that the left covariance matrix of $\boldsymbol{w}_{t1k}$, which describes the dependence among its columns, is equal to the observational covariance matrix $\boldsymbol{V}_{1k}$. Next, we assume that the innovations of the three dynamics (level, gradient, and short-term fluctuations) of the evolution equations are independent.

This then implies that the right covariance matrix is a 3×3 diagonal matrix that we denote by

$$\boldsymbol{U}_{1k}^{\mu} = \begin{pmatrix} \psi_{1k1}^{\mu} & 0 & 0 \\ 0 & \psi_{1k2}^{\mu} & 0 \\ 0 & 0 & \psi_{1k3}^{\mu} \end{pmatrix}.$$

Thus, we write $\boldsymbol{w}_{t1k} \sim MN(\mathbf{0}, \boldsymbol{U}_{1k}^{\mu}, \boldsymbol{V}_{1k})$. Let vec be the operator that transforms a matrix into a column vector by stacking its columns, and let $\otimes$ be the Kronecker product. Then, $\text{vec}(\boldsymbol{w}_{t1k}) \sim N_{3q}(\mathbf{0}, \boldsymbol{V}_{1k} \otimes \boldsymbol{U}_{1k}^{\mu})$. Thus, ψ_{1k1}^{μ}, ψ_{1k2}^{μ}, and ψ_{1k3}^{μ} are signal-to-noise ratio parameters.

Finally, we assume that the q elements of the vector of short-term fluctuations γ_{t1k}^{μ} share the same autoregressive coefficient ϕ_{1k}^{μ}. Thus, the state evolution matrix in Equation (7.9) is

$$\mathbf{G}_{1k}^{\mu} = \begin{pmatrix} 1 & 1 & 0 \\ 0 & 1 & 0 \\ 0 & 0 & \phi_{1k}^{\mu} \end{pmatrix}. \tag{7.10}$$

Similarly, to model the temporal evolution of the latent processes associated with the latent multiscale coefficient $\boldsymbol{\theta}_{tlj}$, we stack $\boldsymbol{\alpha}_{tlj}^{\theta}$, $\boldsymbol{\beta}_{tlj}^{\theta}$, and $\boldsymbol{\gamma}_{tlj}^{\theta}$ as rows of a $3 \times (qd_{lj})$ matrix $\boldsymbol{\Theta}_{tlj}$ defined as

$$\boldsymbol{\Theta}_{tlj} = \begin{pmatrix} (\boldsymbol{\alpha}_{tlj}^{\theta})' \\ (\boldsymbol{\beta}_{tlj}^{\theta})' \\ (\boldsymbol{\gamma}_{tlj}^{\theta})' \end{pmatrix}. \tag{7.11}$$

Then, we assume that $\boldsymbol{\Theta}_{tlj}$ evolves through time according to the matrix-variate system evolution equation

$$\boldsymbol{\Theta}_{tlj} = \mathbf{G}_{lj}^{\theta} \boldsymbol{\Theta}_{t-1,lj} + \boldsymbol{\xi}_{tlj}, \tag{7.12}$$

where the error matrix $\boldsymbol{\xi}_{tlj}$ follows a *singular* matrix normal distribution (SMN) (Díaz-García and Gutiérrez-Jáimez, 2006) such that $\boldsymbol{\xi}_{tlj} \sim SMN(\mathbf{0}, \boldsymbol{U}_{lj}^{\theta}, \boldsymbol{\Omega}_{lj})$. The choices of the left and right covariance matrices for $\boldsymbol{\xi}_{tlj}$ follow reasoning similar to the choices for $\boldsymbol{w}_{t1k}$ explained above. Specifically, these choices assume a signal-to-noise ratio parameterization with the dependence among the columns of $\boldsymbol{\xi}_{tlj}$ being proportional to the dependence among the elements of the error vector $\boldsymbol{e}_{tlj}$ in the corresponding observation equation (7.7). This implies that the left covariance matrix of $\boldsymbol{\xi}_{tlj}$, which describes the dependence among its columns, is equal to the observational covariance matrix $\boldsymbol{\Omega}_{lj}$. Because $\boldsymbol{\Omega}_{lj}$ is a singular semi-positive definite covariance matrix with zero eigenvalue with multiplicity q, the multiscale coefficients satisfy q aggregation constraints that guarantee the coherence of the mean process across the different scales of resolution.

In addition, we again assume that the innovations of the three dynamics (level, gradient, and short-term fluctuations) of the evolution equations are independent, implying that the right covariance matrix of $\boldsymbol{\xi}_{tlj}$ is a 3×3 diagonal matrix that we denote by

$$\boldsymbol{U}_{lj}^{\theta} = \begin{pmatrix} \psi_{lj1}^{\theta} & 0 & 0 \\ 0 & \psi_{lj2}^{\theta} & 0 \\ 0 & 0 & \psi_{lj3}^{\theta} \end{pmatrix},$$

where ψ_{lj1}^{θ}, ψ_{lj2}^{θ}, and ψ_{lj3}^{θ} are signal-to-noise ratio parameters for the three states of the evolution equations.

Finally, we assume that the q elements of the short-term stationary latent process γ_{tlj}^{θ} follow AR(1) processes with the same autoregressive coefficient ϕ_{lj}^{θ}. Thus, the state evolution matrix in Equation (7.12) is

$$\mathbf{G}_{lj}^{\theta} = \begin{pmatrix} 1 & 1 & 0 \\ 0 & 1 & 0 \\ 0 & 0 & \phi_{lj}^{\theta} \end{pmatrix}. \tag{7.13}$$

In what follows, it will be useful to rewrite the observation equations (7.6) and (7.7) in terms of $\boldsymbol{H}_{t1k}$ and $\boldsymbol{\Theta}_{tlj}$, respectively. To that end, let $\boldsymbol{f}' = (1, 0, 1)$. Then the observation equation (7.6) can be rewritten as

$$\boldsymbol{y}_{t1k} = \boldsymbol{H}_{t1k}' \boldsymbol{f} + \boldsymbol{v}_{t1k} \tag{7.14}$$

and the observation equation (7.7) can be rewritten as

$$\boldsymbol{\theta}_{tlj}^{e} = \boldsymbol{\Theta}_{tlj}' \boldsymbol{f} + \boldsymbol{e}_{tlj}. \tag{7.15}$$

The versions of the observation equations given in Equations (7.14) and (7.15) are used to obtain the forward filter and backward sampler algorithm presented in Section 7.4.2.

7.3.2 Priors for the Model Parameters and Latent States

We complete the model specification with conjugate priors for the parameters related to the different subregions at the several levels of resolution. For each signal-to-noise ratio parameter ψ we assign an independent inverse gamma prior $\psi|\eta, \zeta \sim IG(0.5\eta, 0.5\zeta)$ with density $p(\psi|\eta, \zeta) \propto \psi^{-0.5\eta-1} e^{-0.5\zeta/\psi}$. For each autoregressive coefficient ϕ, to ensure stationarity of the system dynamics, we assign a Gaussian prior truncated on the interval $(-1, 1)$. Here, the truncated Gaussian density is of the form $p(\phi|a, b) \propto e^{-0.5\{(\phi-a)/b\}^2} I(\phi)_{(-1,1)}$, where a and b are location and scale parameters and $I(\phi)_{(-1,1)}$ is an indicator function equal to 1 if $\phi \in (-1, 1)$ and 0 otherwise.

To set priors for the latent states at time $t = 0$, let D_0 be the prior information available at time $t = 0$ and recursively define the information up to time t by $D_t = D_{t-1} \cup \{\boldsymbol{y}_t\}$. The prior for the latent states of the multiscale coefficients is $\boldsymbol{\Theta}_{0lj}|D_0 \sim SMN(\mathbf{M}_{0lj}, \mathbf{C}_{0lj}, \boldsymbol{\Omega}_{lj})$. The right covariance matrix $\mathbf{C}_{0lj}$ is assumed to be diagonal with the first two diagonal elements being the prior values of the signal-to-noise ratio parameters for the level and gradient latent states. In addition, note that the marginal covariance matrix of the stationary vector AR(1) process associated with the latent multiscale coefficient $\boldsymbol{\theta}_{tlj}$ is $\{\psi^{\theta}_{lj3}/(1-(\phi^{\theta}_{lj})^2)\}\boldsymbol{\Omega}_{lj}$. Therefore, to preserve the stationarity of the AR(1) component, we set the third diagonal element of $\mathbf{C}_{0lj}$ to be $\psi^{\theta}_{lj3}/(1-(\phi^{\theta}_{lj})^2)$. Similarly, the prior at time 0 for the latent parameters related to the mean process at the coarsest level is $\boldsymbol{H}_{01k}|D_0 \sim MN(\mathbf{M}_{01k}, \mathbf{C}_{01k}, \boldsymbol{V}_{1k})$. The right covariance matrix $\mathbf{C}_{01k}$ is assumed to be diagonal and defined analogously to the right covariance matrix $\mathbf{C}_{0lj}$.

7.4 MCMC for Posterior Exploration

To explore the posterior distribution of the model parameters, we have developed a parallelizable MCMC algorithm with fast and tractable computations. Specifically, we have developed a hybrid Gibbs sampler algorithm with Gibbs steps for the signal-to-noise ratio parameters, Metropolis-Hastings steps for the autoregressive coefficients, and a multivariate singular forward filter backward sampler for the latent states.

The presentation that follows in this section assumes that the observational covariance matrices at the finest resolution level $\boldsymbol{V}_{Lj}$ are known. We then compute the observational covariance matrices at the other levels of resolution by aggregation. Note that the knowledge of the observational covariance matrices implies that the empirical multiscale coefficients are uniquely defined. Thus, the assumption of known covariance matrices is fundamental for the scalability of our multiscale modeling framework. In practice, the observational covariance matrices are usually not known. In that case, we suggest the estimation of the observational covariance matrices with an empirical Bayes procedure, such as the procedure we propose in Section 7.6.1 for the application to tropospheric temperatures.

7.4.1 Simulation of Signal-to-Noise Ratio Parameters and Autoregressive Coefficients

The simulation steps for the signal-to-noise ratio parameters and autoregressive coefficients are as follows.

1. *Simulation of* ψ^{μ}_{1k1}. We simulate the signal-to-noise ratio parameter ψ^{μ}_{1k1}, $k = 1, \ldots n_1$, associated with the latent level process at the coarsest resolution from its full conditional distribution $\psi^{\mu}_{1k1}|\boldsymbol{\alpha}^{\mu}_{1:T,1k}, \boldsymbol{\beta}^{\mu}_{1:T,1k}, \boldsymbol{V}_{1k}, D_T \sim IG(0.5\eta^*_{1k,1}, 0.5\zeta^*_{1k,1})$, where

$$\eta^*_{1k,1} = \eta_{1k,1} + Tq$$

and

$$\zeta^*_{1k,1} = \zeta_{1k,1} + \sum_{t=1}^{T} (\boldsymbol{\alpha}^{\mu}_{t1k} - [\boldsymbol{\alpha}^{\mu}_{t-1,1k} + \boldsymbol{\beta}^{\mu}_{t-1,1k}])' \boldsymbol{V}^{-1}_{1k} (\boldsymbol{\alpha}^{\mu}_{t1k} - [\boldsymbol{\alpha}^{\mu}_{t-1,1k} + \boldsymbol{\beta}^{\mu}_{t-1,1k}]).$$

2. *Simulation of* ψ^{μ}_{1k2}. We simulate the signal-to-noise ratio parameter ψ^{μ}_{1k2}, $k = 1, \ldots n_1$, associated with the latent gradient process at the coarsest resolution from its full conditional distribution $\psi^{\mu}_{1k2}|\boldsymbol{\beta}^{\mu}_{1:T,1k}, \boldsymbol{V}_{1k}, D_T \sim IG(0.5\eta^*_{1k,2}, 0.5\zeta^*_{1k,2})$, where

$$\eta^*_{1k,2} = \eta_{1k,2} + Tq$$

and

$$\zeta^*_{1k,2} = \zeta_{1k,2} + \sum_{t=1}^{T} (\boldsymbol{\beta}^{\mu}_{t1k} - \boldsymbol{\beta}^{\mu}_{t-1,1k})' \boldsymbol{V}^{-1}_{1k} (\boldsymbol{\beta}^{\mu}_{t1k} - \boldsymbol{\beta}^{\mu}_{t-1,1k}).$$

3. *Simulation of* ψ^{μ}_{1k3}. We simulate the signal-to-noise ratio parameter ψ^{μ}_{1k3}, $k = 1, \ldots n_1$, associated with the latent AR(1) process at the coarsest level of resultion from its full conditional distribution $\psi^{\mu}_{1k3}|\boldsymbol{\gamma}^{\mu}_{1:T,1k}, \boldsymbol{V}_{1k}, \phi^{\mu}_{1k}, D_T \sim IG(0.5\eta^*_{1k,3}, 0.5\zeta^*_{1k,3})$, where

$$\eta^*_{1k,3} = \eta_{1k,3} + Tq$$

and

$$\zeta^*_{1k,3} = \zeta_{1k,3} + \sum_{t=1}^{T} (\boldsymbol{\gamma}^{\mu}_{t1k} - \phi^{\mu}_{1k}\boldsymbol{\gamma}^{\mu}_{t-1,1k})' \boldsymbol{V}^{-1}_{1k} (\boldsymbol{\gamma}^{\mu}_{t1k} - \phi^{\mu}_{1k}\boldsymbol{\gamma}^{\mu}_{t-1,1k}).$$

4. *Simulation of* ϕ^{μ}_{1k}. To simulate the autoregressive coefficient ϕ^{μ}_{1k}, we build an efficient Metropolis-Hastings step. Recall that $|\phi^{\mu}_{1k}| < 1$. Thus, the density of the full conditional distribution of ϕ^{μ}_{1k} can be decomposed as the product of two factors $p(\phi^{\mu}_{1k}|\psi^{\mu}_{1k3}, \boldsymbol{\gamma}^{\mu}_{0:T,1k}, \boldsymbol{V}_{1k}, D_T) \propto h(\phi^{\mu}_{1k})g(\phi^{\mu}_{1k})$, where $h(\phi^{\mu}_{1k})$ is the kernel of a truncated normal density and $g(\phi^{\mu}_{1k})$ is the density at time 0 of the coarsest resolution mean process $p(\boldsymbol{H}_{01k}|D_0)$. Thus, we use a Metropolis-Hastings step that proposes a new draw ${\phi^{\mu}_{1k}}^{(prop)}$ from the truncated normal distribution with density

$$h(\phi^{\mu}_{1k}) = p(\phi^{\mu}_{1k}|a^*_{1k}, b^*_{1k}) \propto \exp\{-0.5((\phi^{\mu}_{1k} - a^*_{1k})/b^*_{1k})^2\} I(\phi)_{(-1,1)},$$

where

$$b^{*2}_{1k} = \left[\frac{1}{b^2_{1k}} + \frac{1}{\psi^{\mu}_{1k3}} \sum_{t=1}^{T} (\boldsymbol{\gamma}^{\mu}_{t-1,1k})' \boldsymbol{V}^{-1}_{1k} (\boldsymbol{\gamma}^{\mu}_{t-1,1k})\right]^{-1}$$

and

$$a_{1k}^* = b_{1k}^{*2}\left[\frac{a_{1k}}{b_{1k}^2} + \frac{1}{\psi_{1k3}^{\mu}}\sum_{t=1}^{T}(\gamma_{t-1,1k}^{\mu})'\boldsymbol{V}_{1k}^{-1}(\gamma_{t1k}^{\mu})\right].$$

Let

$$\log(g(\phi_{1k}^{\mu})) = c + \frac{q}{2}\log(1-(\phi_{1k}^{\mu})^2) + \frac{(\phi_{1k}^{\mu})^2}{2\psi_{1k3}^{\mu}}[(\gamma_{01k}^{\mu})'\boldsymbol{V}_{1k}^{-1}(\gamma_{01k}^{\mu})],$$

where c is a constant with respect to ϕ_{1k}^{μ}. Then, we accept the proposal $\phi_{1k}^{\mu(prop)}$ with probability equal to $\min(1,\alpha)$, where

$$\log(\alpha) = \log(g(\phi_{1k}^{\mu(prop)})) - \log(g(\phi_{1k}^{\mu(old)})).$$

5. *Simulation of* ψ_{lj1}^{θ}. We simulate the signal-to-noise ratio parameter ψ_{lj1}^{θ}, $l = 1,\ldots L$, $j = 1,\ldots n_l$, associated with the latent level state from its full conditional distribution $\psi_{lj1}^{\theta}|\boldsymbol{\alpha}_{1:T,lj}^{\theta}, \boldsymbol{\beta}_{1:T,lj}^{\theta}, \boldsymbol{\Omega}_{lj}, D_T \sim IG(0.5\eta_{lj,1}^*, 0.5\zeta_{lj,1}^*)$, where

$$\eta_{lj,1}^* = \eta_{lj,1} + Tq(d_{lj}-1)$$

and

$$\zeta_{lj,1}^* = \zeta_{lj,1} + \sum_{t=1}^{T}(\boldsymbol{\alpha}_{tlj}^{\theta} - [\boldsymbol{\alpha}_{t-1,lj}^{\theta} + \boldsymbol{\beta}_{t-1,lj}^{\theta}])'\boldsymbol{\Omega}_{lj}^{-}(\boldsymbol{\alpha}_{tlj}^{\theta} - [\boldsymbol{\alpha}_{t-1,1k}^{\theta} + \boldsymbol{\beta}_{t-1,lj}^{\theta}]).$$

6. *Simulation of* ψ_{lj2}^{θ}. We simulate the signal-to-noise ratio parameter ψ_{lj2}^{θ}, $l = 1,\ldots L$, $j = 1,\ldots n_l$, associated with the time varying gradient state from its full conditional $\psi_{lj2}^{\theta}|\boldsymbol{\beta}_{1:T,lj}^{\theta}, \boldsymbol{\Omega}_{lj}, D_T \sim IG(0.5\eta_{lj,2}^*, 0.5\zeta_{lj,2}^*)$, where

$$\eta_{lj,2}^* = \eta_{lj,2} + Tq(d_{lj}-1)$$

and

$$\zeta_{1k,2}^* = \zeta_{lj,2} + \sum_{t=1}^{T}(\boldsymbol{\beta}_{tlj}^{\theta} - \boldsymbol{\beta}_{t-1,lj}^{\theta})'\boldsymbol{\Omega}_{lj}^{-}(\boldsymbol{\beta}_{tlj}^{\theta} - \boldsymbol{\beta}_{t-1,lj}^{\theta}).$$

7. *Simulation of* ψ_{lj3}^{θ}. We simulate the signal-to-noise ratio parameter ψ_{lj3}^{θ}, $l = 1,\ldots L$, $j = 1,\ldots n_l$, associated with the evolution state of the AR(1) component from its full conditional $\psi_{lj3}^{\theta}|\gamma_{1:T,lj}^{\theta}, \boldsymbol{\Omega}_{lj}, D_T \sim IG(0.5\eta_{lj,3}^*, 0.5\zeta_{1k,3}^*)$, where

$$\eta_{lj,3}^* = \eta_{lj,3} + Tq$$

and

$$\zeta^*_{lj,3} = \zeta_{lj,3} + \sum_{t=1}^{T} (\gamma^\theta_{tlj} - \phi^\theta_{lj}\gamma^\theta_{t-1,lj})'\mathbf{\Omega}^-_{lj}(\gamma^\theta_{tlj} - \phi^\theta_{lj}\gamma^\theta_{t-1,lj}).$$

8. *Simulation of ϕ^θ_{lj}.* We simulate the autoregressive coefficient ϕ^θ_{lj}, $l = 1,\dots L$, $j = 1,\dots n_l$, with a Metropolis-Hastings step. First, we simulate a new proposal from the truncated normal distribution with density $h(\phi^\theta_{lj}) = p(\phi^\theta_{lj}|a^*, b^*) \propto \exp\{-0.5((\phi^\mu_{lj} - a^*_{lj})/b^*_{lj})^2\}I_{\{-1,1\}}$ where

$$b^{*2}_{lj} = \left[\frac{1}{b^2_{lj}} + \frac{1}{\psi^\theta_{lj3}}\sum_{t=1}^{T}(\gamma^\theta_{t-1,lj})'\mathbf{\Omega}^-_{lj}(\gamma^\theta_{t-1,lj})\right]^{-1}$$

and

$$a^*_{lj} = b^{*2}_{lj}\left[\frac{a_{lj}}{b^2_{lj}} + \frac{1}{\psi^\theta_{lj3}}\sum_{t=1}^{T}(\gamma^\theta_{t-1,lj})'\mathbf{\Omega}^-_{lj}(\gamma^\theta_{tlj})\right].$$

Then, we accept the proposal $\phi^{\theta(prop)}_{lj}$ with probability equal to $\min(1,\alpha)$, where $\log(\alpha) = \log(g(\phi^{\theta\,(prop)}_{lj})) - \log(g(\phi^{\theta\,(old)}_{lj}))$ and

$$\log(g(\phi^\theta_{lj})) = c + \frac{q}{2}\log(1 - (\phi^\theta_{lj})^2) + \frac{\phi^{\theta\,2}_{lj}}{2\psi^\theta_{lj3}}[(\gamma^\theta_{0lj})'\mathbf{\Omega}^-_{lj}(\gamma^\theta_{0lj})],$$

where c is a constant with respect to ϕ^θ_{lj}.

7.4.2 Simulation of the Latent States

We recall that our proposed DMMSTM framework, defined by Equations (7.6)–(7.12), is a multivariate extension of the multiscale spatio-temporal model for univariate Gaussian data proposed by Ferreira et al. (2010, 2011). Similarly to the univariate case, in the DMMSTM framework the latent states can be simulated in parallel as stated in Theorem 7.1.

Theorem 7.1 *Given the covariance matrices at the finest level of resolution $\mathbf{V}_{L1:n_L}$, the signal-to-noise ratio parameters $\psi_{1,1:n_1,-},\dots,\psi_{L-1,1:n_{L-1},-}$, and the autoregressive coefficients $\phi^\mu_{1,1:n_1}, \phi^\theta_{1,1:n_1},\dots,\phi^\theta_{L-1,1:n_{L-1}}$, the latent dynamic arrays $\boldsymbol{H}_{0:T,11},\dots,\boldsymbol{H}_{0:T,1n_1}, \mathbf{\Theta}_{0:T,11},\dots,\mathbf{\Theta}_{0:T,1n_1},\dots,\mathbf{\Theta}_{0:T,L-1n_{L-1}}$ are a posteriori conditionally independent and Gaussian distributed.*

Proof *See appendix.*

Therefore, we simulate in parallel the latent states at the coarsest level and the latent states for the multiscale coefficients at the several scales of resolution. To simulate the latent states at the coarsest level, we use the Forward Filter Backward Sampler (FFBS) (Carter and Kohn, 1994; Frühwirth-Schnatter, 1994) for multivariate dynamic linear models (p. 269, Prado et al., 2021). In addition, to sample the latent multiscale coefficient $\boldsymbol{\Theta}_{tlj}$ through time, we have developed a Multivariate Singular Forward Filter Backward Sampler (MSFFBS). The MSFFBS is employed to sample arrays of singular matrix-variate Gaussian DLMs. Let D_{t-1} represent all the information available up to time $t-1$ and recursively define $D_t = D_{t-1} \cup \{\boldsymbol{y}_{t,L,1:n_L}\}$. Further, set the prior at time 0 to be $\boldsymbol{\Theta}_{0lj}|D_0 \sim MN(\mathbf{M}_{0lj}, \mathbf{C}_{0lj}, \boldsymbol{\Omega}_{lj})$. Then, Theorem 7.2 provides the MSFFBS.

Theorem 7.2 (MSFFBS) *Consider the dynamic linear model for the multiscale coefficients defined by the observational Equation (7.7) and the state space Equation (7.12). Then,*

1. *Posterior distribution at time* $t-1$*:* $\boldsymbol{\Theta}_{t-1,lj}|D_{t-1} \sim MN(\boldsymbol{M}_{t-1,lj}, \boldsymbol{C}_{t-1,lj}, \boldsymbol{\Omega}_{lj})$
2. *Prior distribution at time* t*:* $\boldsymbol{\Theta}_{tlj}|D_{t-1} \sim MN(\boldsymbol{A}_{tlj}, \boldsymbol{R}_{tlj}, \boldsymbol{\Omega}_{lj})$, *where* $\boldsymbol{A}_{tlj} = \boldsymbol{G}^{\theta}_{lj}\boldsymbol{M}_{t-1,lj}$ *and* $\boldsymbol{R}_{tlj} = \boldsymbol{G}^{\theta}_{lj}\boldsymbol{C}_{t-1,lj}(\boldsymbol{G}^{\theta}_{lj})' + \boldsymbol{U}^{\theta}_{lj}$.
3. *One-step ahead forecast distribution:* $(\boldsymbol{\theta}^{e}_{tlj})'|D_{t-1} \sim SMN(\boldsymbol{\vartheta}_{tlj}, k_{tlj}, \boldsymbol{\Omega}_{lj})$, *where* $\boldsymbol{\vartheta}_{tlj} = \boldsymbol{f}'\boldsymbol{A}_{tlj}$ *and* $k_{tlj} = (\boldsymbol{f}'\boldsymbol{R}_{tlj}\boldsymbol{f} + 1)$.
4. *Posterior distribution at time* t*:* $\boldsymbol{\Theta}_{tlj}|D_t \sim MN(\boldsymbol{M}_{tlj}, \boldsymbol{C}_{tlj}, \boldsymbol{\Omega}_{lj})$, *where* $\boldsymbol{C}_{tlj} = (\boldsymbol{R}^{-1}_{tlj} + \boldsymbol{f}\boldsymbol{f}')^{-1}$ *and* $\boldsymbol{M}_{tlj} = \boldsymbol{C}_{tlj}[\boldsymbol{R}^{-1}_{tlj}\boldsymbol{A}_{tlj} + \boldsymbol{f}(\boldsymbol{\theta}^{e}_{tlj})']$.
5. *Backward sampling distribution:* $\boldsymbol{\Theta}_{tlj}|\boldsymbol{\Theta}_{(t+1):T,lj}, D_T \sim MN(\boldsymbol{M}^{*}_{tlj}, \boldsymbol{C}^{*}_{tlj}, \boldsymbol{\Omega}_{lj})$, *where* $\boldsymbol{C}^{*}_{tlj} = [\boldsymbol{C}^{-1}_{tlj} + (\boldsymbol{G}^{\theta}_{lj})'(\boldsymbol{U}^{\theta}_{lj})^{-1}(\boldsymbol{G}^{\theta}_{lj})]^{-1}$ *and* $\boldsymbol{M}^{*}_{tlj} = \boldsymbol{C}^{*}_{tlj}[\boldsymbol{C}^{-1}_{tlj}\boldsymbol{M}_{tlj} + (\boldsymbol{G}^{\theta}_{lj})'(\boldsymbol{U}^{\theta}_{lj})^{-1}\boldsymbol{\Theta}_{t+1,lj}]$.

Proof *See appendix.*

The MSFFBS has two main steps: a forward filter and a backward sampler. The forward filter iterates Steps 1 through 4 of the MSFFBS forward from time 0 to time T, recursively computing the quantities of the posterior distribution $\boldsymbol{\Theta}_{tlj}|D_t$. After that, the MSFFBS simulates the latent process at the last time point T from the distribution $\boldsymbol{\Theta}_{Tlj}|D_T \sim MN(\mathbf{M}_{Tlj}, \mathbf{C}_{Tlj}, \boldsymbol{\Omega}_{lj})$. Finally, the backward sampler iterates Step 5 of the MSFFBS, recursively simulating the latent states related to the multiscale coefficients backward in time from time $T-1$ to time 1 using the distribution $\boldsymbol{\Theta}_{tlj}|\boldsymbol{\Theta}_{t+1,lj} \sim MN(\mathbf{M}^{*}_{tlj}, \mathbf{C}^{*}_{tlj}, \boldsymbol{\Omega}_{lj})$ for $t = T-1, \ldots, 1$.

Recall that, because the mean vector at subregion B_{lj} is a weighted average of the mean vectors at the descendants $\mathcal{D}_{lj}$, there are q aggregation constraints that are encoded in the conditional distribution of the empirical multiscale coefficient $\boldsymbol{\theta}^{e}_{tlj}|\boldsymbol{\theta}_{tlj}, \boldsymbol{\Omega}_{lj}$. These aggregation constraints imply that the conditional distribution of $\boldsymbol{P}_{lj}\boldsymbol{\theta}^{e}_{lj}$ is a degenerate distribution with $P[\boldsymbol{P}_{lj}\boldsymbol{\theta}^{e}_{lj} = \mathbf{0}] = 1$. As explained in Section 7.2, these q aggregation constraints are implicitly encoded in the respective singular Gaussian distribution $\boldsymbol{\theta}^{e}_{tlj}|\boldsymbol{\theta}_{tlj}, \boldsymbol{\Omega}_{lj}$ because its covariance matrix $\boldsymbol{\Omega}_{lj}$ is singular with q zero eigenvalues. In addition, the temporal evolution of the latent states related to the latent multiscale coefficient $\boldsymbol{\theta}_{tlj}$ given in Equation (7.12) has a matrix of innovations $\boldsymbol{\xi}_{tlj}$ with left covariance matrix equal to the observational covariance matrix $\boldsymbol{\Omega}_{lj}$. Thus, the same aggregation constraints of the empirical multiscale coefficient $\boldsymbol{\theta}^e_{tlj}$ are satisfied *a priori* and *a posteriori* by the latent multiscale coefficient $\boldsymbol{\theta}_{tlj}$. That is, $P(\boldsymbol{P}_{lj}\boldsymbol{\theta}_{tlj} = \mathbf{0}) = 1$. Therefore, the framework we propose here respects the q aggregation constraints and, thus, estimates the mean process at each scale of resolution in a coherent manner.

A sample from the posterior distribution of the mean process at each scale of resolution can be easily obtained. Specifically, the MCMC algorithm described in this section provides a sample from the joint posterior distribution of the latent states related to the mean process at the coarsest level of resolution and to the latent multiscale coefficients at the several levels of resolution. From this sample, a sample from the posterior distribution of the mean process at each scale of resolution can be obtained by the recursive application of the disaggregation equation $\boldsymbol{\mu}_{t,\mathcal{D}_{lj}} = \boldsymbol{f}'\boldsymbol{\Theta}_{tlj} + \boldsymbol{\nu}_{lj}\boldsymbol{\mu}_{tlj}$. Note that an analogous disaggregation equation may be used for the latent temporal level field and gradient field. For example, the disaggregation equation for the temporal level field is $\boldsymbol{\alpha}^{\mu}_{t,\mathcal{D}_{lj}} = \boldsymbol{\alpha}^{\theta}_{t,lj} + \boldsymbol{\nu}_{lj}\boldsymbol{\alpha}^{\mu}_{t,lj}$. Finally, note that the framework we propose provides coherent posterior samples of the mean process and related level and gradient fields at the several scales of resolution.

7.4.3 Computational Cost

The computational cost of the MCMC algorithm that we propose depends on the number of time points T, the number of spatial data points n_L, the number of variables q, the number of latent states s per descendants and variables of interest, and the geometry of the multiscale structure as defined by the number of descendants d_{lj}, $l = 1, \ldots, L-1$, $j = 1, \ldots, n_l$. As in this chapter we consider a linear growth model with a stationary component, the number of latent states per descendant and variable of interest is $s = 3$. With the empirical Bayes approach that we use to estimate the covariance matrices, we need to invert the estimates of the covariance matrices $\boldsymbol{V}_{1k}$ and $\boldsymbol{\Omega}_{lj}$ only once before the MCMC algorithm. This makes our MCMC algorithm very efficient.

Let us first consider the cost of simulating the hyperparameters, that is, the signal-to-noise ratio parameters and the autoregressive coefficients. The cost of simulating in one iteration of the MCMC algorithm the hyperparameters related to the jth region at the lth scale is dominated by sums of T terms that involve multiplications of the $\boldsymbol{\Omega}^-_{lj}$ matrix that has dimension $(qd_{lj}) \times (qd_{lj})$ by vectors of length qd_{lj}. Thus, the total cost of simulating the hyperparameters in one iteration is $O(Tq^2 \sum_{l=1}^{L-1} \sum_{j=1}^{n_l} d^2_{lj})$. Typically, d_{lj} is small and, thus, $\sum_{l=1}^{L-1} \sum_{j=1}^{n_l} d^2_{lj} = O(n_L)$. Finally, the number of variables of interest q is usually fixed, implying that the total cost of simulating the hyperparameters in one iteration is $O(Tn_L)$.

Let us now consider the cost of simulating the latent states. In a serial implementation of the MSFFBS in each iteration of the MCMC algorithm, there are three operations that are the most costly. One of these operations is the inversion of $s \times s$ matrices at each time point with a total cost of $O(Tn_L s^3)$. Because s is usually small, the total cost of this operation is $O(Tn_L)$. The other two costly operations are multiplications of matrices in the simulation of $\boldsymbol{\Theta}_{tlj}$ in the backward sampling step of the MSFFBS. The first of these operations multiplies an $s \times s$ matrix to an $s \times (qd_{lj})$ matrix, with a total cost of $O(Ts^2q\sum_{l=1}^{L-1}\sum_{j=1}^{n_l} d_{lj})$. The second of these operations multiplies an $s \times (qd_{lj})$ matrix to a $(qd_{lj}) \times (qd_{lj})$ matrix, with a total cost of $O(Tsq^2\sum_{l=1}^{L-1}\sum_{j=1}^{n_l} d_{lj}^2)$. Finally, as s and q are fixed, the total cost of simulating the latent states in one iteration is $O(Tn_L)$. Therefore, the total cost of one iteration in a serial implementation of our MCMC algorithm is $O(Tn_L)$ and grows linearly with dataset size.

Importantly, our algorithm is highly parallelizable. Specifically, the computations for each multiscale coefficient can be performed independently throughout the whole MCMC algorithm in separate cores. In that case, communication among cores happens only in the preparatory stage before the MCMC algorithm and in the final stage after the MCMC algorithm finishes running. At that point, the samples from each multiscale coefficient are combined to obtain samples of the mean process at different scales of resolution. Therefore, in a super computer with many more cores than the number of multiscale coefficients, the cost of a parallel implementation of our MCMC algorithm would be $O(T)$.

7.5 Model Selection

Our DMMSTM framework allows many different model specifications. In addition, different model specifications may encode different competing scientific hypotheses. Thus, to evaluate which scientific hypothesis is better supported by the observed data, we may use Bayesian model selection. For example, in Section 7.3 we have considered a DMMSTM where each multiscale coefficient has a non-stationary temporal trend and short-term stationary fluctuations. However, there is a dispute amongst climate scientists regarding the nature of the dynamical system of temperatures (Mills, 2007). To shed light on this discussion, in addition to the DMMSTM presented in Section 7.3, in what follows we are also going to consider a DMMSTM without non-stationary trends but instead with purely stationary AR(1) processes to model the temporal evolutions of the coarsest resolution temperatures and the multiscale coefficients. To compare the different model specifications, we use the conditional Bayes factor (Vivar and Ferreira, 2009; Elkhouly and Ferreira, 2021).

As explained in Elkhouly and Ferreira (2021), when considering dynamic multiscale spatio-temporal models, two types of model comparison can be performed with the conditional Bayes factor. The first type compares competing models based on the quality of the prediction of observations at the finest resolution level. This first type is useful for comparing particular specifications of DMMSTM either against each other or against other competing models previously published in the literature. Thus, this first type of model comparison provides a criterion for each model based on the model's global performance when predicting observations at the finest level of resolution.

The second type of model comparison provides a more nuanced approach. Specifically, the second type compares distinct models for the temporal evolution of the mean level at the coarsest level of resolution as well as for the temporal evolution of each multiscale coefficient. Hence, the second type of model selection provides a more detailed analysis of which model is appropriate for each subregion at each scale of resolution. In what follows, we present these two types of model selection.

The first type of model selection considers the conditional predictive density under each model $\mathcal{M}$ at the finest level of resolution L. The conditional predictive density uses a training sample given by the data observed up to a pre-specified time point t^*. Then, we compute the joint predictive density of the observations at the finest level of resolution L from time t^*+1 to time T conditional on the information up to time t^*. This conditional joint predictive density, denoted by $p_{\mathcal{M}}(\boldsymbol{y}_{t^*+1,L},\ldots \boldsymbol{y}_{TL}|D_{t^*})$, is then used to compare the competing models. Note that we can factorize this conditional joint predictive density as the product of one-step-ahead predictive densities $p_{\mathcal{M}}(\boldsymbol{y}_{t^*+1,L},\ldots \boldsymbol{y}_{TL}|D_{t^*}) = \prod_{t=t^*+1}^{T} p_{\mathcal{M}}(\boldsymbol{y}_{tL}|D_{t-1})$. In addition, each one-step ahead predictive density can be written as

$$p_{\mathcal{M}}(\boldsymbol{y}_{tL}|D_{t-1}) = \int p_{\mathcal{M}}(\boldsymbol{y}_{tL}|\boldsymbol{\mu}_{tL})p_{\mathcal{M}}(\boldsymbol{\mu}_{tL}|D_{t-1})d\boldsymbol{\mu}_{tL}. \tag{7.16}$$

This one-step-ahead predictive density may be approximated with Monte Carlo integration in the following way. Let $\boldsymbol{\Psi}$ be the vector of hyperparameters. First, we run the MCMC algorithm given in Section 7.4 using data from time 1 to time $t-1$. After convergence of the MCMC algorithm, we have G draws from the joint posterior distribution $\boldsymbol{\Theta}_{1:(t-1),lj}, \boldsymbol{H}_{1:(t-1),k}, \boldsymbol{\Psi}|D_{t-1}$. Denote the gth posterior draw by $(\boldsymbol{\Theta}^{(g)}_{1:(t-1),lj}, \boldsymbol{H}^{(g)}_{1:(t-1),k}, \boldsymbol{\Psi}^{(g)})$. Then, use the evolution equations (7.9) and (7.12) to simulate draws from the joint predictive distribution $\boldsymbol{\Theta}_{tlj}, \boldsymbol{H}_{t1}|D_{t-1}$. We then obtain draws of $\boldsymbol{\mu}_{t1j}$ and $\boldsymbol{\theta}_{tlj}$ using the equalities

$$\boldsymbol{\mu}_{t1j} = \boldsymbol{H}'_{t1j}\boldsymbol{f}$$

and

$$\boldsymbol{\theta}_{tlj} = \boldsymbol{\Theta}'_{tlj}\boldsymbol{f}.$$

Further, we recursively use the aggregation equation

$$\boldsymbol{\mu}_{t,\mathcal{D}_{lj}} = \boldsymbol{\theta}_{tlj} + \boldsymbol{\nu}_{lj}\boldsymbol{\mu}_{tlj}$$

to obtain draws from the distribution $\boldsymbol{\mu}_{tL}|D_{t-1}$. With these draws, we approximate the one-step ahead predictive density given in Equation (7.16) with the Rao-Blackwellized approximation

$$\widehat{p}_{\mathcal{M}}(\boldsymbol{Y}_{tL}|D_{t-1}) = G^{-1}\sum_{g=1}^{G} p_{\mathcal{M}}(\boldsymbol{y}_{tL}|\boldsymbol{\mu}_{tL}^{(g)}).$$

After that, the conditional joint predictive density can be approximated with

$$\widehat{p}_{\mathcal{M}}(\boldsymbol{y}_{t^*+1,L}, \ldots \boldsymbol{y}_{TL}|D_{t^*}) = \prod_{t=t^*+1}^{T} \widehat{p}_{\mathcal{M}}(\boldsymbol{y}_{tL}|D_{t-1}).$$

Finally, two competing models $\mathcal{M}_1$ and $\mathcal{M}_2$ can be compared with the conditional Bayes factor (CBF)

$$CBF_{\mathcal{M}_1\mathcal{M}_2} = \frac{\widehat{p}_{\mathcal{M}_1}(\boldsymbol{y}_{t^*+1,L}, \ldots, \boldsymbol{y}_{TL}|D_{t^*})}{\widehat{p}_{\mathcal{M}_2}(\boldsymbol{y}_{t^*+1,L}, \ldots \boldsymbol{y}_{TL}|D_{t^*})}. \tag{7.17}$$

This conditional Bayes factor provides a comparison of the global performance of model $\mathcal{M}_1$ versus model $\mathcal{M}_2$.

The second type of model comparison provides more detailed information about the support that each model receives from the data at each subregion and scale of resolution. Specifically, for any two models, we can calculate separate CBFs for different subregions at the coarsest resolution and for the multiscale coefficients at the multiple resolutions. This feature may be useful because we may be interested in whether a winning model is also the best model in multiple parts of the multiscale system, or if different parts of the multiscale system support different models.

To perform the second type of model comparison, we consider the one-step ahead predictive density of the empirical multiscale coefficient $\boldsymbol{\theta}_{tlj}^{e}$ which is given by

$$\begin{aligned} p_{\mathcal{M}}(\boldsymbol{\theta}_{tlj}^{e}|D_{t-1}) &= \int p_{\mathcal{M}}(\boldsymbol{\theta}_{tlj}^{e}|\boldsymbol{\Theta}_{tlj})p_{\mathcal{M}}(\boldsymbol{\Theta}_{tlj}|\boldsymbol{\Theta}_{t-1,lj}, \boldsymbol{\Psi}) \\ &\quad p_{\mathcal{M}}(\boldsymbol{\Theta}_{1:(t-1),lj}, \boldsymbol{\Psi}|D_{t-1})d\boldsymbol{\Theta}_{1:(t-1),lj}d\boldsymbol{\Theta}_{tlj}d\boldsymbol{\Psi} \end{aligned}$$

and the one-step ahead predictive density of the observations in the kth subregion at the coarsest resolution

$$\begin{aligned} p_{\mathcal{M}}(\boldsymbol{y}_{t1k}|D_{t-1}) &= \int p_{\mathcal{M}}(\boldsymbol{y}_{t1k}|\boldsymbol{H}_{t1k})p_{\mathcal{M}}(\boldsymbol{H}_{t1k}|\boldsymbol{H}_{t-1,1k}, \boldsymbol{\Psi}) \\ &\quad p_{\mathcal{M}}(\boldsymbol{H}_{1:(t-1),1k}, \boldsymbol{\Psi}|D_{t-1})d\boldsymbol{H}_{1:(t-1),1k}d\boldsymbol{H}_{t1k}d\boldsymbol{\Psi}. \end{aligned}$$

Approximations to each of these one-step-ahead predictive densities can be computed using the output of an MCMC algorithm run on the data from times 1 to $t-1$. We can then use the evolution equations (7.9) and (7.12) to simulate draws from the joint predictive distribution $\boldsymbol{\Theta}_{tlj}, \boldsymbol{H}_{t1}|D_{t-1}$. After that, we obtain draws of $\boldsymbol{\mu}_{t1j}$ and $\boldsymbol{\theta}_{tlj}$ using the equalities $\boldsymbol{\mu}_{t1j} = \boldsymbol{H}'_{t1j}\boldsymbol{f}$ and $\boldsymbol{\theta}_{tlj} = \boldsymbol{\Theta}'_{tlj}\boldsymbol{f}$. We can then compute the Rao-Blackwellized estimates of the one-step-ahead predictive densities

$$\widehat{p}_{\mathcal{M}}(\boldsymbol{\theta}^e_{tlj}|D_{t-1}) = \frac{1}{G}\sum_{g=1}^{G} p_{\mathcal{M}}(\boldsymbol{\theta}^e_{tlj}|\boldsymbol{\theta}^{(g)}_{tlj})$$

and

$$\widehat{p}_{\mathcal{M}}(\boldsymbol{y}_{t1j}|D_{t-1}) = \frac{1}{G}\sum_{g=1}^{G} p_{\mathcal{M}}(\boldsymbol{y}_{t1j}|\boldsymbol{\mu}^{(g)}_{t1j}).$$

Then, the conditional joint predictive density from time $t^* + 1$ to time T of the observations in subregion j at the coarsest resolution level 1 can be approximated with

$$\widehat{p}_{\mathcal{M}}(\boldsymbol{y}_{t^*+1,1j}, \ldots, \boldsymbol{y}_{T1j}|D_{t^*}) = \prod_{t=t^*+1}^{T} \widehat{p}_{\mathcal{M}}(\boldsymbol{y}_{t1j}|D_{t-1}).$$

Then, models $\mathcal{M}_1$ and $\mathcal{M}_2$ can be compared for subregion j at the coarsest resolution level 1 with the conditional Bayes factor

$$\frac{\widehat{p}_{\mathcal{M}_1}(\boldsymbol{y}_{t^*+1,1j}, \ldots, \boldsymbol{y}_{T1j}|D_{t^*})}{\widehat{p}_{\mathcal{M}_2}(\boldsymbol{y}_{t^*+1,1j}, \ldots \boldsymbol{y}_{T1j}|D_{t^*})}.$$

Similarly, the conditional joint predictive density from time $t^* + 1$ to time T of the empirical multiscale coefficient in subregion j of resolution level l can be approximated with $\widehat{p}_{\mathcal{M}}(\boldsymbol{\theta}^e_{t^*+1,lj}, \ldots, \boldsymbol{\theta}^e_{Tlj}|D_{t^*}) = \prod_{t=t^*+1}^{T} \widehat{p}_{\mathcal{M}}(\boldsymbol{\theta}^e_{tlj}|D_{t-1})$. Finally, to compare two competing models $\mathcal{M}_1$ and $\mathcal{M}_2$ for the temporal evolution of $\boldsymbol{\theta}_{tlj}$ we use the ratio of their conditional joint predictive densities

$$\frac{\widehat{p}_{\mathcal{M}_1}(\boldsymbol{\theta}^e_{t^*+1,lj}, \ldots, \boldsymbol{\theta}^e_{Tlj}|D_{t^*})}{\widehat{p}_{\mathcal{M}_2}(\boldsymbol{\theta}^e_{t^*+1,lj}, \ldots, \boldsymbol{\theta}^e_{Tlj}|D_{t^*})}.$$

We are going to illustrate these flexible model selection features with the application to tropospheric temperatures presented in the next section.

7.6 Application: Tropospheric Temperatures

In this section, we use our DMMSTM framework to model spatio-temporal data on temperature anomalies at the upper and middle tropospheric altitudes. Specifically, we analyze the annual summer average tropospheric temperatures provided by the NCEP/NCAR Reanalysis dataset and available at https://psl.noaa.gov/data/gridded/data.ncep.reanalysis.html. Specifically, we consider five tropospheric altitudes pressure levels (250, 300, 400, 500 and 600 hPa) for the years 1951 to 2016. We focus on a spatial domain that covers the United States and some parts of Canada and Mexico (-127.5°:-70° Longitude and 25°:52.5° Latitude on a 2.5°×2.5° grid, Figure 7.1a). To reduce the latitude effect, we consider the anomalies of the annual average daily temperature at 18.00 UTC for the summer months of June, July, and August. Therefore, for each of the 288 spatial subregions and for each altitude pressure level, we compute the temperature anomalies relative to the average of the study period 1951-2016.

We employ our DMMSTM framework to estimate the underlying mean process through the years 1951 to 2016. We consider four different spatial resolutions connected hierarchically through parent-children relationships. We use a quad tree to connect each pair of successive levels of the fourth, third, and second resolutions. The fourth spatial resolution – the finest resolution – contains the actual observations of air temperature anomalies at the 288 subregions. The third resolution consists of 72 subregions, where each subregion has four descendant subregions at the finest resolution. Next, the second resolution consists of 18 main subregions, where each subregion has four descendant subregions at the third resolution. Panels (a), (b), and (c) of Figure 7.1 present these three resolutions of our domain of interest. Finally, each subregion at the first resolution – the coarsest resolution – has only two descendant subregions at the second resolution in a dyadic tree structure. Therefore, the coarsest resolution (Figure 7.1(d)) contains nine subregions that represent the main geographic sections of North America. These nine regions are north, center and south for latitude sections as well as west (Pacific), center and east (Atlantic) for longitude sections.

7.6.1 Empirical Bayes Estimation of the Covariance Matrices

To apply our DMMSTM framework, we need to estimate the covariance matrices for the subregions at the finest level of resolution. First, we use empirical Bayes to estimate the covariance matrix at each of the subregions at the finest resolution. After that, we obtain the covariance matrices at the successive scales of resolution through the Aggregation Equation (7.2).

To estimate the covariance matrices at the finest level of resolution, we assume each vector of realizations for a subregion at the finest resolution marginally follows a multivariate DLM defined by Equations (7.5) and (7.9). Hence, we use the MARSS package (Holmes et al., 2012) which maximizes the likelihood using the Kalman filter/smoother incorporated with an EM algorithm. For the setting of the MARSS function, we set the covariance of the evolution equations to be a diagonal matrix and assume an unstructured observational covariance matrix. The corresponding estimated observational covariances are the covariances for the locations at the finest resolution $\boldsymbol{V}_{Lj}$. In addition, each of the four descendants of the same node at the third resolution seems to share the same observational covariance structure. Therefore, we use the average of the estimated observational covariance matrices of those descendants as a more robust estimate for their corresponding observational covariance matrix at the finest level.

Our framework is able to accommodate changes in the covariance matrix among different descendants at different resolutions. Figure 7.3(a) shows a panel of the covariance matrix of the five air temperature anomalies for 72 subregions of Figure 7.1(b). We observe spatial variability of the estimated covariance of realizations which may be because of topography. Closer to the Atlantic and the Pacific oceans, there is a weak interdependence among the considered tropospheric temperature anomalies. Moreover, both the strength and direction of the interdependence change spatially. On the north-west corner of the spatial domain, there is a negative association between the highest tropospheric altitude pressure level (250 hPa) and the other altitudes. This negative association diminishes slightly as we move toward the east or toward the south as noticeable in the correlation maps (Figure 7.3(b)). In addition, the interdependence among the lower tropospheric pressure level altitudes (850-1000 hPa) and the other pressure level altitudes vary spatially. Moreover, there is a clustering between the groups of middle and upper tropospheric altitude pressure levels in the south-west regions. Such a spatial variation in both the strength and direction of the association among the considered variables is easily accommodated in our proposed framework.

To aggregate from a specific resolution to a coarser resolution, we use a simple arithmetic average via the equal weights version of $\boldsymbol{P}_{lj}$. For signal-to-noise ratio parameters, we use the inverse gamma prior $IG(0.5\eta, 0.5\zeta)$. In this parametrization of the IG, the prior mean is $0.5\zeta/(0.5\eta - 1)$. Note that the signal-to-noise ratio parameters control the possible amount of variability in the non-stationary trends. With this in mind, we set an IG prior with $\eta = 8.2$ and $\zeta = 0.018$ for the signal-to-noise parameters associated with the latent level and gradient states. With these choices, the IG prior has its first four moments and the prior mean is 0.003. Thus, the change per year in the latent states due to the random innovations is with high probability between $\pm 2\sqrt{0.003} \approx \pm 0.11$ times the respective covariance matrix. Lastly, we set a truncated normal prior for the AR(1) parameter, ϕ, with prior mean $a = 0.5$ and prior standard deviation $b = 0.2$ which allows for moderate persistence in the short-term stationary fluctuations.

7.6.2 Results

To analyze the troposphere temperature data with our DMMSTM framework, we have run MCMC chains for a total of 20,000 iterations. To check the convergence of the MCMC algorithm, we have used the Gelman-Rubin convergence diagnostic (Gelman and Rubin, 1992) implemented in the R package coda (Plummer et al., 2006). In addition, we have visually inspected for convergence the trace plots for all the DMMSTM unknown quantities. The MCMC

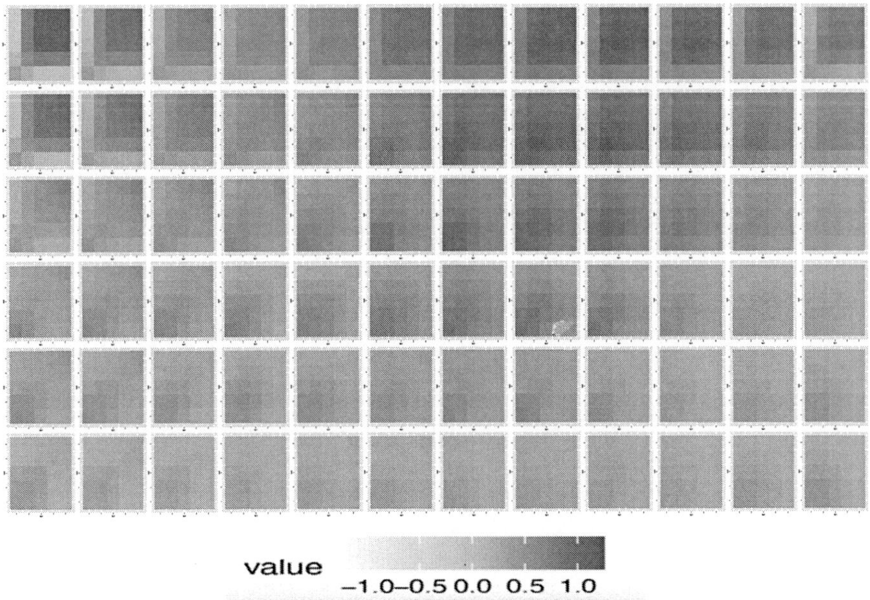

(a) Covariance matrices

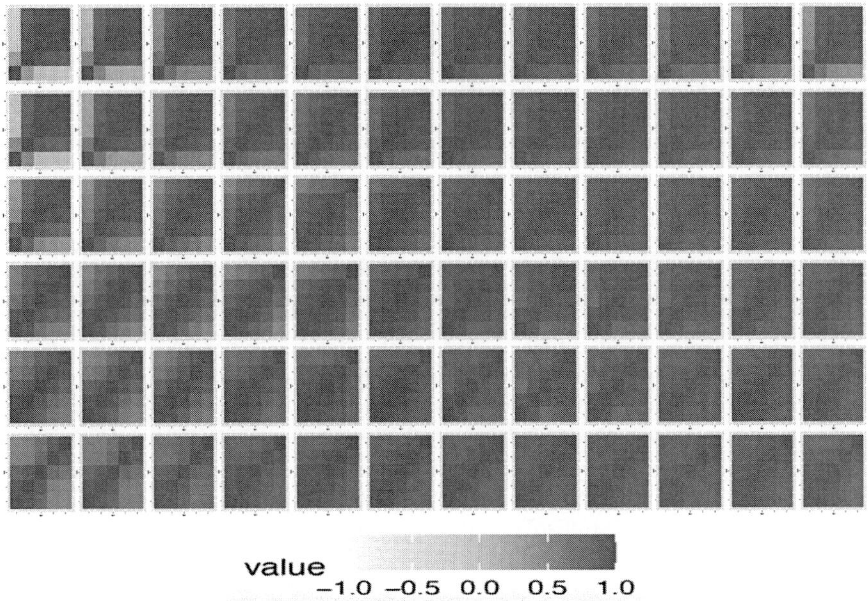

(b) Correlation matrices

FIGURE 7.3
The spatial distribution of the covariance and correlation structures. a) Estimated covariance matrices for the 72 blocks of the study region. b) Estimated correlation matrices for the 72 blocks of the study region.

algorithm converged after 10,000 iterations; thus, we have discarded the first 10,000 iterations as burn-in. To deal with computer memory constraints, we have thinned the MCMC sample where we kept one draw from each 10 iterations. In what follows, we report results based on these 1,000 MCMC draws.

As explained in Section 7.3, the model we consider contains a possible long-term trend with level and gradient components, as well as a short-term transitory fluctuations term. The inclusion of the short-term dynamics is very important: not including the short-term dynamics in this application would lead to a confounding of the short-term transitory fluctuations and the long-term trend, with the variances of the evolution of the long-term trend being overestimated and the model being over-adaptive. In contrast, including the short-term transitory fluctuations term allows the long-term trend to be estimated with higher precision. Note that the magnitude of the autoregressive coefficient indicates the level of persistence of the short-term transitory fluctuations term. Here, we have estimated low to moderate persistence of the short-term transitory fluctuations for the multiscale coefficients as well as for the mean process at the coarsest level. Specifically, for the multiscale vectors connecting the finest resolution to the third level of resolution, the estimates of the autoregressive coefficients are around 0.2 indicating weakly persistent short-term fluctuations. In addition, for the multiscale coefficients connecting the second and third resolutions, about half of the autoregressive coefficient estimates are around 0.5. Finally, for both the mean process at the coarsest level and the multiscale coefficients connecting the coarsest level to the second resolutions, the estimated autoregressive coefficients are around 0.2 to 0.3. Therefore, the estimated autoregressive coefficients indicate that the short-term fluctuations have low to moderate persistence.

Figures 7.4 to 7.8 present the estimated mean level of temperature anomalies in Kelvin at pressure levels 250, 300, 400, 500, and 600 hPa from 1951 to 2016. These figures are in grayscale, with darker gray meaning higher positive temperature anomalies and lighter gray meaning lower negative temperature anomalies. These figures show that the estimated mean level of tropospheric temperature anomalies varies spatially, temporally, and through the different altitude pressure levels over the same region. One of the most prominent features in these figures is the existence of temporal trends in the troposphere above the Great Plains. Specifically, in that region, the estimated mean level of temperature anomalies was positive during the 1950s but with a decreasing trend, then they turned negative, and then the trend became increasing and the estimated mean level anomalies turned positive again during the second half of the study period.

Another interesting feature shown in Figures 7.4 to 7.8 is that the lower the altitude, the later the switch in the trend direction. For the 250 hPa altitude pressure, the estimated mean anomalies began increasing around 1985 and 1986. Later in 1987 and 1988, the estimated mean anomalies corresponding to the altitude pressure levels 300 and 400 hPa started increasing. Finally, the

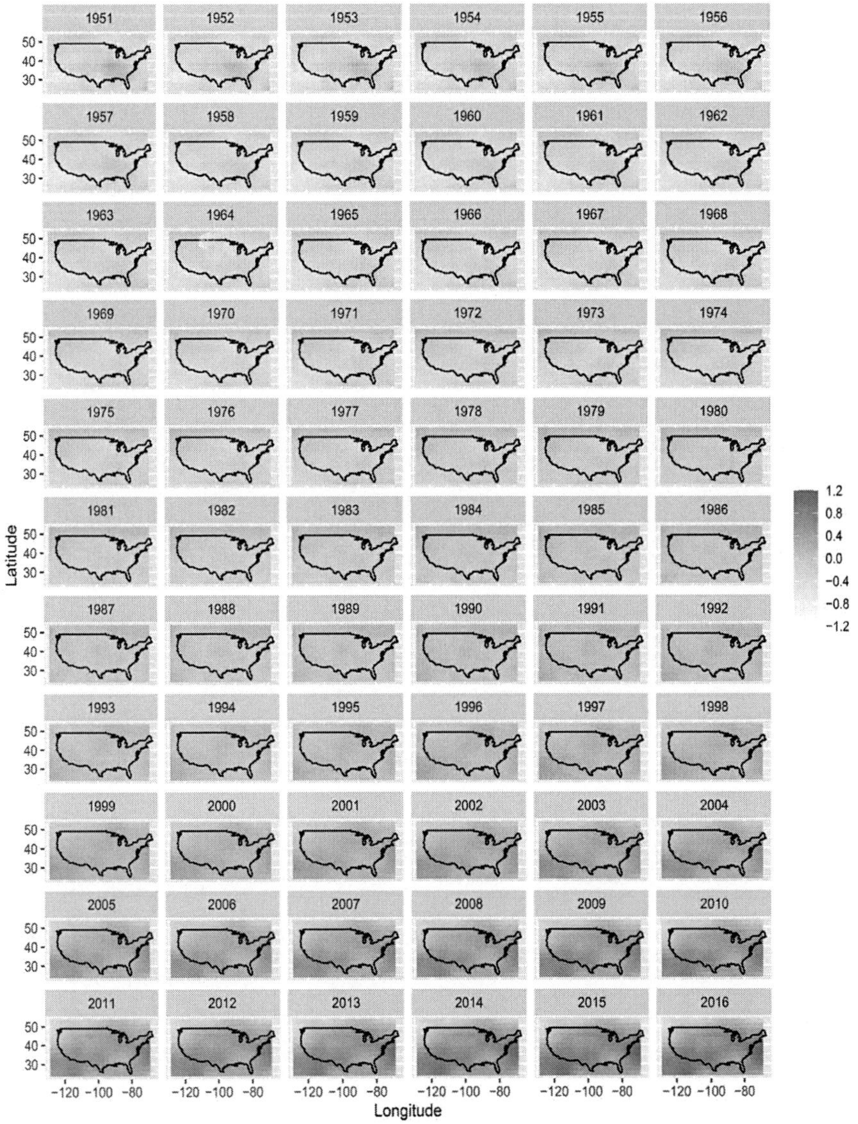

FIGURE 7.4
Estimated mean level of temperature anomalies at altitude pressure level 250 hPa from 1951 to 2016.

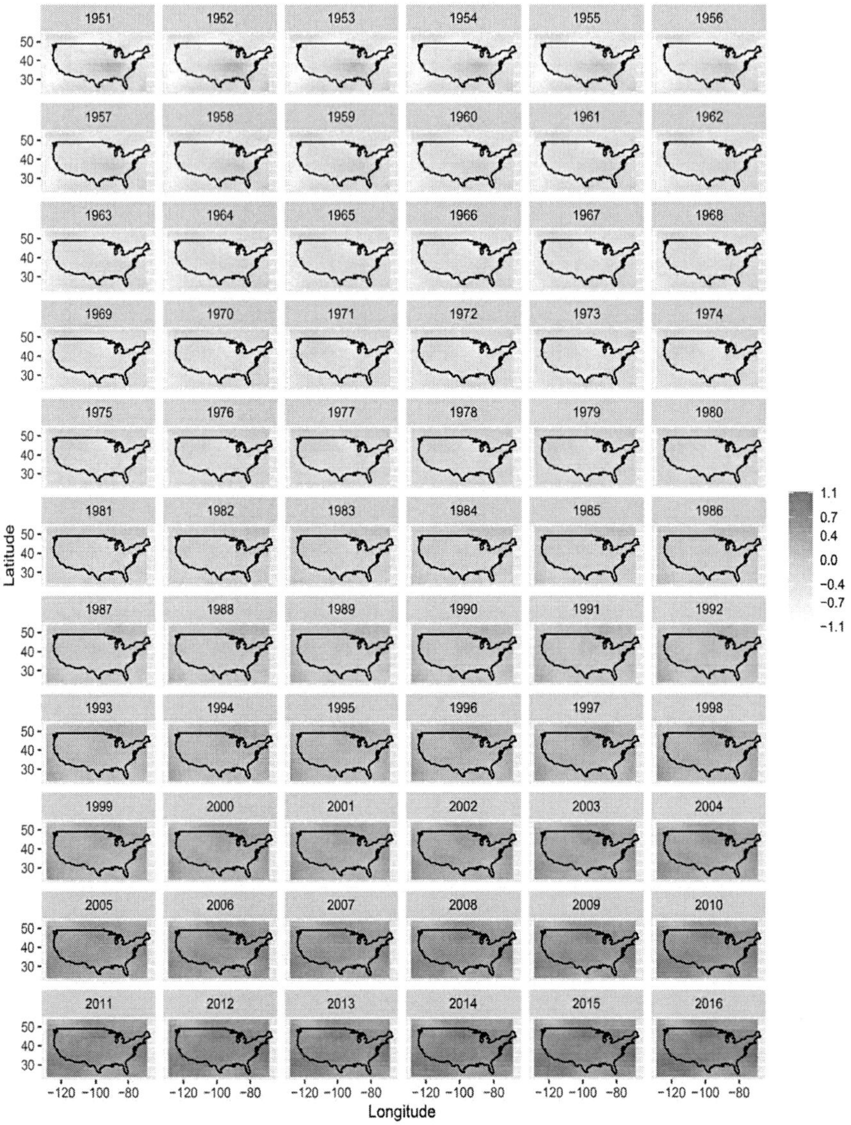

FIGURE 7.5
Estimated mean level of temperature anomalies at altitude pressure level 300 hPa from 1951 to 2016.

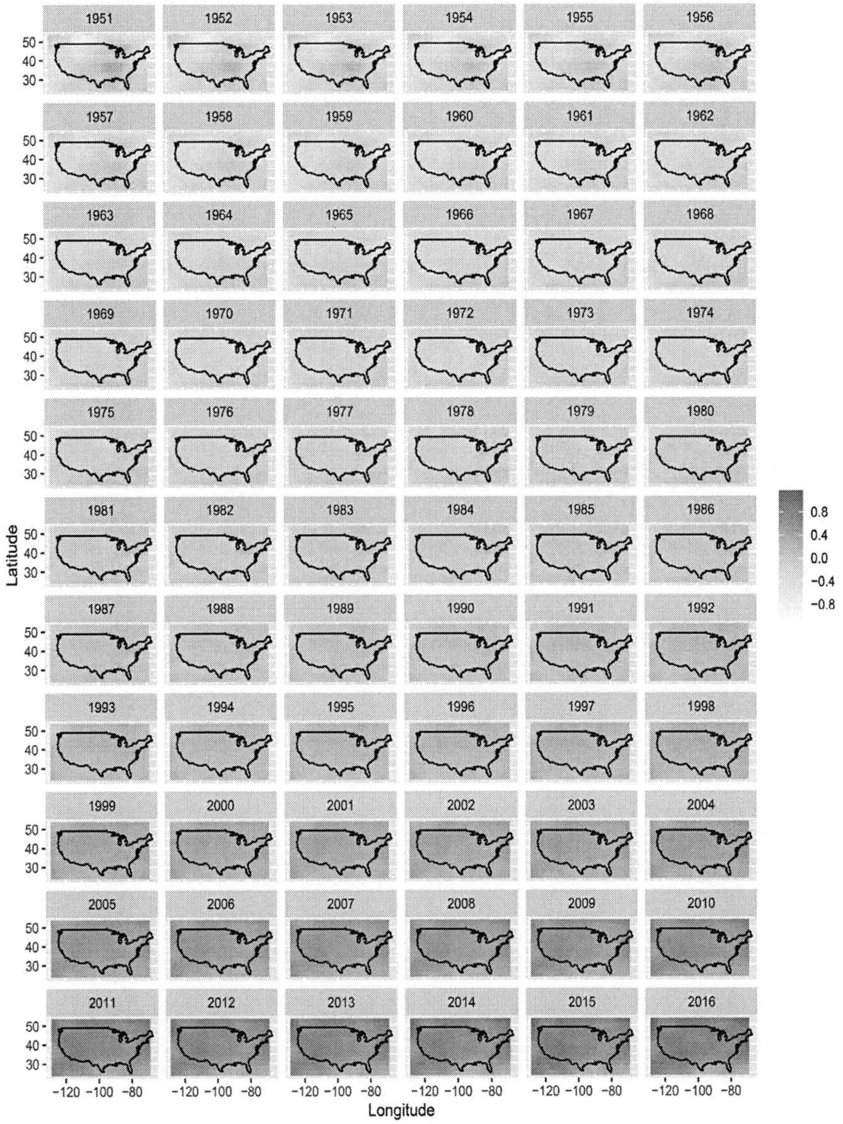

FIGURE 7.6
Estimated mean level of temperature anomalies at altitude pressure level 400 hPa from 1951 to 2016.

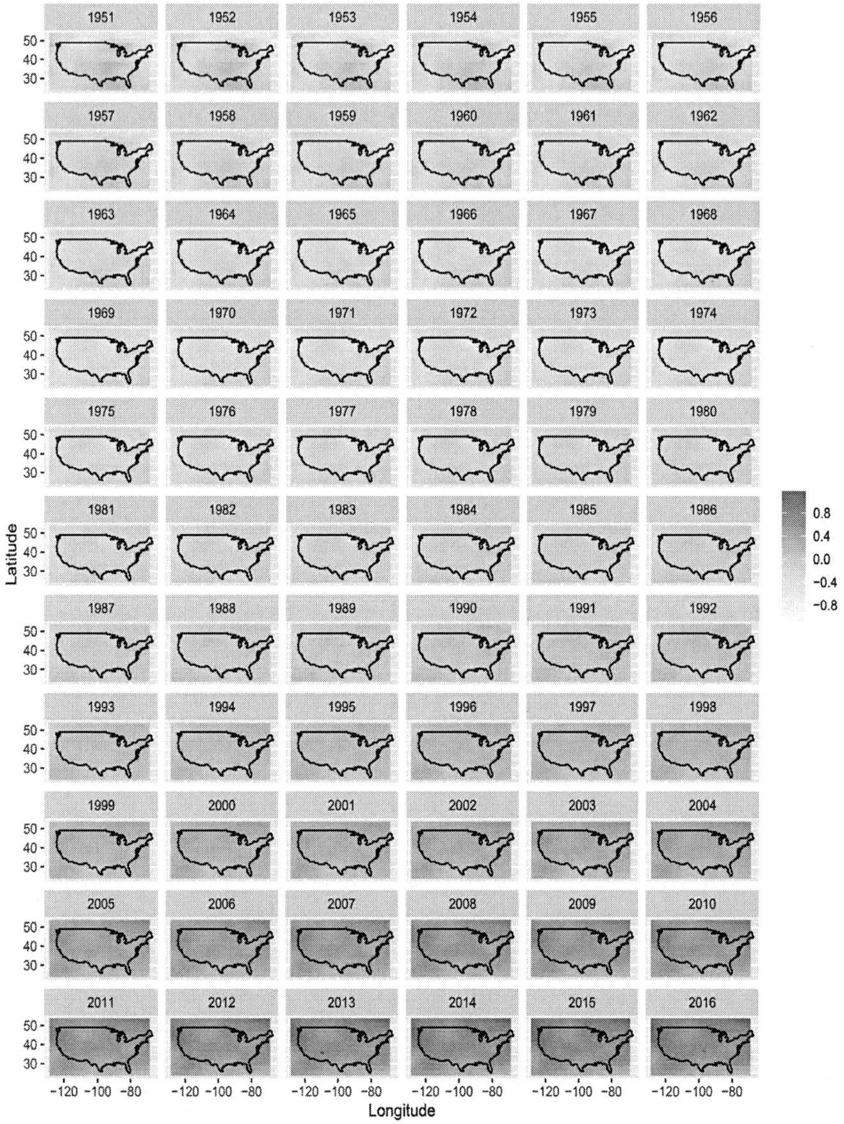

FIGURE 7.7
Estimated mean level of temperature anomalies at altitude pressure level 500 hPa from 1951 to 2016.

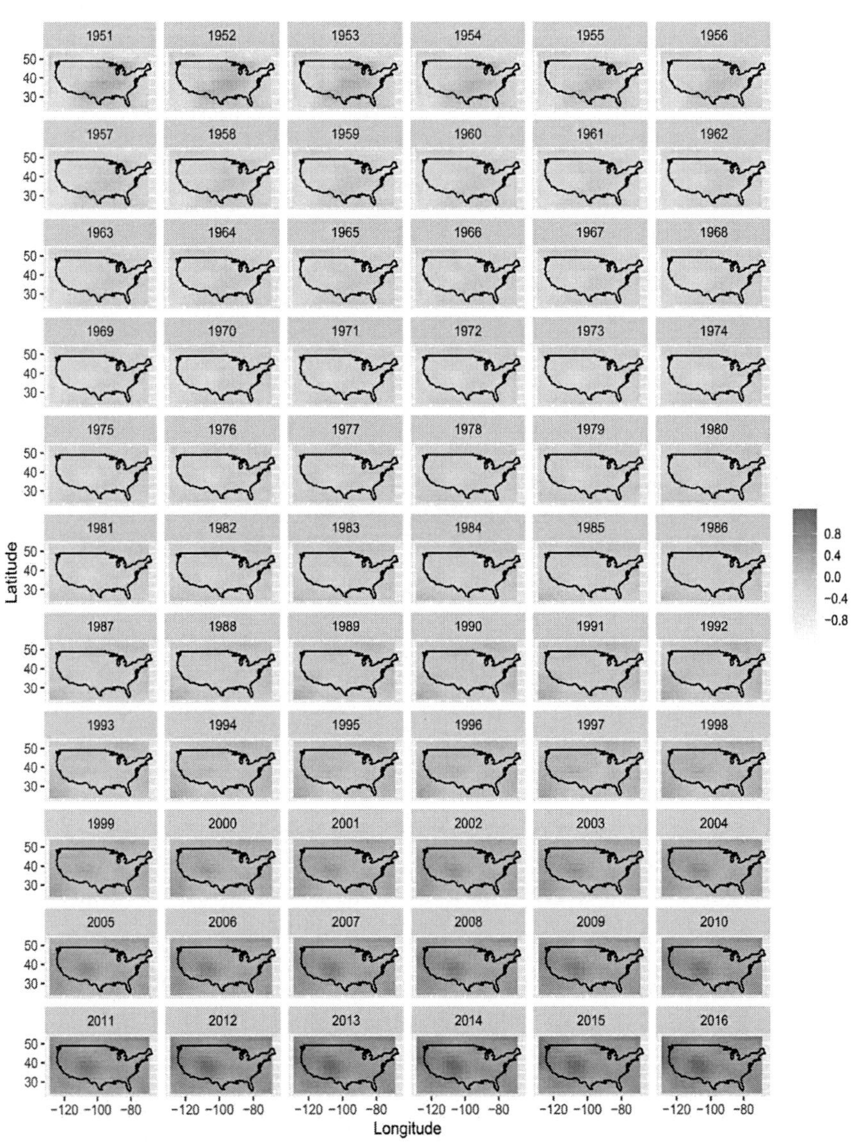

FIGURE 7.8
Estimated mean level of temperature anomalies at altitude pressure level 600 hPa from 1951 to 2016.

estimated mean anomalies for mid-tropospheric pressures 500 and 600 hPa started increasing above the Great Plains around the first half of the 1990s. Another region with a switch in the trend is the west-north and north region at an altitude pressure level of 250 hPa, where the estimated mean anomalies had been increasing since the 1950s, then reverted trend to decrease since around 2004 (Figure 7.4). Finally, at the other regions and considered tropospheric altitudes, the estimated means of temperature anomalies have been increasing persistently through time.

Figures 7.9 to 7.13 present the estimated annual gradient fields of temperature anomalies in Kelvin at altitude pressure levels 250, 300, 400, 500, and 600 hPa from 1951 to 2016. At the beginning of the study period in the 1950s, most of the study region at the five considered altitude pressure levels had negative gradient and, thus, was experiencing decreasing trends in temperature anomalies. The gradient fields of these decreasing trends got progressively attenuated until the end of the 1960s, when the trends started to change into increasing trends. Figures 7.9–7.13 confirm the previous conclusion based on Figures 7.4–7.8 that the lower the altitude, the later was the reversal in the trend in the tropospheric temperatures from decreasing to increasing. Then, for most of the study region at each of the five considered altitudes, the estimated gradient field became positive and progressively increased until the end of the study period. At the end of the study period in 2016, the average gradients for altitude pressure levels 250, 300, 400, 500, and 600 hPa were 0.018, 0.028, 0.028, 0.026, and 0.028, respectively. Noteworthy is the fact that the estimated average gradients over the study region at altitude pressure levels 300, 400, 500, and 600 hPa were about the same in 2016 implying a trend of increase in average temperatures at those altitudes of about 2.8 K per century.

There has been some discussion amongst climate scientists regarding whether the dynamical system of temperatures has a nonstationary trend or just stationary fluctuations (Mills, 2007). Figures 7.4–7.13 are based on a DMMSTM with a local linear trend and short-term stationary fluctuations. To shed some light on the climate discussion, we perform model comparisons of this DMMSTM versus a DMMSTM that contains only AR(1) short-term stationary fluctuations and does not have long-term trends. To compare the performance of these spatiotemporal models, we use the conditional Bayes factor (Vivar and Ferreira, 2009; Elkhouly and Ferreira, 2021) specialized to the DMMSTM framework as explained in Section 7.5.

Because different spatiotemporal models may be better at different resolutions, subregions, and time periods, instead of considering only a single global measure of model performance, we consider the conditional Bayes factor applied to each latent component of the multiscale factorization. That is, we consider the CBF applied to the mean process of each subregion at the coarsest level of resolution, as well as the CBF for the multiscale coefficient that connects each subregion of each level of resolution to its descendants at the subsequent finer level. In addition, to take into account the possibility that different models may be better at different time periods, we consider the

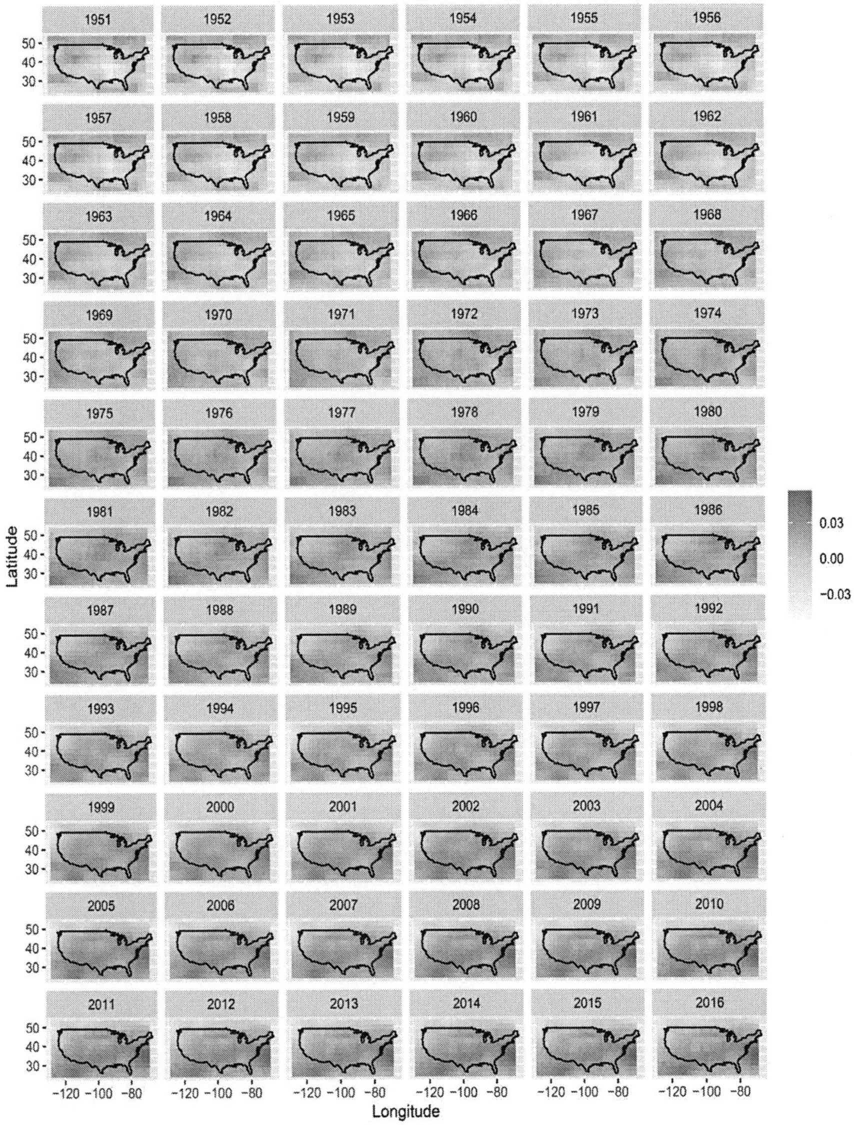

FIGURE 7.9
Estimated annual gradients in the level of temperature anomalies at altitude pressure level 250 hPa from 1951 to 2016.

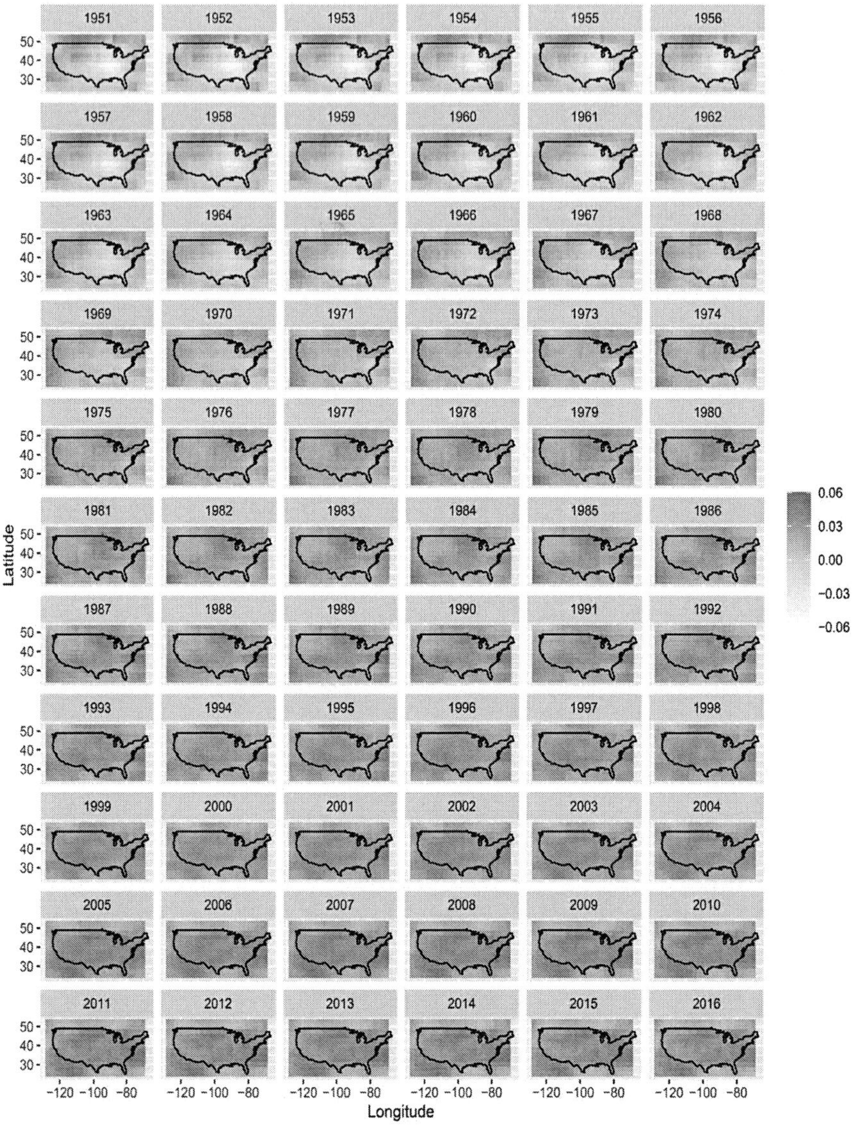

FIGURE 7.10
Estimated annual gradients in the level of temperature anomalies at altitude pressure level 300 hPa from 1951 to 2016.

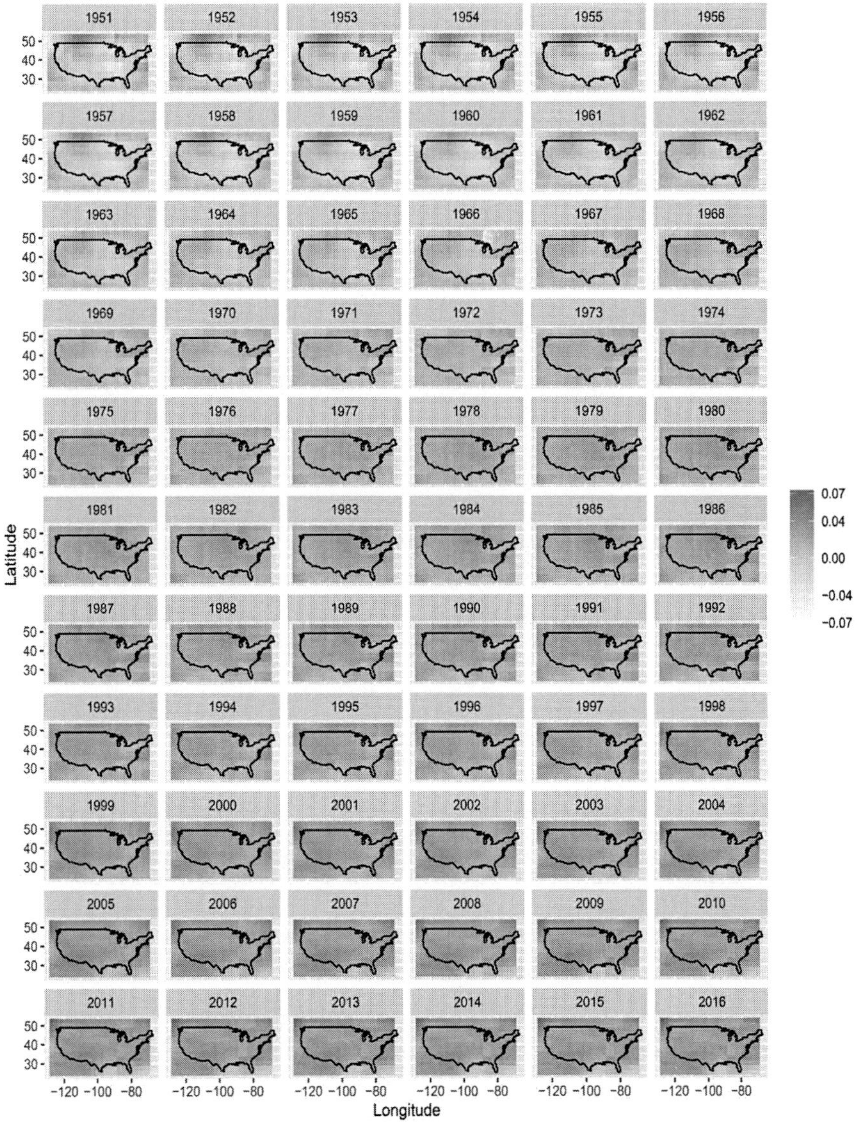

FIGURE 7.11
Estimated annual gradients in the level of temperature anomalies at altitude pressure level 400 hPa from 1951 to 2016.

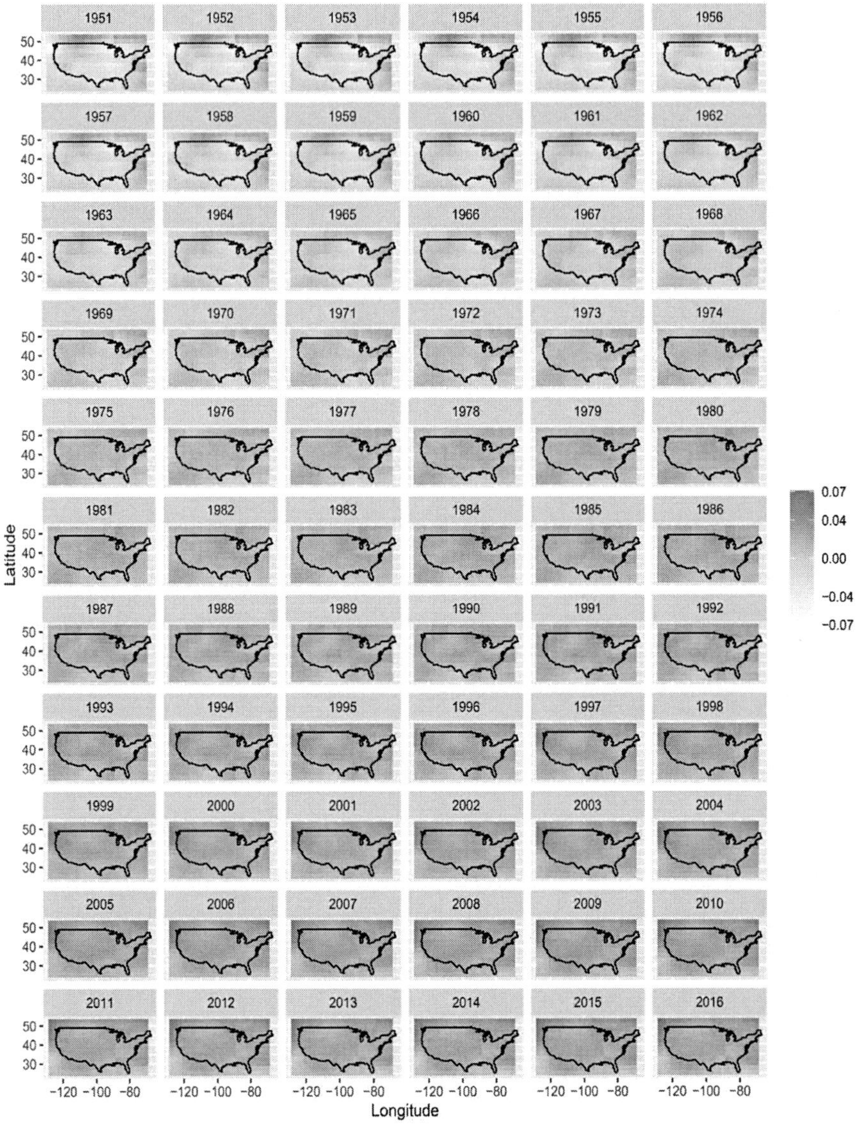

FIGURE 7.12
Estimated annual gradients in the level of temperature anomalies at altitude pressure level 500 hPa from 1951 to 2016.

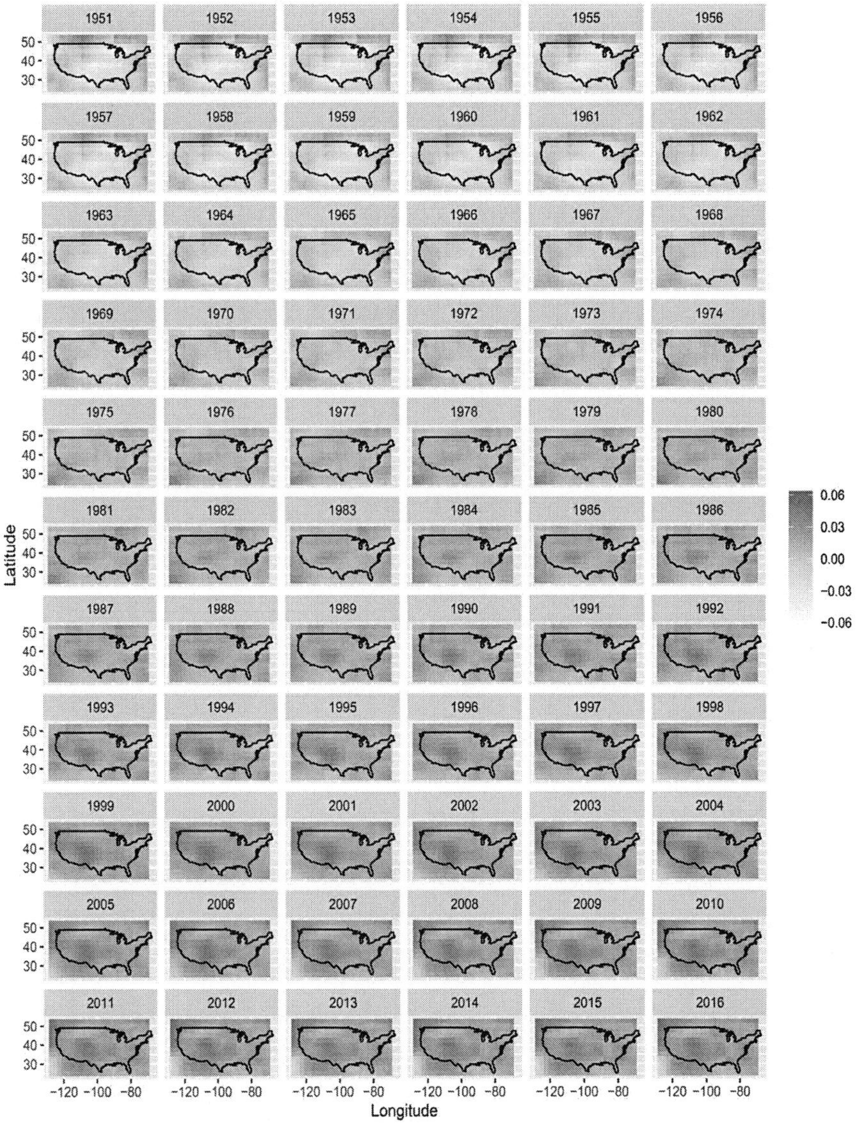

FIGURE 7.13
Estimated annual gradients in the level of temperature anomalies at altitude pressure level 600 hPa from 1951 to 2016.

CBF calculated for each time period on a 5-year moving window from 1972 to 2016.

To shed light on the dynamical system of temperatures, we focus on the analysis of the mean process of each subregion at the aggregated coarsest resolution. Because the coarsest resolution partitions the study region in 9 subregions, our focus on these subregions provides information about the existence of trends in tropospheric temperature at a synoptic scale. Figure 7.14 presents, for the mean process of each subregion at the aggregated coarsest resolution over five-year windows, the logarithm of the conditional Bayes factor of the linear growth plus AR(1) model versus the stationary AR(1) model. With the exception of one subregion in the period 1980–1984, all other subregions and time periods have positive log CBF, thus favoring for each subregion at the synaptic scale the DMMSTM with nonstationary long-term trend.

As pointed out by Elkhouly and Ferreira (2021), the multiscale coefficients may be interpreted as spatial directional derivatives. Thus, comparing the DMMSTMs with and without local linear trend applied to the multiscale coefficients sheds light on whether there are long-term trends in the spatial directional derivatives. That is, such model comparison provides information, in a specific subregion at a given spatial scale, about whether temporal changes in the spatial differences among descendant subregions are short-term transitory or if they are persistent with long-term trends. With this in mind, next we examine the log CBFs of the DMMSTM with local linear trend and stationary AR(1) variations versus the DMMSTM with only stationary AR(1) variations for each multiscale coefficient at each resolution level.

Figure 7.15 presents, for the linear growth plus AR(1) model versus the stationary AR(1) model, the logarithm of the conditional Bayes factor for multiscale coefficients that connect each subregion at the coarsest level to its descendants at the second level of resolution over five-year windows. From the period 1973–1976 to the period 1995–1999 there is no clear and consistent evidence in favor of the model with local linear trend or the model without local linear trend. However, starting in the period 2005–2009 to the period 2012–2016 there is evidence in favor of the model with local linear trend. Hence, there is evidence that toward the end of study period, there is a trend in the multiscale coefficients that connect each subregion at the coarsest level to its descendants at the second level of resolution. Therefore, in the last decade or so of the study period, there is evidence of temporal changes in the spatial differences among sibling subregions at the second scale of resolution.

Figure 7.16 presents the logarithm of the conditional Bayes factor for the multiscale coefficients that connect each subregion at the second level of resolution to its descendants at the third level of resolution over five-year windows. Throughout the entire study period, there are more 5-year periods supporting the model with local linear trend than the model without local linear trend. However, many 5-year periods have mixed results with some subregions supporting the model with trend and other subregions supporting the model without trend. This may be an indication that, for the multiscale co-

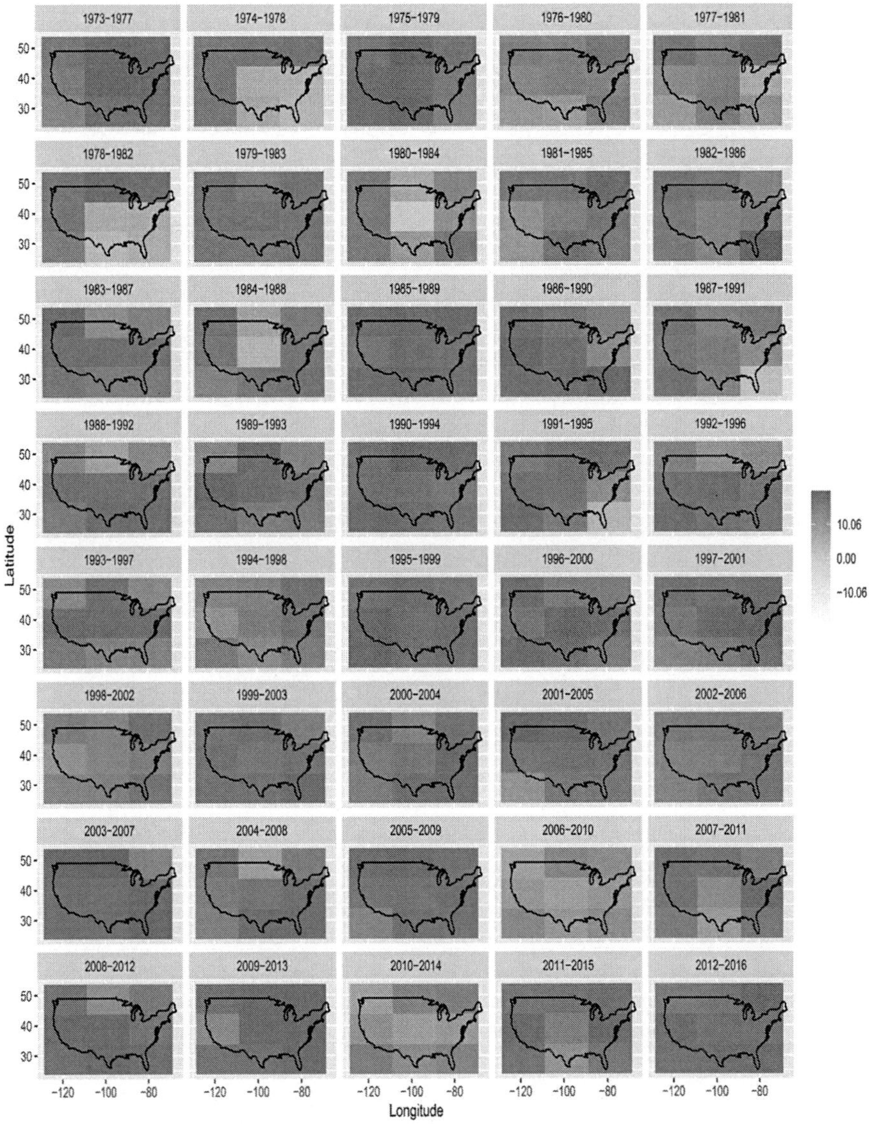

FIGURE 7.14
Log CBF for the mean process of each subregion at the aggregated coarsest resolution over five-year windows – Local linear growth plus AR(1) model versus stationary AR(1) model.

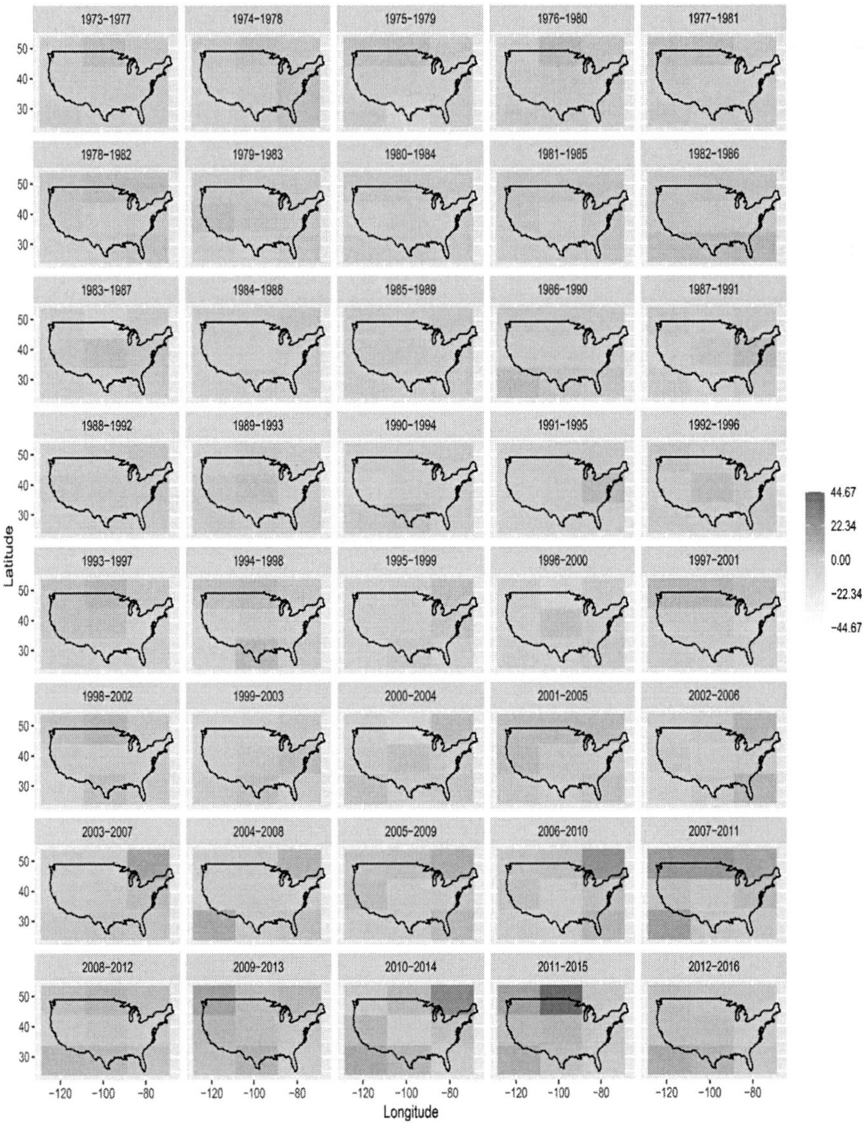

FIGURE 7.15
Log CBF for multiscale coefficients that connect each subregion at the coarsest level to its descendants at the second level of resolution over five-year windows – Local linear growth plus AR(1) model versus stationary AR(1) model.

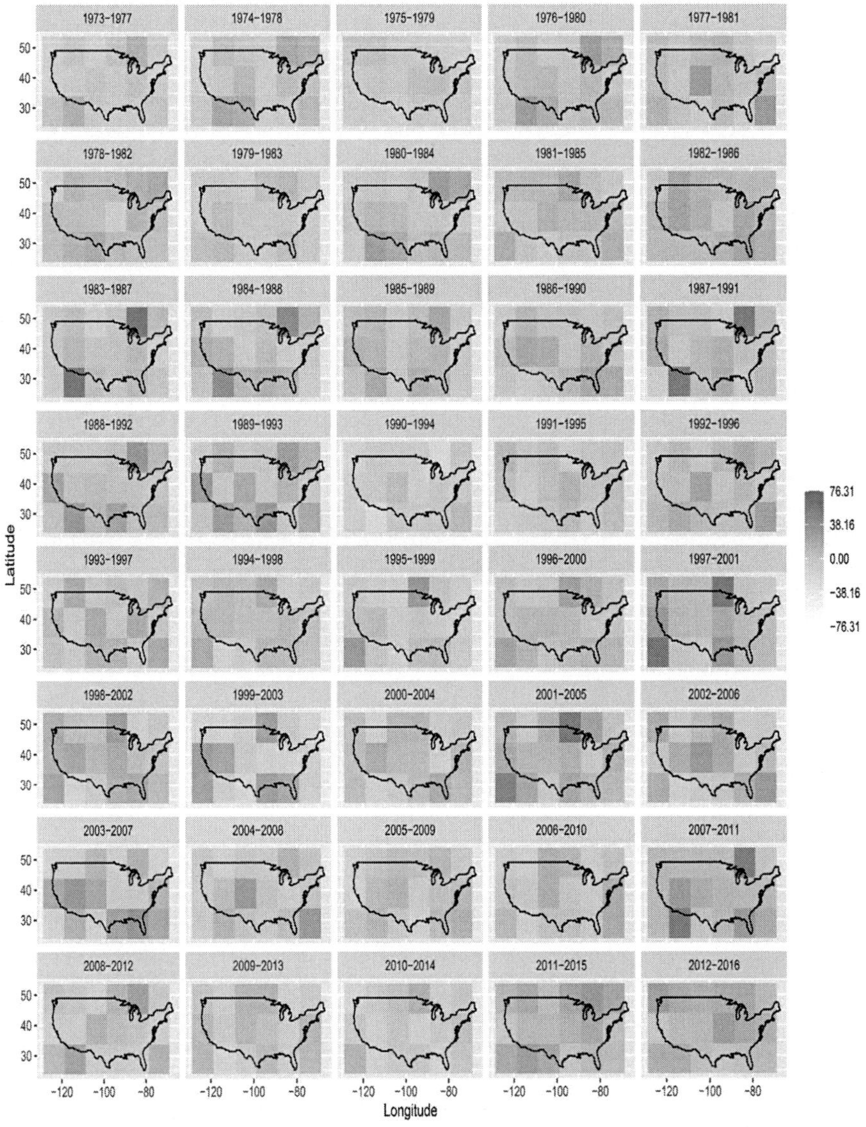

FIGURE 7.16
Log CBF for multiscale coefficients that connect each subregion at the second scale of resolution to its descendants at the third scale of resolution over five-year windows – Local linear growth plus AR(1) model versus stationary AR(1) model.

efficients that connect each subregion at the second level of resolution to its descendants at the third level of resolution, a model that is a mixture of the models with and without local linear trend may provide better probabilistic predictions.

Figure 7.17 presents similar maps for the multiscale coefficients that connect each subregion at the third scale of resolution to its descendants at the fourth (finest) scale of resolution. In this case, the vast majority of the log CBFs is negative indicating that the data strongly and consistently support the stationary AR(1) model without trend. This provides strong evidence that differences in temperature anomalies among sibling subregions at the finest level of resolution follow short-term stationary processes.

Given the above results, in this application it does not seem to make sense to compute one global conditional Bayes factor to compare a DMMSTM with local linear trend and stationary AR(1) processes versus a DMMSTM with only stationary AR(1) processes for all aggregated coarsest mean processes and multiscale coefficients at all resolution levels jointly. That is because we have found that while the DMMSTM with local linear trend is strongly supported for the aggregated coarsest mean processes, the DMMSTM with only stationary components is strongly supported for the multiscale coefficients that connect each subregion at the third scale of resolution to its descendants at the finest scale of resolution. Therefore, in this application different parts of the multiscale system of tropospheric temperatures seem to need different temporal evolution models.

7.7 Discussion

We have proposed in this chapter a Bayesian framework to analyze multivariate multiscale spatio-temporal heteroskedastic Gaussian data. The framework we propose here is a multivariate extension of the univariate framework proposed by Ferreira et al. (2010, 2011). Our current framework is also an extension of the multivariate framework of Elkhouly and Ferreira (2021). There are two main differences between our work here and the class of models proposed in Elkhouly and Ferreira (2021). First, while Elkhouly and Ferreira (2021) assume a multiscale structure where at each level of resolution the number of descendants is the same, here we allow the number of descendants to vary. Second, and maybe more importantly, while Elkhouly and Ferreira (2021) assume that the covariance matrix is fixed, here we allow for heteroskedasticity where the covariance matrices vary spatially.

The DMMSTM framework that we propose here has six salient features. First, the DMMSTM analyzes multiple variables jointly using a multiscale approach. Second, our framework accommodates sudden spatial changes of the mean vectors among adjacent regions. Third, our framework allows for

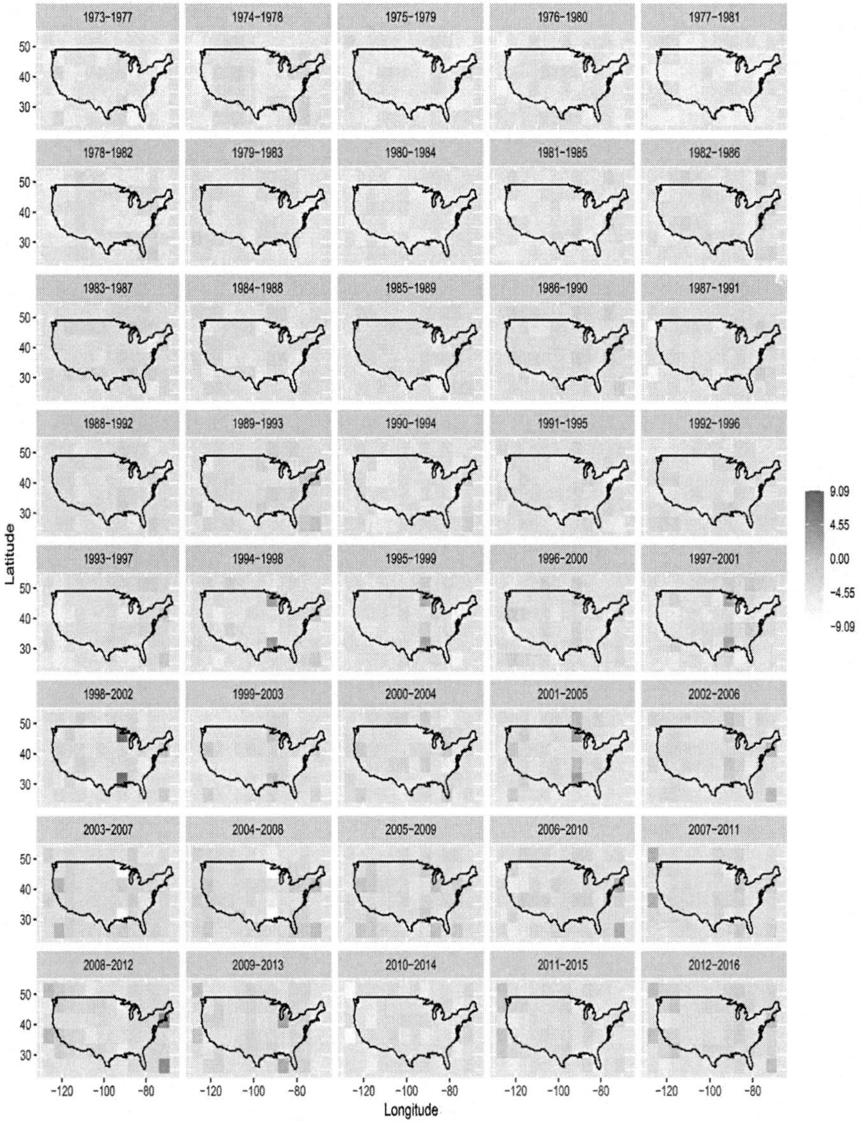

FIGURE 7.17
Log CBF for multiscale coefficients that connect each subregion at the third scale of resolution to its descendants at the fourth (finest) scale of resolution over five-year windows – Local linear growth plus AR(1) model versus stationary AR(1) model.

spatially varying covariance matrices. Fourth, we model the temporal evolutions with a state-space approach that allows for possible non-stationary and stationary evolutions. Fifth, the DMMSTM decomposes multivariate spatio-temporal data into conditionally independent multiscale coefficient processes. Sixth, our framework allows the estimation of multivariate mean and gradient fields at multiple scales of resolution in a coherent manner. That is, the estimated mean level fields are compatible across different scales of resolution.

We have developed an MCMC algorithm to explore the posterior distribution of the system dynamics as well as the parameters. Specifically, to explore the posterior distribution of the latent evolution states of the multivariate multiscale coefficients, we have developed a multivariate singular FFBS algorithm. Further, the DMMSTM framework implements the aggregation constraints on the multiscale coefficients via tractable and direct computations. Furthermore, most of the full conditional distributions for the parameters are available for sampling, leading to fast computations. Therefore, our DMMSTM framework uses an MCMC algorithm that is parallelizable, scalable, and fast.

We have employed the DMMSTM to analyze the summer tropospheric temperatures over North America from the NCEP/NCAR Reanalysis-I dataset for the period from 1951 to 2016. Our results indicate the existence of long-term trends in temperatures at upper and middle tropospheric altitudes over North America. Importantly, the average estimated gradient over North America at altitude pressure levels 300, 400, 500, and 600 hPa at the end of the study period in 2016 imply a trend of increase in average summer temperatures at those altitudes of about 2.8 Kelvin per century.

7.8 Appendix

7.8.1 The Singular Gaussian Distribution

Definition: We say that an n-dimensional random vector $\boldsymbol{X}$ follows a singular Gaussian distribution, denoted by $\mathbf{X}_n \sim N(\boldsymbol{\mu},, \boldsymbol{\Sigma})$, with mean vector $\boldsymbol{\mu}$ and positive semi-definite covariance matrix $\boldsymbol{\Sigma}$ if rank$(\boldsymbol{\Sigma}) = r < n$. In that case, the probability density function of $\boldsymbol{X}$ is given by

$$p(\mathbf{X}|\boldsymbol{\mu}, \boldsymbol{\Sigma}) = \frac{1}{(2\pi)^{r/2}(\prod_{i=1}^{r} \lambda_i^{1/2})} \exp\left[-\frac{1}{2}\left\{(\mathbf{X}-\boldsymbol{\mu})'\boldsymbol{\Sigma}^{-}(\mathbf{X}-\boldsymbol{\mu})\right\}\right],$$

where $\boldsymbol{\Sigma}^{-}$ is the Moore-Penrose inverse of $\boldsymbol{\Sigma}$ and $\lambda_1, \ldots, \lambda_r$ are the non-zero eigenvalues of $\boldsymbol{\Sigma}$. This singular Gaussian distribution implies $n - r$ implicit linear constraints on the vector $\boldsymbol{X}$.

7.8.2 The Singular Gaussian Matrix-Variate Distribution

Definition: We say that an $n \times m$ random matrix $\boldsymbol{X}$ follows a singular Gaussian matrix-variate distribution (Díaz-García and Gutiérrez-Jáimez, 2006), denoted by $\mathbf{X}_{n\times m} \sim MN(\mathbf{M}_{n\times m}, \boldsymbol{\mho}_{n\times n}, \boldsymbol{\Sigma}_{m\times m})$, with mean matrix $\mathbf{M}_{n\times m}$, row covariance matrix $\boldsymbol{\mho}$, and column covariance matrix $\boldsymbol{\Sigma}$ such that $\text{rank}(\boldsymbol{\Sigma}) = r < m$ or $\text{rank}(\boldsymbol{\mho}) = k < n$, if the probability density function is given by

$$
\begin{aligned}
p(\mathbf{X}|\mathbf{M}, \boldsymbol{\mho}, \boldsymbol{\Sigma}) &= \frac{1}{(2\pi)^{rk/2}(\prod_{i=1}^{r} \lambda_i{}^{k/2})(\prod_{j=1}^{k} \delta_i{}^{r/2})} \\
&\quad \times \exp\left[-\frac{1}{2}\text{tr}\left\{(\mathbf{X}-\mathbf{M})'\boldsymbol{\mho}^{-}(\mathbf{X}-\mathbf{M})\boldsymbol{\Sigma}^{-}\right\}\right],
\end{aligned}
$$

where $\boldsymbol{\mho}^{-}$ is the Moore-Penrose inverse of $\boldsymbol{\mho}$, $\boldsymbol{\Sigma}^{-}$ is the Moore-Penrose inverse of $\boldsymbol{\Sigma}$, $\delta_1, \ldots, \delta_k$ are the non-zero eigenvalues of the row covariance matrix $\boldsymbol{\mho}$, and $\lambda_1, \ldots, \lambda_r$ are the non-zero eigenvalues of the column covariance matrix $\boldsymbol{\Sigma}$.

7.8.3 Proof of Theorem 7.1: Posterior Conditional Independence

By Bayes theorem, given the different covariance matrices at the finest resolution $\boldsymbol{V}_{L1:n_L}$, the different signal-to-noise ratio parameters for the three possible dynamics $\psi_{11:n_1,-}, \cdots, \psi_{L-11:n_{L-1},-}$ and the AR(1) coefficients $\phi_{11:n_1}, \ldots, \phi_{L-11:n_{L-1}}$, the posterior distribution of the latent dynamics $\boldsymbol{H}_{0:T,11}, \ldots \boldsymbol{H}_{0:T,1n_1}, \boldsymbol{\Theta}_{0:T,11}, \ldots \boldsymbol{\Theta}_{0:T,1n_1}, \ldots \boldsymbol{\Theta}_{0:T,L-1n_{L-1}}$ is proportional to the product of the factorized likelihood given in Equation (7.3) for each time point times the priors for $\boldsymbol{H}_{01k}$ and $\boldsymbol{\Theta}_{0lj}$ times the priors that are implied by the system Equations (7.9) and (7.12).

$$
\prod_{t=1}^{T}\left\{\prod_{k=1}^{n_1} p(\boldsymbol{y}_{t1k}|\boldsymbol{H}_{t1k}, \boldsymbol{V}_{1k}) \prod_{l=1}^{L-1}\prod_{j=1}^{n_l} p(\boldsymbol{\theta}^{e}_{lj}|\boldsymbol{\Theta}_{tlj}, \boldsymbol{\Omega}_{lj})\right\}
$$

$$
\times \left\{\prod_{k=1}^{n_1} p(\boldsymbol{H}_{01k}|\mathbf{M}_{01k}\mathbf{C}_{01k}, \boldsymbol{V}_{1k}) \prod_{l=1}^{L-1}\prod_{j=1}^{n_l} p(\boldsymbol{\Theta}_{0lj}|\mathbf{M}_{0lj}, \mathbf{C}_{0lj}, \boldsymbol{\Omega}_{lj})\right\}
$$

$$
\times \prod_{t=1}^{T}\left\{\prod_{k=1}^{n_1} p(\boldsymbol{H}_{t1k}|\boldsymbol{H}_{t-1,1k}\mathbf{C}_{1k}, \boldsymbol{V}_{1k}) \prod_{l=1}^{L-1}\prod_{j=1}^{n_l} p(\boldsymbol{\Theta}_{tlj}|\boldsymbol{\Theta}_{t-1,lj}, \mathbf{C}_{lj}, \boldsymbol{\Omega}_{lj})\right\}
$$

After re-arranging the terms in the expression above, we obtain

$$\prod_{t=1}^{T}\prod_{k=1}^{n_1}\left[p(\boldsymbol{y}_{t1k}|\boldsymbol{H}_{t1k},\boldsymbol{V}_{1k})p(\boldsymbol{H}_{01k}|\mathbf{M}_{01k}\mathbf{C}_{01k},\boldsymbol{V}_{1k})p(\boldsymbol{H}_{t1k}|\boldsymbol{H}_{t-1,1k}\mathbf{C}_{1k},\boldsymbol{V}_{1k})\right]$$

$$\times\prod_{l=1}^{L-1}\prod_{j=1}^{n_l}\left[p(\boldsymbol{\theta}^e_{lj}|\boldsymbol{\Theta}_{tlj},\boldsymbol{\Omega}_{lj})p(\boldsymbol{\Theta}_{tlj}|\boldsymbol{\Theta}_{t-1,lj},\mathbf{C}_{lj},\boldsymbol{\Omega}_{lj})p(\boldsymbol{\Theta}_{0lj}|\mathbf{M}_{0lj},\mathbf{C}_{0lj},\boldsymbol{\Omega}_{lj})\right]$$

Therefore, the posterior distribution can be factorized into two parts: the first part refers to $\boldsymbol{H}_{0:T,1k}$ for $k = 1,\dots n_1$ for the latent dynamics at the coarsest level of resolution; the second part refers to the latent dynamic of the multiscale coefficients $\boldsymbol{\Theta}_{0:T,lj}$ for $l = 1,\dots L-1$ and $j = 1,\dots n_l$. Therefore, $\boldsymbol{H}_{0:T,11},\dots \boldsymbol{H}_{0:T,1n_1}$, $\boldsymbol{\Theta}_{0:T,11}$, $\dots \boldsymbol{\Theta}_{0:T,1n_1}$, $\dots \boldsymbol{\Theta}_{0:T,L-1n_{L-1}}$ are conditionally independent a posteriori.

7.8.4 Proof of Theorem 7.2: The Multivariate Singular FFBS

First, let us prove the results of the forward filter. The proof is by induction. Assume that the posterior at time $t-1$ is $\boldsymbol{\Theta}_{t-1,lj}|D_{t-1} \sim SMN(\mathbf{M}_{t-1,lj},\mathbf{C}_{t-1,lj},\boldsymbol{\Omega}_{lj})$. Then, using Equation (7.12) and properties of the matrix-variate Gaussian distribution, the prior of the latent state at time t is $\boldsymbol{\Theta}_{tlj}|D_{t-1} \sim SMN(\mathbf{A}_{tlj},\ \mathbf{R}_{tlj},\boldsymbol{\Omega}_{lj})$. Recall that at time t we have the observed empirical multiscale coefficients vector with distribution $\boldsymbol{\theta}^e_{tlj}|\boldsymbol{\Theta}_{tlj} \sim SMN(\boldsymbol{f}'\boldsymbol{\Theta}_{tlj},1,\boldsymbol{\Omega}_{lj})$. Thus, by Bayes Theorem, the posterior distribution of the forward filter at time t is

$$\begin{aligned}
p(\boldsymbol{\Theta}_{tlj}|D_t) &= p(\boldsymbol{\Theta}_{tlj}|\boldsymbol{\theta}^e_{tlj},D_{t-1})\\
&\propto p(\boldsymbol{\theta}^e_{tlj}|\boldsymbol{\Theta}_{tlj})p(\boldsymbol{\Theta}_{tlj}|D_{t-1})\\
&\propto \exp\left[-\frac{1}{2}\mathrm{tr}\left\{(\boldsymbol{\Theta}_{tlj}-\mathbf{A}_{tlj})'\mathbf{R}^{-1}_{tlj}(\boldsymbol{\Theta}_{tlj}-\mathbf{A}_{tlj})\boldsymbol{\Omega}^{-}_{lj}\right\}\right]\\
&\quad\times\exp\left[-\frac{1}{2}\mathrm{tr}\left\{(\boldsymbol{\theta}^e_{tlj}-\boldsymbol{f}'\boldsymbol{\Theta}_{tlj})'(\boldsymbol{\theta}^e_{tlj}-\boldsymbol{f}'\boldsymbol{\Theta}_{tlj})\boldsymbol{\Omega}^{-}_{lj}\right\}\right]\\
&\propto \exp\Bigg[-\frac{1}{2}\mathrm{tr}\Bigg\{\Big(\boldsymbol{\Theta}'_{tlj}[\mathbf{R}^{-1}_{tlj}+\boldsymbol{f}\boldsymbol{f}']\boldsymbol{\Theta}_{tlj}\\
&\qquad -2\boldsymbol{\Theta}'_{tlj}[\mathbf{R}^{-1}_{tlj}\mathbf{A}_{tlj}+\boldsymbol{f}\boldsymbol{\theta}^e_{tlj}]\Big)\boldsymbol{\Omega}^{-}_{lj}\Bigg\}\Bigg],
\end{aligned}$$

which is the kernel of a singular matrix-variate Gaussian distribution with left covariance matrix $\boldsymbol{\Omega}_{lj}$, right covariance matrix $\mathbf{C}_{tlj} = (\mathbf{R}^{-1}_{tlj}+\boldsymbol{f}\boldsymbol{f}')^{-1}$, and mean matrix $\mathbf{M}_{tlj} = \mathbf{C}_{tlj}[\mathbf{R}^{-1}_{tlj}\mathbf{A}_{tlj}+\boldsymbol{f}\boldsymbol{\theta}^e_{tlj}]$. This concludes the derivation of the forward filter.

Now, let us prove the results of the backward sampler. After processing the information from times 1 to T, the forward filter provides the distribution

$\mathbf{\Theta}_{Tlj}|D_T \sim SMN(\mathbf{M}_{Tlj}, \mathbf{C}_{Tlj}, \mathbf{\Omega}_{lj})$. Then, the backward sampler starts by sampling $\mathbf{\Theta}_{Tlj}$ from this distribution. After that, the backward sampler recursively samples $\mathbf{\Theta}_{tlj}$ from the conditional distribution $\mathbf{\Theta}_{tlj}|\mathbf{\Theta}_{(t+1):T,lj}, D_T$ for $t = T-1, \ldots, 1$. Because of the Markovian property, $\mathbf{\Theta}_{tlj}|\mathbf{\Theta}_{(t+1):T,lj}, D_T \equiv \mathbf{\Theta}_{tlj}|\mathbf{\Theta}_{t+1,lj}$. Further, note that $\mathbf{\Theta}_{t+1,lj}|\mathbf{\Theta}_{tlj} \sim SMN(\mathbf{G}^{\Theta}_{lj}\mathbf{\Theta}_{tlj},\ \boldsymbol{U}^{\theta}_{lj}, \mathbf{\Omega}_{lj})$. Then, by Bayes Theorem

$$
\begin{aligned}
p(\mathbf{\Theta}_{tlj}|\mathbf{\Theta}_{t+1,lj}, D_T) &\propto p(\mathbf{\Theta}_{t+1,lj}|\mathbf{\Theta}_{tlj})p(\mathbf{\Theta}_{tlj}|D_t)\\
&\propto \exp\left[-\frac{1}{2}\mathrm{tr}\left\{(\mathbf{\Theta}_{tlj} - \mathbf{M}_{tlj})'\mathbf{C}^{-1}_{tlj}(\mathbf{\Theta}_{tlj} - \mathbf{M}_{tlj})\mathbf{\Omega}^{-}_{lj}]\right\}\right]\\
&\quad \times \exp\left[-\frac{1}{2}\mathrm{tr}\left\{(\mathbf{\Theta}_{t+1,lj} - \mathbf{G}^{\Theta}_{lj}\mathbf{\Theta}_{tlj})'\boldsymbol{U}^{-1}_{lj}\right.\right.\\
&\qquad\qquad \left.\left.(\mathbf{\Theta}_{t+1,lj} - \mathbf{G}^{\Theta}_{lj}\mathbf{\Theta}_{tlj})\mathbf{\Omega}^{-}_{lj}]\right\}\right]\\
&\propto \exp\left[-\frac{1}{2}\mathrm{tr}\left\{(\mathbf{\Theta}'_{tlj}[\mathbf{C}^{-1}_{tlj} + (\mathbf{G}^{\Theta}_{lj})'\boldsymbol{U}^{-1}_{lj}\mathbf{G}^{\Theta}_{lj}]\mathbf{\Theta}_{tlj}\right.\right.\\
&\qquad \left.\left.-2\mathbf{\Theta}'_{tlj}[\mathbf{C}^{-1}_{tlj}\mathbf{M}_{tlj} + (\mathbf{G}^{\Theta}_{lj})'\boldsymbol{U}^{-1}_{lj}\mathbf{\Theta}_{t+1,lj}])\mathbf{\Omega}^{-}_{lj}]\right\}\right],
\end{aligned}
$$

which is the kernel of a singular matrix-variate Gaussian distribution with left covariance matrix $\mathbf{\Omega}_{lj}$, right covariance matrix $\mathbf{C}^{*}_{tlj} = [\mathbf{C}^{-1}_{tlj} + (\mathbf{G}^{\Theta}_{lj})'\boldsymbol{U}^{-1}_{lj}\mathbf{G}^{\Theta}_{lj}]^{-1}$, and mean matrix $\mathbf{M}^{*}_{tlj} = \mathbf{C}^{*}_{tlj}[\mathbf{C}^{-1}_{tlj}\mathbf{M}_{tlj} + (\mathbf{G}^{\Theta}_{lj})'\boldsymbol{U}^{-1}_{lj}\mathbf{\Theta}_{t+1,lj}]$.

Bibliography

Anderson, T. W. (1984). *An Introduction to Multivariate Statistical Analysis* (2nd ed.). New York: John Wiley & Sons.

Calder, C. A. (2007). Dynamic factor process convolution models for multivariate space-time data with application to air quality assessment. *Environmental and Ecological Statistics 14*(3), 229–247.

Carter, C. K. and R. Kohn (1994). On Gibbs sampling for state space models. *Biometrika 81*, 541–553.

Carvalho, C. M. and M. West (2007). Dynamic matrix-variate graphical models. *Bayesian Analysis 2*(1), 69–97.

Christy, J. R. and W. B. Norris (2004). What may we conclude about global tropospheric temperature trends? *Geophysical Research Letters 31*(L06211), 1–4.

Díaz-García, J. A. and R. Gutiérrez-Jáimez (2006). Distribution of the generalised inverse of a random matrix and its applications. *Journal of Statistical Planning and Inference 136*(1), 183–192.

Douglass, D. H., B. D. Pearson, and S. F. Singer (2004). Altitude dependence of atmospheric temperature trends: Climate models versus observation. *Geophysical Research Letters 31*(13), 2–5.

Elkhouly, M. (2019). *Bayesian Analysis of Temporal and Spatio-temporal Multivariate Environmental Data*. Ph. D. thesis, Virginia Polytechnic Institute and State University.

Elkhouly, M. and M. A. R. Ferreira (2021). Dynamic multiscale spatiotemporal models for multivariate Gaussian data. *Spatial Statistics 41*, 100475.

Elkhouly, M., S. E. Zick, and M. A. R. Ferreira (2023). Long term temporal trends in synoptic-scale weather conditions favoring significant tornado occurrence over the central United States. *PLoS One 18*(2), e0281312.

Ferreira, M. A. R. (2011). Invited discussion on "Characterizing uncertainty of future climate change projections using hierarchical Bayesian models by Tebaldi, Smith and Sansó". In J. M. Bernardo, M. J. Bayarri, J. O. Berger, A. P. Dawid, D. Heckerman, A. F. M. Smith, and M. West (Eds.), *Bayesian Statistics 9*, pp. 639–658. Oxford: Oxford University Press.

Ferreira, M. A. R. (2021). Bayesian spatial and spatiotemporal models based on multiscale factorizations. *Wiley Interdisciplinary Reviews: Computational Statistics 13*(2), e1509.

Ferreira, M. A. R., A. I. Bertolde, and S. Holan (2010). Analysis of economic data with multi-scale spatio-temporal models. In A. O'Hagan and M. West (Eds.), *Handbook of Applied Bayesian Analysis*, pp. 295–318. Oxford University Press.

Ferreira, M. A. R., S. H. Holan, and A. I. Bertolde (2011). Dynamic multiscale spatio-temporal models for Gaussian areal data. *Journal of the Royal Statistical Society (Series B: Methodological) 73*, 663–688.

Fonseca, T. C. O. and M. A. R. Ferreira (2017). Dynamic multiscale spatiotemporal models for Poisson data. *Journal of the American Statistical Association 112*, 215–234.

Frühwirth-Schnatter, S. (1994). Data augmentation and dynamic linear models. *Journal of Time Series Analysis 15*(2), 183–202.

Gelman, A. and D. B. Rubin (1992). Inference from iterative simulation using multiple sequences. *Statistical Science 7*, 457–511.

Hoegh, A., M. A. R. Ferreira, and S. Leman (2016). Spatiotemporal model fusion: multiscale modelling of civil unrest. *Journal of the Royal Statistical Society: Series C (Applied Statistics) 65*, 529–545.

Holmes, E. E., E. J. Ward, and K. Wills (2012). MARSS: Multivariate Autoregressive State-space Models for Analyzing Time-Series Data. *R Journal 4*(1), 11–19.

Kalnay, E., M. Kanamitsu, R. Kistler, W. Collins, D. Deaven, L. Gandin, M. Iredell, S. Saha, G. White, J. Woollen, et al. (1996). The NCEP/NCAR 40-year reanalysis project. *Bulletin of the American Meteorological Society 77*(3), 437–472.

Karpman, D., M. A. R. Ferreira, and C. K. Wikle (2013). A point process model for tornado report climatology. *Stat 2*(1), 1–8.

Landim, F. and D. Gamerman (2000). Dynamic hierarchical models: an extension to matrix-variate observations. *Computational Statistics & Data Analysis 35*(1), 11–42.

Lopes, H., D. Gamerman, and E. Salazar (2011). Generalized spatial dynamic factor models. *Computational Statistics & Data Analysis 55*, 1319–1330.

Lopes, H. F., E. Salazar, and D. Gamerman (2008). Spatial dynamic factor analysis. *Bayesian Analysis 3*(4), 759–792.

Mills, T. C. (2007). Time series modelling of two millennia of northern hemisphere temperatures: Long memory or shifting trends? *Journal of the Royal Statistical Society – Series A 170*(1), 83–94.

Muirhead, R. J. (1982). *Aspects of Multivariate Statistical Theory.* New York: Wiley.

Perlwitz, J., M. Hoerling, and R. Dole (2015). Arctic tropospheric warming: Causes and linkages to lower latitudes. *Journal of Climate 28*(6), 2154–2167.

Plummer, M., N. Best, K. Cowles, and K. Vines (2006). CODA: convergence diagnosis and output analysis for MCMC. *R News 6*(March), 7–11.

Prado, R., M. A. R. Ferreira, and M. West (2021). *Time Series: Modeling, Computation, and Inference* (2nd ed.). Boca Raton: CRC Press.

Quintana, J. M. and M. West (1987). An analysis of international exchange rates using multivariate DLM's. *Journal of the Royal Statistical Society: Series D (The Statistician) 36*(2-3), 275–281.

Shin, H. and M. A. R. Ferreira (2023). Dynamic ICAR spatiotemporal factor models. *Spatial Statistics 56*, 100763.

Siotani, M., T. Hayakawa, and Y. Fujikoshi (1985). *Modern Multivariate Statistical Analysis.* Columbus: American Sciences Press.

Stein, M. L. (2008). A modeling approach for large spatial datasets. *Journal of the Korean Statistical Society 37*(1), 3–10.

Stein, M. L. (2014). Limitations on low rank approximations for covariance matrices of spatial data. *Spatial Statistics 8*(C), 1–19.

Sun, Y., B. Li, and M. G. Genton (2012). Geostatistics for large datasets. In E. Porcu, J. M. Montero, and M. Schlather (Eds.), *Advances and Challenges in Space-Time Modelling of Natural Events*, pp. 55–77. Springer.

Tebaldi, C., R. L. Smith, and B. Sansó (2011). Characterizing uncertainty of future climate change projections using hierarchical Bayesian models. In J. M. Bernardo, M. J. Bayarri, J. O. Berger, A. P. Dawid, D. Heckerman, A. F. M. Smith, and M. West (Eds.), *Bayesian Statistics 9*, pp. 639–658. Oxford: Oxford University Press.

Tian, B., E. J. Fetzer, B. H. Kahn, J. Teixeira, E. Manning, and T. Hearty (2013). Evaluating CMIP5 models using AIRS tropospheric air temperature and specific humidity climatology. *Journal of Geophysical Research: Atmospheres 118*(1), 114–134.

Vidakovic, B. (1999). *Statistical Modeling by Wavelets.* Wiley.

Vivar, J. C. and M. A. R. Ferreira (2009). Spatio-temporal models for Gaussian areal data. *Journal of Computational and Graphical Statistics 18*, 658–674.

Walsh, S., M. A. R. Ferreira, D. Higdon, and S. E. Zick (2023). A Bayesian hierarchical model framework to quantify uncertainty of tropical cyclone precipitation forecasts. *The Annals of Applied Statistics 17*(3), 1984–2001.

Wentz, F. J. and M. Schabel (1998). Effects of orbital decay on satellite-derived lower-tropospheric temperature trends. *Nature 394*(6694), 661–664.

West, M. and P. J. Harrison (1997). *Bayesian Forecasting and Dynamic Models* (2nd. ed.). New York: Springer.

8

A Model Selection Paradox with Implications to Multiscale Modeling

Marco A. R. Ferreira
Virginia Tech

M. Alejandra Jaramillo
Universidad Militar Nueva Granada

Elizabeth B. Hypólito
Instituto Brasileiro de Geografía e Estatística

We present here a Bayesian model selection paradox that has implications for multiscale modeling. We consider the M-open case when the true model is not among the models being considered in the analysis. In this case, the set of models that minimize the Kullback-Leibler divergence to the true model is the asymptotic carrier. For regular parametric models, as the sample size increases, the posterior probability of the asymptotic carrier converges to one. Thus, if the asymptotic carrier has only one model, then the data will eventually provide overwhelming evidence for that model. But what happens if the asymptotic carrier has multiple models? In the latter case, a scientist may naïvely wish for the posterior distribution to reflect such uncertainty by converging to a uniform distribution on the set of models that belong to the asymptotic carrier. Unfortunately, that is not the asymptotic behavior.

DOI: 10.1201/9781032623443-8

Instead, in that case there is no limiting distribution for the posterior probabilities of models within the asymptotic carrier. We call this the M-open multiple optima paradox. We illustrate this paradoxal behavior and some of its practical implications with four examples: a toy example to introduce the main concepts, two examples related to inference on phylogenetic trees, and one example on wavelet nonparametric regression where the multiscale tree structure is assumed to be unknown.

8.1 Introduction

We present here a Bayesian model selection paradox that has implications for multiscale modeling. We consider the M-open case in model selection, that is, the case when the true model is not among the models being considered in the analysis (Yao et al., 2018; Li and Dunson, 2020). For simplicity of exposition, throughout we assume a uniform prior distribution on the set of considered models. For regular parametric models, as the sample size increases the posterior probability of the model closest to the true model in terms of Kullback-Leibler divergence will converge to one (Li and Dunson, 2020). But what happens if there are multiple models that minimize the Kullback-Leibler divergence from the true model? Then, as the sample size increases, the posterior probability of this set of models converges to one. Thus, the set of models that minimizes the Kullback-Leibler divergence from the true model is the asymptotic carrier (Berk, 1966).

But how do the posterior probabilities get distributed among the models in the asymptotic carrier? A scientist would wish for the posterior distribution within the asymptotic carrier to converge to a uniform distribution. This would provide somehow appropriate information about the uncertainty on the support provided by the data to each model considered in the analysis. Unfortunately, that is not the case. In reality, while the posterior probability of the set of models in the asymptotic carrier converges to one as the sample size increases, there is no limiting distribution for the individual posterior probabilities of the models in the asymptotic carrier. Even more vexing, as the sample size increases, the probability of the event "the posterior distribution within the asymptotic carrier is near a uniform distribution" converges to zero. We call this latter phenomenon the M-open multiple optima (MOMO) paradox.

We illustrate the MOMO paradox with four examples. First, we present a toy example that provides motivation for the study of the paradox and introduces some key concepts. In addition, this toy example allows for an easier understanding of why the paradox happens. The MOMO paradox has important consequences for the practical use of Bayesian model selection. To illustrate this latter point, we present two examples of the occurrence of the

MOMO paradox related to inference on phylogenetic trees. Usually, phylogenetic reconstruction models assume bifurcating trees. The two phylogenetic examples we consider show that the MOMO paradox may occur when the evolutionary tree is not bifurcating. Finally, we show an example of the occurrence of the MOMO paradox in wavelet nonparametric regression. This example points to the issue that the MOMO paradox may have consequences for multiscale modeling and, in particular, may impact the ability to estimate multiscale structures. We end this chapter with a short discussion and concluding remarks.

8.2 Examples

8.2.1 Toy Example

This section presents a toy example that provides motivation for the study of the MOMO paradox. Thus, this toy example helps us introduce key concepts related to the behavior of model posterior probabilities when the true model is not one of the models being considered and two or more models are equally distant from the true model. This toy example is inspired by an example given by Berk (1966), who considered the limiting behavior of posterior distributions when the model is incorrect. Berk (1966) considered the case when the possibly incorrect model is indexed by a parameter vector taking values in a parameter space that is a Borel subset of a complete separable metric space. In that case, as the sample size increases the posterior distribution of the parameter vector concentrates in a set that may contain more than one point. Berk (1966) calls this set the asymptotic carrier. In the case considered by Berk (1966), the asymptotic carrier is the set of parameter values that minimize the Kullback-Leibler divergence between the incorrect model and the true data-generating process.

Our toy example considers data from a true data-generating process that is a mixture of two Gaussian distributions with mixing probability 0.5, means -0.1 and 0.1, and common standard deviation 1. In this toy example, instead of analyzing the data with the true data-generating model, we analyze the data with four competing models that are incorrect. These competing models are univariate Gaussian distributions with a standard deviation of 1 and possible mean values $\mu = -2, -1, 1$, and 2. We denote these four models by $\mathcal{M}_{-2}$, $\mathcal{M}_{-1}$, $\mathcal{M}_1$, and $\mathcal{M}_2$. Note that incorrect models $\mathcal{M}_{-1}$ and $\mathcal{M}_1$ are equally distant from the true data-generating process in terms of Kullback-Leibler divergence. In addition, when compared to $\mathcal{M}_{-2}$ and $\mathcal{M}_2$, models $\mathcal{M}_{-1}$ and $\mathcal{M}_1$ are closer to the true data-generating process. Thus, in this toy example, the set $\{-1, 1\}$ is the asymptotic carrier.

We have simulated one sample of 50,000 independent observations from the mixture of two Gaussian distributions presented above. Figure 8.1 presents

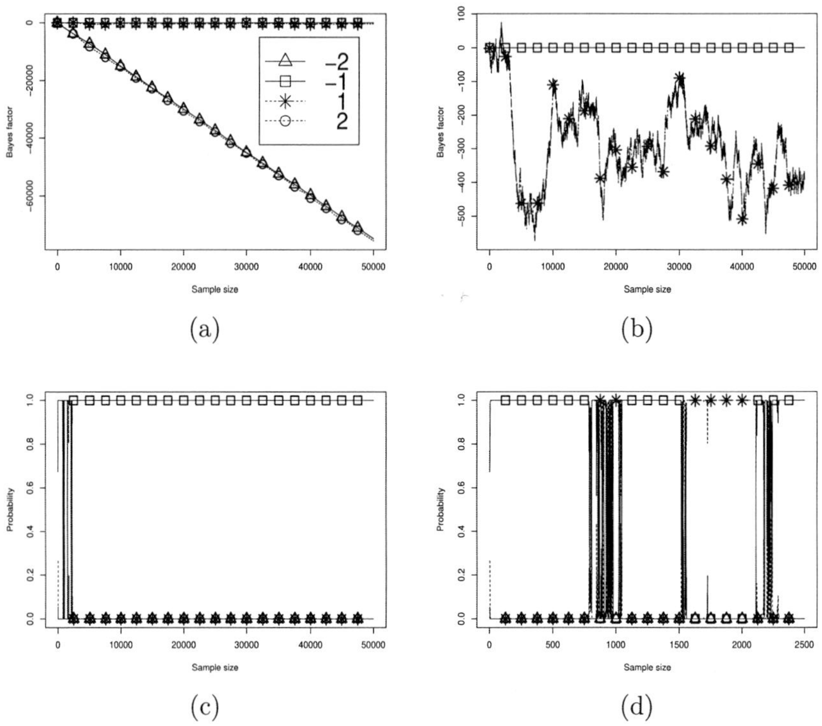

FIGURE 8.1
Toy example. One representative realization. Progression of Bayes factors of models $N(\mu, 1)$ with $\mu = -2, -1, 1$, and 2 versus model $N(-1.1)$ and posterior probabilities as the sample size increases. True generating process is a mixture of two Gaussians with mixing probability 0.5, means -0.1 and 0.1, and common standard deviation 1. (a) Bayes factors considering all four competing models; (b) Bayes factors considering only the models with $\mu = -1$ and $\mu = 1$, which are the models closest to the true model in terms of Kullback-Leibler divergence; (c) Posterior probability of the four competing models as the sample increases from 1 to 50,000; (d) Zoomed in posterior probabilities as the sample increases from 1 to 2,500. In this particular realization, as the sample size increases from 1 to 2,500, the models with $\mu = -1$ and $\mu = 1$ take turns are having posterior probability close to 1. From sample size 2,500 to 50,000, only the model with $\mu = -1$ has posterior probability close to 1. In another realization (e.g., another researcher collecting data and computing posterior probabilities as functions of sample size), the model with $\mu = 1$ may have posterior probability close to 1 for most of the sample sizes.

the progression of the logarithm of Bayes factors and posterior probabilities of incorrect models $\mathcal{M}_{-2}$, $\mathcal{M}_{-1}$, $\mathcal{M}_1$, and $\mathcal{M}_2$ versus incorrect model $\mathcal{M}_{-1}$ as the sample size increases. Thus, in this figure a sample of size n refers to the subsample of the first n observations in the sample of 50,000 simulated observations. Specifically, Panel (a) of Figure 8.1 presents the log-Bayes factors of each of the four competing models versus model $\mathcal{M}_{-1}$. The log-Bayes factor of model $\mathcal{M}_{-1}$ versus itself (provided as a baseline and represented with a solid line and squares) is constant and equal to zero as the sample size increases. In Panel (a), this log-Bayes factor is not distinguishable from the log-Bayes factor of model $\mathcal{M}_1$ versus model $\mathcal{M}_{-1}$ (represented with a dashed line and asterisks). Finally, the log-Bayes factors of models $\mathcal{M}_{-2}$ (represented with a solid line and triangles) and $\mathcal{M}_2$ (represented with a dashed line and circles) versus model $\mathcal{M}_{-1}$ are not distinguishable and decrease linearly with sample size. These behaviors of the latter log-Bayes factors are due to two facts: models $\mathcal{M}_{-2}$ and $\mathcal{M}_2$ are equally distant from the true generating process when compared with each other, but are farther from the true generating process when compared with model $\mathcal{M}_{-1}$.

To provide a more detailed comparison of the incorrect models that are in the asymptotic carrier, Panel (b) of Figure 8.1 focuses on the log-Bayes factors of models $\mathcal{M}_1$ and $\mathcal{M}_{-1}$ versus model $\mathcal{M}_{-1}$. To make sense of this plot, we note that Kass and Raftery (1995) provide guidelines for the interpretation of log-Bayes factors (for the case when the true model is among the models being considered in the analysis) and recommend that values larger than 5 indicate very strong evidence in favor of a model. Unfortunately, our toy example shows that these guidelines do not hold in the case when the asymptotic carrier contains two or more competing models. Specifically, Panel (b) shows that the log-Bayes factor of model $\mathcal{M}_1$ versus model $\mathcal{M}_{-1}$ varies from -573.299 to 74.867, and is rarely between -5 and 5. That is, this log-Bayes factor is with high probability either in the region of very strong evidence in favor of model $\mathcal{M}_1$ or in the region of very strong evidence in favor of model $\mathcal{M}_{-1}$.

Often, uncertainty quantification for model selection is performed by examining the posterior probabilities of the competing models. Assuming uniform prior probabilities for the four competing models, Panel (c) of Figure 8.1 presents the posterior probabilities of these models as the sample size increases from 1 to 50,000. For sample sizes from 2,500 to 50,000, the posterior probability of model $\mathcal{M}_{-1}$ is very close to one while the posterior probabilities of the models with $\mu = -2$, $\mu = 1$, and $\mu = 2$ are close to zero. The behaviors of the posterior probabilities for sample sizes from 1 to 2,500 are less clear. To more closely examine the behavior of the posterior probabilities for smaller sample sizes, Panel (d) of Figure 8.1 zooms in posterior probabilities as the sample size increases from 1 to 2,500. In this case, models $\mathcal{M}_{-1}$ and $\mathcal{M}_1$ take turns are having posterior probability close to 1.

To shed light on the behavior of the log-Bayes factors and posterior probabilities, let us consider the log-Bayes factor of model $\mathcal{M}_1$ versus model $\mathcal{M}_{-1}$

based on the first n observations $x_1, \ldots, x_n$, which is

$$l_n = -0.5\sum_{i=1}^{n}(x_i - 1)^2 + 0.5\sum_{i=1}^{n}(x_i + 1)^2 = 2\sum_{i=1}^{n} x_i. \tag{8.1}$$

Thus, we can write the log-Bayes factor as a random walk:

$$l_n = l_{n-1} + \epsilon_n, \tag{8.2}$$

where $l_0 = 0$, $\epsilon_n = 2x_n$, $E(\epsilon_n) = 0$, and $Var(\epsilon_n) = 4Var(x_n) = 8$ (The expectation and variances are computed with respect to the true mixture model).

Note that when the log-Bayes factor l_n is equal to 0 the posterior probabilities of $\mathcal{M}_1$ and $\mathcal{M}_{-1}$ are equal to each other. That is, when the log-Bayes factor is $l_n = 0$ the posterior distribution on the models in the asymptotic carrier is exactly uniform. Conversely, this posterior distribution is farther from uniform when l_n is farther from zero. Thus, let us consider the probability of the event $\{|l_n| < u\}$ for an arbitrary positive scalar u. Thus, this event corresponds to the posterior distribution on the set of models in the asymptotic carrier being close (as measured by u) to a uniform distribution. By the Central Limit Theorem, for large n the log-Bayes factor l_n will have an approximate Gaussian distribution with mean 0 and variance $8n$. Thus, the probability that the absolute value of l_n is less than u is

$$P(|l_n| < u) \approx \Phi\left(\frac{u}{\sqrt{8n}}\right) - \Phi\left(-\frac{u}{\sqrt{8n}}\right) = 2\Phi\left(\frac{u}{\sqrt{8n}}\right) - 1,$$

where $\Phi(z)$ is the cumulative distribution function of the standard Gaussian distribution computed at z. Thus, as the sample size n grows, the probability that l_n is close to zero becomes progressively smaller. In the limit when $n \to \infty$, the probability that the posterior distribution on the set of models in the asymptotic carrier is close to uniform converges to 0. We call this the MOMO paradox.

Take for example $u = 5$; note that $l_n > 5$ would indicate very strong evidence in favor of model $\mathcal{M}_1$ whereas $l_n < -5$ would indicate very strong evidence in favor of model $\mathcal{M}_{-1}$. With a sample of size 50,000 as considered here, the probability of $|l_n < 5|$ is about 0.0063. Hence, for such a sample we have

$$P[\{P(\mathcal{M}_1|\text{Data}) > 0.993\} \text{ or } \{P(\mathcal{M}_{-1}|\text{Data}) > 0.993\}] = 0.9937.$$

That is, in this toy example the posterior distribution on the models in the asymptotic carrier not only is very far from uniform but also tends to concentrate either in $\mathcal{M}_1$ or $\mathcal{M}_{-1}$

Note that Figure 8.1 provides results only for one realization of this process that represents one scientist obtaining a sample and performing data analysis. If another scientist collected another independent sample and computed Bayes factors and posterior probabilities as functions of sample size, model $\mathcal{M}_1$ may have posterior probability close to 1 for most of the considered sample sizes. In fact, if 10,000 scientists are to obtain independent samples of size 50,000 of our true mixture model, consider in the data analysis only models $\mathcal{M}_{-2}$, $\mathcal{M}_{-1}$, $\mathcal{M}_1$, and $\mathcal{M}_2$, and make a decision based on posterior probabilities, then based on the calculations from the previous paragraph on average only 63 scientists will find a Bayes factor of $\mathcal{M}_1$ versus $\mathcal{M}_{-1}$ between -5 and 5, about half of the scientists will conclude that their data provides strong evidence in favor of model $\mathcal{M}_{-1}$, and about half will conclude that there is strong evidence in favor of model $\mathcal{M}_1$. That is, scientists collecting independent distinct samples from the same true data generating process, but not considering the true generating process when analyzing the data and, instead, considering the same set of incorrect models may arrive at conflicting conclusions about which model is best supported by their respective datasets.

8.2.2 Inference for Phylogenetic Trees

The scientific field of phylogenetic studies the evolutionary history of organisms (Wiley and Lieberman, 2011). The result of a phylogenetic analysis is a phylogenetic tree that represents the inferred evolutionary history and relationships among a group of organisms. Bayesian reconstruction of phylogenetic trees is widely used by phylogeneticists, both by its flexibility and computational power (Rambaut et al., 2018; Bouckaert et al., 2019). For example, Bayesian phylogenetics has been recently used for inferring the evolutionary history of different lineages of the severe acute respiratory syndrome coronavirus 2 (SARS-CoV-2), the virus that causes COVID (Candido et al., 2020; Alteri et al., 2021; Du Plessis et al., 2021). Inspired by the literature on inference for phylogenetic trees, this section presents two examples of the MOMO paradox. The explanation for the MOMO paradox in these two examples first appeared in Hypólito (2005).

The flexibility of Bayesian modeling and computation has allowed many extensions of the usual phylogenetic models, such as for example models for recombination detection (Husmeier and McGuire, 2003), joint estimation of alignment and phylogeny (Redelings and Suchard, 2005), stochastic modeling of horizontal gene transfer (Suchard, 2005), and models that allow not only bifurcations but also polytomies (Lewis et al., 2005). While this flexibility has led to progressively more realistic phylogenetic models, a substantial part of the current popularity of Bayesian methods derives from its computational power: Bayesian inference coupled with Markov chain Monte Carlo provides uncertainty assessments simultaneously with the estimates of parameters, and thus allows analyses of large molecular sequences databases in a realistic time frame (Huelsenbeck and Ronquist, 2001; Ronquist and Huelsenbeck, 2003).

Nevertheless, Bayesian phylogenetic reconstruction has been criticized as providing unreliable posterior probabilities of trees (Suzuki et al., 2002; Simmons et al., 2004; Pickett and Randle, 2005). One of the most influential criticisms was delivered by Suzuki et al. (2002), who showed a paradoxical

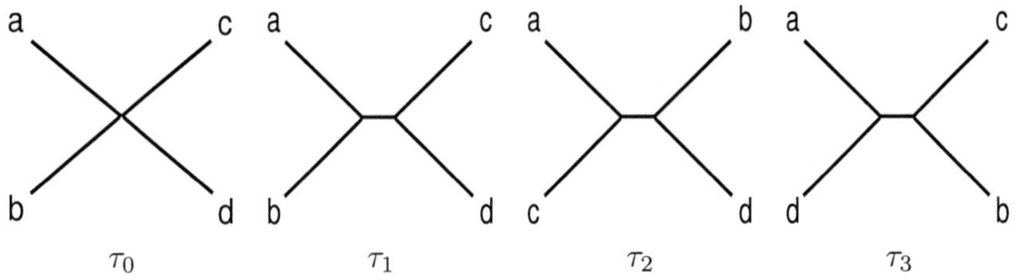

FIGURE 8.2
Possible unrooted topologies for 4-taxa DNA sequences. The star tree is denoted by τ_0 and the three possible unrooted bifurcating trees are denoted by τ_1, τ_2, and τ_3.

behavior of the posterior probabilities of trees that has been called by Lewis et al. (2005) the star tree paradox. Here, we show that the star tree paradox is an example of the MOMO paradox.

Suzuki et al. (2002) performed comparative studies between bootstrap support and Bayesian posterior probabilities for two types of simulated long molecular sequences, that we refer to as the mixture of topologies case and the star tree case. Suzuki et al. (2002) considered four taxa, their three possible unrooted bifurcating trees and the star tree. Figure 8.2 depicts the star tree, denoted by τ_0, and the three unrooted bifurcating trees, denoted by τ_1, τ_2, and τ_3. For the mixture of topologies case, they simulated molecular sequences from each of the unrooted bifurcating trees τ_1, τ_2 and τ_3, concatenated the sequences, and applied Bayesian phylogenetic reconstruction assuming one unique true bifurcating topology. For the star tree case, they simulated molecular sequences from the star tree τ_0, and applied Bayesian phylogenetic reconstruction assuming one unique true bifurcating topology τ_1, τ_2 or τ_3. Intuitively, they expected the Bayesian posterior probability for each topology to be about 1/3, and in that case the analysis would be inconclusive as to what was the true topology. Instead, they found that very often the posterior probability of one of the topologies would be fairly close to one, and thus the analysis would erroneously point to one of the bifurcating evolutionary histories as the truth.

The results found by Suzuki et al. (2002) spurred many researchers to further examine the credibility of Bayesian phylogenies using simulated and empirical data (Alfaro et al., 2003; Huelsenbeck and Rannala, 2004; Yang and Rannala, 2005). On one hand, Huelsenbeck and Rannala (2004) showed that Bayesian phylogenetic reconstruction is more sensitive to underspecification of the nucleotide substitution model than to overspecification of the model. On the other hand, Yang and Rannala (2005) showed that the bootstrap proportion is not always conservative, while Bayesian posterior probabilities may underestimate or overestimate clade support depending on the prior distribution of the internal branch lengths. While these studies have shown that

Bayesian methods do not intrinsically overestimate clade support, they have not explained the paradoxical behavior found by Suzuki et al. (2002), that is, that Bayesian phylogenetic reconstruction of bifurcating trees based on *long* molecular sequences actually simulated from a mixture of topologies or from a star tree very often erroneously point to one of the bifurcating evolutionary histories as having posterior probability close to one.

In this section, we show that for the conditions considered by Suzuki et al. (2002), in the limit when the length of the molecular sequence converges to infinity the Bayesian analysis tends to choose one of the bifurcating topologies as strongly supported by the data. We note that this paradoxical behavior is not an intrinsic limitation of Bayesian phylogenetic reconstruction but, instead, results from the not inclusion among the candidate models of an appropriate topological model to analyze the simulated data. That is, the conditions considered by Suzuki et al. (2002) lead to the MOMO paradox.

In the next two sections we explain the paradoxical behavior of Bayesian posterior probabilities for each of the two cases considered by Suzuki et al. (2002). Specifically, Section 8.2.2.1 provides an explanation of the paradox for the mixture of topologies case and Section 8.2.2.2 provides an explanation for the star tree case. For both the mixture of topologies case and the star tree case, the paradoxical behavior is related to the simulation of molecular sequences from a topological structure that is not included within the class of topological structures considered in the phylogenetic reconstruction. That is, the paradoxical behavior happens because of the misspecification of the class of topological models considered in the phylogenetic analysis.

8.2.2.1 Mixture of Topologies Case

To focus on the most relevant aspect of the paradox, which is the discrepancy between the topological structure used to simulate the data and the class of topological structures considered in the phylogenetic reconstruction, here we assume that the molecular substitution model and its parameters are known. When considering four taxa, there are three possible unrooted bifurcating topologies τ_1, τ_2, and τ_3 as depicted in Figure 8.2. The choice among one of these three possible trees can be seen as a problem of model selection where each bifurcating topology corresponds to a model. In this section, as the extension to comparisons among three topologies is straightforward, for ease of exposition we consider comparisons between two topologies τ_1 and τ_2.

The mixture of topologies case considered by Suzuki et al. (2002) is equivalent to assuming that two molecular sequences of length n are generated from a molecular substitution model, each sequence from a different topology τ_1 or τ_2. We assume that each nucleotide of each of these sequences follows a probability distribution $P(.|\tau)$ indexed by a parameter τ. Thus, the mixture of topologies case considered by Suzuki et al. (2002) is equivalent to assuming that two independent random samples of the same size n are generated from the distribution $P(.|\tau)$, each of them with a different topology τ:

$$x_1, \ldots, x_n|\tau = \tau_1 \overset{iid}{\sim} P(x|\tau = \tau_1) \tag{8.3}$$

$$y_1, \ldots, y_n|\tau = \tau_2 \overset{iid}{\sim} P(y|\tau = \tau_2). \tag{8.4}$$

Analogously to Suzuki et al. (2002), we concatenate the two sequences and (erroneusly) assume that the resulting concatenated sequence was generated from the molecular substitution model with a unique topology. That is, we concatenate the two samples and assume that the concatenated sample $(x_1, \ldots, x_n, y_1, \ldots, y_n)$ is a random sample from the distribution $P(.|\tau)$ with a unique value of τ. Thus, under this inappropriate assumption, the likelihood function of τ will be

$$p(\vec{x}_n, \vec{y}_n|\tau) = \prod_{i=1}^{n} p(x_i|\tau)p(y_i|\tau), \tag{8.5}$$

where $\vec{x}_n = (x_1, \ldots, x_n)$, $\vec{y}_n = (y_1, \ldots, y_n)$, and $p(.|\tau)$ is the probability function corresponding to the probability distribution $P(.|\tau)$.

Based on the concatenated sequence, we may make one of three possible decisions: (1) decide that the concatenated sequence was generated from τ_1, (2) decide that the concatenated sequence was generated from τ_2, or (3) decide that the topology is not resolved. Similarly to Suzuki et al. (2002), let us compare model $\mathcal{M}_1 : \tau = \tau_1$ versus model $\mathcal{M}_2 : \tau = \tau_2$, using the dataset simulated as in Equations (8.3) and (8.4), but with the misspecified model given by Equation (8.5). The posterior probability of model $\mathcal{M}_1 : \tau = \tau_1$ is:

$$p(\tau = \tau_1|\vec{x}_n, \vec{y}_n) = \frac{p(\vec{x}_n, \vec{y}_n|\tau = \tau_1)p(\tau = \tau_1)}{p(\vec{x}_n, \vec{y}_n|\tau = \tau_1)p(\tau = \tau_1) + p(\vec{x}_n, \vec{y}_n|\tau = \tau_2)p(\tau = \tau_2).}$$

Assuming as usual that the two models are equally likely a priori, that is $p(\tau = \tau_1) = p(\tau = \tau_2)$, the posterior odds ratio will be equal to the likelihood ratio:

$$\frac{p(\tau = \tau_1|\vec{x}_n, \vec{y}_n)}{p(\tau = \tau_2|\vec{x}_n, \vec{y}_n)} = \frac{p(\vec{x}_n, \vec{y}_n|\tau = \tau_1)}{p(\vec{x}_n, \vec{y}_n|\tau = \tau_2)}.$$

Note that we can rewrite the likelihood ratio as

$$\frac{p(\vec{x}_n, \vec{y}_n|\tau_1)}{p(\vec{x}_n, \vec{y}_n|\tau_2)} = \frac{\prod_{i=1}^{n} p(x_i|\tau_1)p(y_i|\tau_1)}{\prod_{i=1}^{n} p(x_i|\tau_2)p(y_i|\tau_2)} = \frac{\prod_{i=1}^{n} p(x_i|\tau_1)/p(x_i|\tau_2)}{\prod_{i=1}^{n} p(y_i|\tau_2)/p(y_i|\tau_1)}. \tag{8.6}$$

We can compute the log likelihood ratio sequentially, adding at each time two sites coming from each of the different topologies. Define l_n as the logarithm of the likelihood ratio after including n pairs of sites. Thus

$$\begin{aligned} l_n &= \log\left[\frac{p(\vec{x}_n, \vec{y}_n|\tau_1)}{p(\vec{x}_n, \vec{y}_n|\tau_2)}\right] \\ &= \sum_{i=1}^{n}\{[\log p(x_i|\tau_1) - \log p(x_i|\tau_2)] - [\log p(y_i|\tau_2) - \log p(y_i|\tau_1)]\}. \end{aligned}$$

Let g_i be the contribution of the ith pair of sites, that is,

$$g_i = \log\left[\frac{p(x_i|\tau_1)/p(x_i|\tau_2)}{p(y_i|\tau_2)/p(y_i|\tau_1)}\right]. \tag{8.7}$$

Note that, by symmetry, the expected value of the evidence given by x_i in favor of $\mathcal{M}_1$ against $\mathcal{M}_2$ is equal to the evidence given by y_i in favor of $\mathcal{M}_2$ against $\mathcal{M}_1$, that is

$$E\left[\log p(x_i|\tau_1) - \log p(x_i|\tau_2)\right] = E\left[\log p(y_i|\tau_2) - \log p(y_i|\tau_1)\right],$$

which implies that $E(g_n) = 0$. That is, because the n-th pair of sites is composed of one site coming from one topology and another site coming from the other topology, the contribution g_n of the pair has expected value 0.

Thus, we can write the logarithm of the likelihood ratio l_n in a recursive manner as

$$l_n = l_{n-1} + g_n,$$

where $l_0 = 0$ and $E(g_n) = 0$. Denote the variance of g_n by σ_g^2. Therefore, l_n is a random walk and by the Central Limit Theorem $l_n \overset{a}{\sim} N(0, n\sigma_g^2)$ for large n (e.g., see Feller, 1968, pg 357).

It is common practice to decide that the concatenated sequence was generated from a particular topology if the posterior probability of that topology is greater than a given threshold. In particular, Suzuki et al. (2002) assumed a threshold equal to 95%. If each of the posterior probabilities is smaller than the threshold, then we decide that the topology is not resolved. Note that the condition for deciding that the topology is not resolved can also be expressed in terms of the log posterior odds ratio, that is, the logarithm of the ratio between the posterior probabilities of each topology. Thus, one decides that the topology is not resolved if the log posterior odds ratio is in absolute value smaller than a given threshold.

The paradoxical behavior pointed out by Suzuki et al. (2002) is related to the behavior of the likelihood ratio given in Equation (8.6) when the sample size goes to infinity. Suzuki et al. (2002) assumed that the data would give the same evidence in favor of model $\mathcal{M}_1$ and model $\mathcal{M}_2$, that is, that as the sample size goes to infinity the likelihood ratio (8.6) should converge to 1. But as we show below this is not the case.

The choice between topology τ_1 and topology τ_2 may be framed within a decision theory framework (Parmigiani and Inoue, 2009). Such framework considers decisions, costs associated with those decisions that depend on the state of nature, and probabilities associated with each possible state of nature. Let us consider the following three possible decisions:

- d_1: decide for $\mathcal{M}_1$, that is, τ_1 is the true topology;
- d_2: decide for $\mathcal{M}_2$, that is, τ_2 is the true topology;
- d_3: decide that the topology is not resolved.

Assume that if decision d_1 is taken and τ_1 is the true topology then the loss is 0, while if τ_2 is the true topology then the loss is $c_2 > 0$. Moreover, if decision d_2 is taken and τ_2 is the true topology then the loss is 0, while if τ_1 is the true topology then the loss is c_2. That is, choosing the correct topology has cost 0 while choosing the incorrect topology has cost $c_2 > 0$. Finally, if decision d_3 is taken then the loss is $c_1 > 0$ regardless of which one is the true topology. Typically, c_1 is much smaller than c_2; this reflects the fact that deciding that the topology is not resolved and thus more research is necessary is less problematic than stating that a wrong topology is the true topology.

Define $\epsilon = c_1/c_2$. Then, it is easy to show that the resulting decision rule is of the type considered by Suzuki et al. (2002): if $P(\tau = \tau_1|\vec{x}_n, \vec{y}_n) > 1 - \epsilon$ then decide that there is evidence in favor of $\mathcal{M}_1$ against $\mathcal{M}_2$; if $P(\tau = \tau_2|\vec{x}_n, \vec{y}_n) > 1 - \epsilon$ then decide that there is evidence in favor of $\mathcal{M}_2$ against $\mathcal{M}_1$; if $P(\tau = \tau_1|\vec{x}_n, \vec{y}_n) \leq 1 - \epsilon$ and $P(\tau = \tau_2|\vec{x}_n, \vec{y}_n) \leq 1 - \epsilon$ then decide that the topology is not resolved.

Let us consider the probability p_n of the decision "the topology is not resolved". When n is large:

$$
\begin{aligned}
p_n &= P\left[\{P(\tau = \tau_1|\vec{x}_n, \vec{y}_n) \leq 1 - \epsilon\} \text{ and } \{P(\tau = \tau_2|\vec{x}_n, \vec{y}_n) \leq 1 - \epsilon\}\right] \\
&= P\left[|l_n| \leq \log(1-\epsilon) - \log\epsilon\right] \\
&\approx 2\left[\Phi\left(\frac{\log(1-\epsilon) - \log\epsilon}{\sigma_g\sqrt{n}}\right) - 0.5\right],
\end{aligned}
$$

where the last approximate equality holds by the Central Limit Theorem and $\Phi(.)$ is the standard Gaussian cumulative distribution function.

Therefore, as the sample size increases to infinity, the probability p_n of the decision "the topology is not resolved" converges to

$$\lim_{n\to\infty} p_n = 2\left[\Phi(0) - 0.5\right] = 0. \tag{8.8}$$

From this we conclude that as the sample size goes to infinity the probability of choosing one of the two topologies will converge to one. Note that not only a Bayesian test but also a likelihood ratio test would have the same undesirable behavior. That is, the probability of the log likelihood ratio being smaller than a pre-specified threshold converges to zero as the length n of the set of molecular sequences goes to infinity. That explains the result found by Suzuki et al. (2002) for the case of a mixture of topologies.

We note that the structure of the mixture of topologies case is somehow different from the structure of the toy example presented in Section 8.2.1. In the toy example, the observations were independently simulated from a mixture of two Gaussians. The simulation in the toy example was performed by first simulating for each observation a latent variable that indicated to which of the Gaussian components the observation would be simulated from. Thus, in the toy example the latent variables that indicate the Gaussian component are independent and the observations are exchangeable. In contrast, the mixture of topologies case considered in this section corresponds to partitioning the genetic sequence in three parts of equal length, and within each partition simulating independent observations from a distinct topology. Thus, in the mixture of topologies case we could think that there are latent variables that are constant within each partition but are different across partitions. Hence, in this case observations near each other tend to have been generated from the same topology. Therefore, in the mixture of topologies case the observations are not exchangeable.

Finally, we note that to eliminate the MOMO paradox in the mixture of topologies case, we would need to include in the Bayesian model selection analysis a more appropriate model. Such a model would need to incorporate the possibility of the observations to come from different topologies, as well as the tendency of observations near each other to have been generated from the same topology. As an example of a model that would allow for both possibilities, a researcher could consider a hidden Markov model where latent variables indicating topology would follow a Markov process and the genetic sequence would arise conditional on these latent variables (Frühwirth-Schnatter, 2006; Cui et al., 2015).

8.2.2.2 Star Tree Case

The explanation of the MOMO paradox in the star tree case and in the toy example presented in Section 8.2.1 are similar. This is because, as in the toy example, the observations in the star tree case are exchangeable. As a result, in both examples, the log likelihood ratio computed as we accumulate observations will be a random walk. These examples are somehow different from the mixture of topologies case. Specifically, to show the MOMO paradox in the mixture of topologies case (for two topologies, as we considered in the previous section) we need to consider the log likelihood ratio computed as we accumulate joint observations from each topology. As we have seen in the previous section, this also leads to the resulting log likelihood ratio following a random walk in the mixture of topologies case.

Let us now consider the star tree case. In this case, Suzuki et al. (2002) simulated molecular sequences from the star tree τ_0 and applied Bayesian phylogenetic reconstruction assuming that the true topology was one of the bifurcating topology τ_1, τ_2, or τ_3. Let x_i be the nucleotide at site i, with probability function $p(x_i|\tau)$, indexed by the topology τ. Then

$$x_1, \ldots, x_n|\tau = \tau_0 \overset{iid}{\sim} P(x|\tau = \tau_0).$$

Based on this sample, Suzuki et al. (2002) compared the models $\mathcal{M}_1 : \tau = \tau_1$, $\mathcal{M}_2 : \tau = \tau_2$, and $\mathcal{M}_3 : \tau = \tau_3$. As in Suzuki et al. (2002), let us erroneously assume that the true topology τ belongs to the set of bifurcating topologies $\{\tau_1, \tau_2, \tau_3\}$ and let us assign uniform prior for τ in this set, that is, $P(\tau = \tau_k) = 1/3, k = 1, 2, 3$.

By symmetry, the Kullback-Leibler divergence between the true model with $\tau = \tau_0$ and each of the models $\mathcal{M}_1$, $\mathcal{M}_2$, and $\mathcal{M}_3$ is the same. More precisely, when $\tau = \tau_0$ the Kullback-Leibler divergence between $p(x_i|\tau = \tau_k)$ and $p(x_i|\tau = \tau_0)$ is constant for $k = 1, 2, 3$, that is,

$$E\left[\log \frac{p(x_i|\tau = \tau_1)}{p(x_i|\tau = \tau_0)}\right] = E\left[\log \frac{p(x_i|\tau = \tau_2)}{p(x_i|\tau = \tau_0)}\right] = E\left[\log \frac{p(x_i|\tau = \tau_3)}{p(x_i|\tau = \tau_0)}\right], \quad (8.9)$$

where the expectations are taken with respect to $p(x_i|\tau = \tau_0)$. Therefore, models $\mathcal{M}_1$, $\mathcal{M}_2$, and $\mathcal{M}_3$ are in the asymptotic carrier.

The likelihood function for a topology τ will be $L(\tau|\vec{x}_n) = \prod_{i=1}^{n} p(x_i|\tau)$. As a uniform prior is assumed for τ in the set of bifurcating topologies and the parameters of the molecular substitution model are assumed to be known, the ratio of posterior probabilities is equal to the likelihood ratio. Thus, let us consider a bivariate vector that has as elements the log likelihood ratio of topology τ_2 versus topology τ_1 and the log likelihood ratio of topology τ_3 versus topology τ_1. Thus, the vector of log likelihood ratios is

$$\boldsymbol{l}_n = \begin{pmatrix} \sum_{i=1}^{n} [\log p(x_i|\tau = \tau_2) - \log p(x_i|\tau = \tau_1)] \\ \sum_{i=1}^{n} [\log p(x_i|\tau = \tau_3) - \log p(x_i|\tau = \tau_1)] \end{pmatrix}.$$

Consider the vector of contributions of the nth observation to the log likelihood ratio

$$\boldsymbol{h}_n = \begin{pmatrix} \log p(x_i|\tau = \tau_2) - \log p(x_i|\tau = \tau_1) \\ \log p(x_i|\tau = \tau_3) - \log p(x_i|\tau = \tau_1) \end{pmatrix}.$$

Using the result in Equation (8.9), it is easy to show that $E[\boldsymbol{h}_n] = (0, 0)'$. Thus, we can rewrite the log likelihood ratio recursively as the bivariate random walk $\boldsymbol{l}_n = \boldsymbol{l}_{n-1} + \boldsymbol{h}_n$. Therefore, as the sample size increases to infinity, the probability of the decision "the topology is not resolved" converges to zero. This undesirable behavior of the log likelihood ratio and of the corresponding posterior probabilities in the star tree case is an example of the MOMO paradox.

8.2.3 Wavelets Example

In this section, we present an example of the MOMO paradox in the context of wavelet nonparametric regression. Wavelets have been successfully used in time series analysis (Percival and Walden, 2006), image processing (Starck et al., 2015), and analysis of functional magnetic resonance imaging data (Sanyal and Ferreira, 2012, 2017). First, we simulate a regression function at $n = 2^L$ equally spaced points on the interval $[0, 1]$ with a periodic boundary condition using a Gaussian process with Matérn covariance function. After that, we simulate observations at the $n = 2^L$ equally spaced points adding iid Gaussian white noise to the true regression function. Then, we fit wavelet nonparametric regression to the observations. Because of the periodic boundary condition, we can shift (translate) the wavelet basis in 2^L different ways. Thus, we have 2^L different wavelet models, one for each possible shift.

It may be tempting to compute the posterior probability for each of these wavelet models, and to estimate the true function using Bayesian model averaging (BMA). However, because the true function follows a continuous Gaussian process, the Kullback-Leibler divergence between each of the 2^L wavelet models and the true model is the same. As a result, all 2^L wavelet models are in the asymptotic carrier. That is, the asymptotic carrier contains the same number of models as the number of observations. Therefore, in this example the asymptotic carrier grows with the number of observations.

Here, we assume observations y_i at equally spaced points $t_i = (i-1)2^{-L}$, $i = 1, \ldots, n$, where we recall that $n = 2^L$. We consider a periodic true function $f(.)$ on the interval $(0, 1)$ arising from a Gaussian process. Denote the value of f at t_i by $f_i = f(t_i)$, $i = 1, \ldots, n$. Then, we simulate the observations from the model

$$y_i = f_i + \epsilon_i,$$

where $\boldsymbol{\epsilon} = (\epsilon_1, \ldots, \epsilon_n)' \sim N(\mathbf{0}, \tau^2 \boldsymbol{I})$ is a vector of measurement errors, and τ^2 is the measurement error variance, also called the nugget in the spatial statistics literature. Here, $\boldsymbol{f} = (f_1, \ldots, f_n)'$ follows a Gaussian process with Matérn covariance function (Banerjee et al., 2014)

$$C(h) = \frac{\sigma^2}{2^{\kappa-1}\Gamma(\kappa)}(\phi h)^{\kappa} K_{\kappa}(\phi h),$$

where K_κ is the modified Bessel function of order κ, ϕ is a spatial decay parameter, κ controls the smoothness of the Gaussian process, and σ^2 is the partial sill. Figure 8.3 presents observations simulated with $L = 8$, $n = 256$, partial sill $\sigma^2 = 5$, smoothness parameter $\kappa = 5$, spatial decay parameter $\phi = 1/60$, and nugget $\tau^2 = 1.5$.

In practice we would not know what type of process generated the true regression function. Thus, here we estimate the function $f(.)$ with wavelet nonparametric regression. In this case, the regression function is written in terms of a wavelet basis. In such wavelet analysis f_i, $i = 1, \ldots, n$, is assumed to be written in terms of a discrete wavelet expansion (e.g., see Ferreira and Lee, 2007)

$$f_i = c_0 \phi_i + \sum_{l=0}^{L-1} \sum_{k=0}^{2^l - 1} d_{lk} \psi_{lki},$$

where ϕ_i is the value of a discretized scaling function computed at t_i and ψ_{lki} is the value of a discretized wavelet function at scale l and location k computed at t_i. See Vidakovic (1999) and Ferreira and Lee (2007) for details.

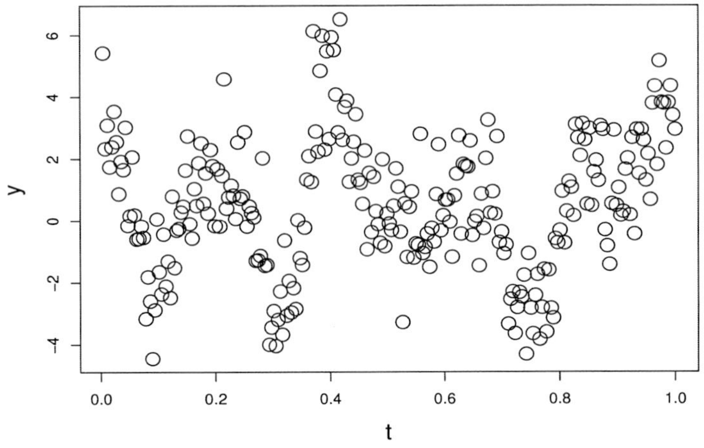

FIGURE 8.3
Wavelets example – Observed data.

Let $\boldsymbol{y} = (y_1, \ldots, y_n)'$ be the vector of observations, $\boldsymbol{d} = (c_0, d_{00}, d_{10}, d_{11}, \ldots, d_{L-1,2^{L-1}-1})$ be the vector of wavelet coefficients, and

$$\boldsymbol{W} = \begin{bmatrix} \phi_1 & \psi_{001} & \psi_{101} & \psi_{111} & \cdots & \psi_{L-1,2^{L-1}-1,1} \\ \phi_2 & \psi_{002} & \psi_{102} & \psi_{112} & \cdots & \psi_{L-1,2^{L-1}-1,2} \\ \vdots & \vdots & \vdots & \vdots & & \vdots \\ \phi_n & \psi_{00n} & \psi_{10n} & \psi_{11n} & \cdots & \psi_{L-1,2^{L-1}-1,n} \end{bmatrix}$$

be the orthogonal matrix that contains the discretized scaling and wavelet functions computed at $t_1, \ldots, t_n$. Then, we write the wavelet regression model in matrix notation as

$$\boldsymbol{y} = \boldsymbol{W}\boldsymbol{d} + \boldsymbol{\epsilon},$$

where $\boldsymbol{\epsilon} \sim N(0, \tau^2)$.

Because $\boldsymbol{W}$ is orthogonal, the least squares estimate of $\boldsymbol{d}$, known in the wavelets literature as the vector of empirical wavelet coefficients, may be computed as $\widehat{\boldsymbol{d}} = \boldsymbol{W}'\boldsymbol{y}$ with $O(n^2)$ operations. However, in practice $\widehat{\boldsymbol{d}}$ is computed with the discrete wavelet transform (DWT) implemented with a fast cascade algorithm that costs $O(n)$ operations (Mallat, 1999). Here, we compute $\widehat{\boldsymbol{d}}$ with the DWT implemented in the R package wavethresh (Nason, 2022).

Thus, the wavelet regression model may be rewritten in terms of the vector of empirical wavelet coefficients as

$$\widehat{\boldsymbol{d}} = \boldsymbol{d} + \boldsymbol{\epsilon}^*,$$

where, because $\boldsymbol{W}$ is orthogonal, $\boldsymbol{\epsilon}^* \sim N(\boldsymbol{0}, \tau^2\boldsymbol{I})$.

As is typical in Bayesian wavelet analysis, we assume for each wavelet coefficient the spike-and-slab prior (Abramovich et al., 1998; Clyde et al., 1998; Ferreira and Lee, 2007).

$$d_{lk} \sim w_l N(0, c_l \tau^2) + (1 - w_l)\delta(0),$$

where w_l is the prior probability that a wavelet coefficient at level l is different than zero, $c_l \tau^2$ is the prior variance of a wavelet coefficient at level l, and $\delta(0)$ is the Dirac delta function that represents a point mass at zero. Similarly to Abramovich et al. (1998), we assume that $c_l = 2^{-\alpha l} C_1$ and $w_l = \min(2^{-\beta l} C_2, 1)$, $l = 1, \ldots, L-1$. We estimate $(\alpha, \beta, C_1, C_2)$ with maximum likelihood estimation. After that, we take an empirical Bayes approach that fixes these parameters at their MLEs to compute estimates of the wavelet coefficients.

The presentation up to now has assumed one original wavelet basis on the interval $(0, 1)$. But because of the periodic boundary condition, we can shift (translate) the wavelet basis in $n = 2^L$ different ways. Thus, we have $n = 2^L$ different wavelet models, one for each possible shifted basis. Fitting a wavelet regression model to the observation vector $\boldsymbol{y}$ with a periodic boundary condition and with a wavelet basis shifted to the right by $h2^{-L}$, $h = 0, \ldots, 2^L - 1$, is equivalent to fitting such a model with the original wavelet basis to a shifted observation vector $\boldsymbol{P}^h \boldsymbol{y}$ where h is the shift and $\boldsymbol{P}$ is the $n \times n$ permutation matrix

$$\boldsymbol{P} = \begin{bmatrix} 0 & 1 & 0 & 0 & \ldots & 0 & 0 \\ 0 & 0 & 1 & 0 & \ldots & 0 & 0 \\ \vdots & \vdots & \vdots & \vdots & & \vdots & \vdots \\ 0 & 0 & 0 & 0 & \ldots & 1 & 0 \\ 0 & 0 & 0 & 0 & \ldots & 0 & 1 \\ 1 & 0 & 0 & 0 & \ldots & 0 & 0 \end{bmatrix}.$$

Let us denote by $\mathcal{M}_h$ the wavelet regression model with the wavelet basis shifted to the right by $h2^{-L}$. We assume a uniform prior such that $P(\mathcal{M}_h) = 2^{-L}$, $h = 0, \ldots, 2^L - 1$. To fit each of these wavelet regression models, we use Daubechies least asymmetric wavelets with six vanishing moments (Daubechies, 1992).

Because the number of unknown parameters is the same for all the considered wavelet models, we compute posterior model probabilities with an empirical Bayes approach. First, we estimate the variance of the measurement error with the median absolute deviation estimator (Donoho and Johnstone, 1995; Donoho et al., 1995) given by $\widehat{\tau}^2 = \{\text{median}_j(|\hat{d}_{L-1,j}|)/0.6745\}^2$. Second, we estimate the parameters of the spike-and-slab prior with maximum likelihood estimation. Finally, we assume that these hyperparameters are fixed at their estimates when we compute the posterior probabilities of the different wavelet models.

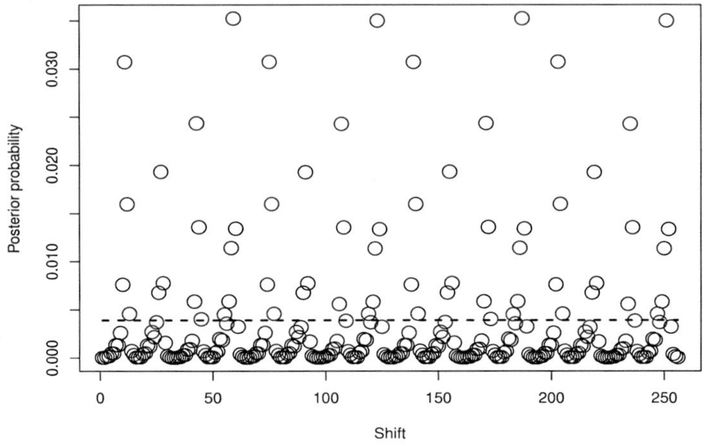

FIGURE 8.4
Wavelet example – Posterior probabilities of different shifted wavelet bases (circles) and horizontal line at 1/256 (dashed line).

Figure 8.4 presents the posterior probabilities of the $2^8 = 256$ wavelet regression models based on different shifted wavelet bases. For reference, a horizontal dashed line indicates the value 1/256, which would be the probability of all shifted wavelet bases if the posterior distribution on the model space was uniform. Thus, we can see from the figure that the posterior distribution is considerably far from a uniform distribution. In particular, 21.5% of the models have more than 80% of the posterior mass. Hence, the posterior distribution concentrates on a much smaller number of models than the 256 possible models. In addition, we note that these 256 possible models are equally distant from the true model. Thus, this is again an example of the MOMO paradox where the posterior distribution is far from the uniform distribution.

To estimate the true function, Bayesian model averaging would compute a weighted average of the function estimates based on the 256 shifted basis models using the posterior probabilities as weights. But because of the MOMO paradox, for the example presented in this section the posterior probabilities are not a reliable representation of uncertainty. Alternatively, we may consider translation-invariant denoising (Coifman and Donoho, 1995), that is, we may use an ensemble estimate of the true function computed as an equal-weight average of the function estimates based on the 256 shifted basis models.

Figure 8.5 presents the true function (solid line), the Bayesian model average (BMA) estimate based on the posterior probabilities (dotted line), and the ensemble estimate based on an equal-weight average (dashed line). Both the

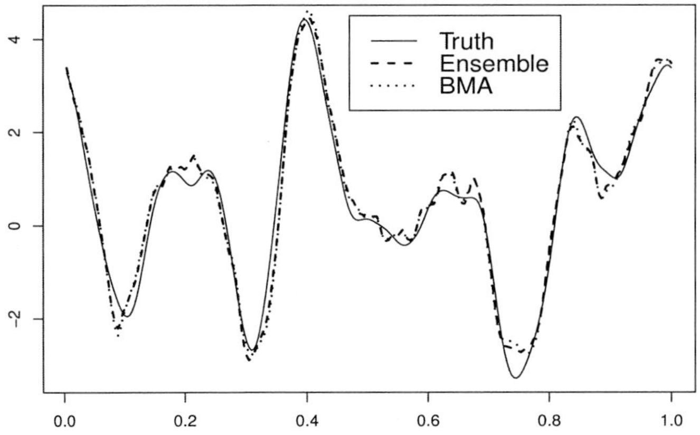

FIGURE 8.5
Wavelet example – True function (solid line), ensemble estimate (dashed line), and Bayesian model averaging estimate (dotted line).

BMA and the ensemble estimates perform well at estimating the true function. This good performance is a consequence of the good properties of wavelet nonparametric regression. To evaluate which of the two estimates perform best, we compare their mean squared errors (MSEs). While the MSE of the BMA estimate is 0.108, the MSE of the ensemble estimate is 0.093 thus providing a reduction of 13.6% in MSE. Therefore, in this wavelet nonparametric regression setting with presence of the MOMO paradox, instead of BMA, it is more advantageous to use the equal-weight ensemble estimate.

This example of the MOMO paradox in wavelet nonparametric regression is distinct from the examples presented in Sections 8.2.1 and 8.2.2. In those sections, the number of possible models was fixed and the number of observations was large. In contrast, in the wavelet nonparametric regression example presented in this section, the number of possible models increases with sample size. An interesting possible future research direction would be a theoretical investigation of the asymptotic properties of the posterior model probabilities in the wavelet nonparametric regression with possibly shifted wavelet basis.

8.3 Discussion

We have presented four examples of the M-open multiple optima paradox, the MOMO paradox. Specifically, we consider Bayesian model selection in the M-open case when the set of models being entertained does not include the true model. In that case, asymptotically the posterior distribution concentrates on a set of models called the asymptotic carrier. If the asymptotic carrier has only one model, then the posterior probability of that model converges to one. However, if the asymptotic carrier has more than one model, then the situation is not as clear. In the latter case, a scientist would wish for the posterior distribution to reflect the existing uncertainty and converge to a uniform distribution on the models in the asymptotic carrier. Unfortunately, that is not the case for the set of examples that we consider. On the contrary, for large sample sizes the posterior distribution on the models in the asymptotic carrier tends to be far from a uniform distribution. We call this behavior the MOMO paradox.

The researcher should consider whether the MOMO paradox may be present or not in their problem. For example, consider the case when the researcher believes that a spatio-temporal process under study is smooth, but because of computational speed the researcher is considering dynamic multiscale spatio-temporal models (DMSTM) (Ferreira et al., 2010, 2011; Hoegh et al., 2016; Fonseca and Ferreira, 2017; Ferreira, 2021; Elkhouly and Ferreira, 2021) with unknown multiscale structure (Fonseca and Ferreira, 2024). Because DMSTMs have spatial covariance functions that are step functions – thus not smooth – the researcher may want to consider fitting DMSTMs with all possible multiscale structures and use some sort of averaging of model fits and predictions. Because the researcher believes the process under study is smooth, by construction none of the DMSTMs are going to be the true model and there is a substantial chance that multiple DMSTMs may be in the asymptotic carrier. Thus, in this case the MOMO paradox may be present. Thus, the example presented in Section 8.2.3 suggests that, instead of Bayesian model averaging, it may be more advantageous to use ensembles computed with equal-weight averages as proposed in Chapter 9 (Fonseca and Ferreira, 2024).

Despite the MOMO paradox, we think that general Bayesian model selection based on posterior probabilities is useful and important for the advancement of science. We share the point of view espoused by Rouder et al. (2009) that discoveries of invariances frequently lead to advances in science. Important examples of invariances are Kepler's laws of planetary motion, which inspired Newton to discover his Theory of Mechanics (Hawking, 2002). As beautifully explained by Rouder et al. (2009), invariances may often be described in mathematical terms as null models and, unlike frequentist statistics, Bayesian model selection allows for the data to provide evidence in favor of null models. Therefore, Bayesian model selection permits researchers to discover invariances in their scientific fields.

An example where Bayesian model selection has been successful in finding invariances is in genome-wide association studies (GWAS) (Williams et al., 2022, 2023; Xu et al., 2023). In GWAS, the objective is to find genetic variants that may influence phenotypes of interest such as the occurrence of cancer, yield of crops, resilience of crops to salinity, and propensity to addiction. Thus, invariance of a phenotype to a specific genetic variant means that, when considering all other genetic variants, the probability distribution of the phenotype does not change for different variations of the specific genetic variant. In addition, GWAS data analysis is usually performed with linear mixed models (LMM) (Yu et al., 2006; Kang et al., 2008, 2010). In the LMM framework, testing for invariance of a phenotype to a genetic variant is equivalent to testing if the regression coefficient of the genetic variant is equal to zero (corresponding to invariance) or not. One of the main aspects of GWAS is that the number of genetic variants is in the order of 100,000s to 1,000,000s while the sample size is in the order of 1,000s to 10,000s. Importantly, usually this is a situation where sparsity prevails (Dobra et al., 2004; McCullagh and Polson, 2018) in that the phenotype is invariant to most of the genetic variants. In this context, Bayesian model selection and Bayesian false discovery control (Müller et al., 2007) are useful, where a hierarchical model specification helps to find the relatively few genetic variants that influence the phenotype (Williams et al., 2022, 2023; Xu et al., 2023).

A fair question is whether invariances and models are the truth. As G. E. P. Box famously wrote "All models are wrong, some are useful" (Box, 1979). Thus, models are not the truth, but models may capture important aspects of the truth. In addition, we know that if the models being considered are reasonable approximations and the MOMO paradox is not present, then the asymptotic carrier will have only one model that is going to be the closest to the truth. And all we really need is for that one model to be useful to solve the problems at hand. For example, we know from Einstein's Theory of Relativity that Newton's Theory of Mechanics is not the truth but an approximation. However, despite being an approximation, calculations based on Newton's Theory of Mechanics have allowed humanity to send humans to the moon.

Bibliography

Abramovich, F., T. Sapatinas, and B. W. Silverman (1998). Wavelet thresholding via a Bayesian approach. *Journal of the Royal Statistical Society 60*, 725–749.

Alfaro, M. E., S. Zoller, and F. Lutzoni (2003). Bayes or bootstrap? A simulation study comparing the performance of Bayesian Markov chain Monte Carlo sampling and bootstrapping in assessing phylogenetic confidence. *Molecular Biology and Evolution 20*, 255–266.

Alteri, C., V. Cento, A. Piralla, V. Costabile, M. Tallarita, L. Colagrossi, S. Renica, F. Giardina, F. Novazzi, S. Gaiarsa, et al. (2021). Genomic epidemiology of SARS-CoV-2 reveals multiple lineages and early spread of SARS-CoV-2 infections in Lombardy, Italy. *Nature Communications 12*(1), 434.

Banerjee, S., B. P. Carlin, and A. E. Gelfand (2014). *Hierarchical Modeling and Analysis for Spatial Data* (2nd ed.). Boca Raton, FL: CRC Press.

Berk, R. H. (1966). Limiting behavior of posterior distributions when the model is incorrect. *The Annals of Mathematical Statistics 37*, 51–58.

Bouckaert, R., T. G. Vaughan, J. Barido-Sottani, S. Duchêne, M. Fourment, A. Gavryushkina, J. Heled, G. Jones, D. Kühnert, N. De Maio, et al. (2019). BEAST 2.5: An advanced software platform for Bayesian evolutionary analysis. *PLoS Computational Biology 15*(4), e1006650.

Box, G. E. P. (1979). Robustness in the strategy of scientific model building. In R. L. Launer and G. N. Wilkinson (Eds.), *Robustness in Statistics*, pp. 201–236. Elsevier.

Candido, D. S., I. M. Claro, J. G. De Jesus, W. M. Souza, F. R. Moreira, S. Dellicour, T. A. Mellan, L. Du Plessis, R. H. Pereira, F. C. Sales, et al. (2020). Evolution and epidemic spread of SARS-CoV-2 in Brazil. *Science 369*(6508), 1255–1260.

Clyde, M., G. Parmigiani, and B. Vidakovic (1998). Multiple shrinkage and subset selection in wavelets. *Biometrika 85*, 391–402.

Coifman, R. R. and D. L. Donoho (1995). Translation-invariant de-noising. In A. Antoniadis and G. Oppenheim (Eds.), *Wavelets and Statistics*, pp. 125–150. New York, NY: Springer New York.

Cui, S., S. Guha, M. A. R. Ferreira, and A. N. Tegge (2015). hmmseq: A hidden Markov model for detecting differentially expressed genes from RNA-seq data. *The Annals of Applied Statistics 9*(2), 901–925.

Daubechies, I. (1992). *Ten Lectures on Wavelets.* Philadelphia, PA: SIAM.

Dobra, A., C. Hans, B. Jones, J. R. Nevins, G. Yao, and M. West (2004). Sparse graphical models for exploring gene expression data. *Journal of Multivariate Analysis 90*(1), 196–212.

Donoho, D. L. and I. M. Johnstone (1995). Adapting to unknown smoothness via wavelet shrinkage. *Journal of the American Statistical Association 90*, 1200–1224.

Donoho, D. L., I. M. Johnstone, G. Kerryacharian, and D. Picard (1995). Wavelet shrinkage: Asymptopia? *Journal of the Royal Statistical Society, Series B 57*, 301–369.

Du Plessis, L., J. T. McCrone, A. E. Zarebski, V. Hill, C. Ruis, B. Gutierrez, J. Raghwani, J. Ashworth, R. Colquhoun, T. R. Connor, et al. (2021). Establishment and lineage dynamics of the SARS-CoV-2 epidemic in the UK. *Science 371*(6530), 708–712.

Elkhouly, M. and M. A. R. Ferreira (2021). Dynamic multiscale spatiotemporal models for multivariate Gaussian data. *Spatial Statistics 41*, 100475.

Feller, W. (1968). *An Introduction to Probability Theory and its Applications*, Volume 1, 3rd ed. Wiley.

Ferreira, M. A. R. (2021). Bayesian spatial and spatiotemporal models based on multiscale factorizations. *Wiley Interdisciplinary Reviews: Computational Statistics 13*(2), e1509.

Ferreira, M. A. R., A. I. Bertolde, and S. Holan (2010). Analysis of economic data with multi-scale spatio-temporal models. In A. O'Hagan and M. West (Eds.), *Handbook of Applied Bayesian Analysis*, pp. 295–318. Oxford University Press.

Ferreira, M. A. R., S. H. Holan, and A. I. Bertolde (2011). Dynamic multiscale spatio-temporal models for Gaussian areal data. *Journal of the Royal Statistical Society (Series B: Methodological) 73*, 663–688.

Ferreira, M. A. R. and H. K. H. Lee (2007). *Multiscale Modeling: A Bayesian Perspective.* Springer Series in Statistics. New York: Springer.

Fonseca, T. C. O. and M. A. R. Ferreira (2017). Dynamic multiscale spatiotemporal models for Poisson data. *Journal of the American Statistical Association 112*, 215–234.

Fonseca, T. C. O. and M. A. R. Ferreira (2024). Ensembles of dynamic multiscale spatio-temporal models. In M. A. R. Ferreira (Ed.), *Modeling Spatio-Temporal Data: Markov Random Fields, Objective Bayes, and Multiscale Models*, pp. 221–245. Chapman & Hall/CRC.

Frühwirth-Schnatter, S. (2006). *Finite mixture and Markov switching models.* Springer.

Hawking, S. (2002). *On the Shoulders of Giants: The Great Works of Physics and Astronomy.* Philadelphia: Running Press.

Hoegh, A., M. A. R. Ferreira, and S. Leman (2016). Spatiotemporal model fusion: multiscale modelling of civil unrest. *Journal of the Royal Statistical Society: Series C (Applied Statistics) 65*, 529–545.

Huelsenbeck, J. P. and B. Rannala (2004). Frequentist properties of Bayesian posterior probabilities of phylogenetic trees under simple and complex substitution models. *Systematic Biology 53*, 904–913.

Huelsenbeck, J. P. and F. Ronquist (2001). MrBayes: Bayesian inference of phylogenetic trees. *Bioinformatics 17*, 754–755.

Husmeier, D. and G. McGuire (2003). Detecting recombination in 4-taxa DNA sequence alignments with Bayesian hidden Markov models and Markov chain Monte Carlo. *Molecular Biology and Evolution 20*, 315–337.

Hypólito, E. B. (2005). Uma resposta Bayesiana ao paradoxo de Suzuki (in Portuguese). Master's thesis, Departamento de Métodos Estatísticos, Universidade Federal do Rio de Janeiro.

Kang, H. M., J. H. Sul, S. K. Service, N. A. Zaitlen, S. Y. Kong, N. B. Freimer, C. Sabatti, and E. Eskin (2010). Variance component model to account for sample structure in genome-wide association studies. *Nature Genetics 42*(4), 348–354.

Kang, H. M., N. A. Zaitlen, C. M. Wade, A. Kirby, D. Heckerman, M. J. Daly, and E. Eskin (2008). Efficient control of population structure in model organism association mapping. *Genetics 178*(3), 1709–1723.

Kass, R. E. and A. E. Raftery (1995). Bayes factors. *Journal of the American Statistical Association 90*, 773–795.

Lewis, P. O., M. T. Holder, and K. E. Holsinger (2005). Polytomies and Bayesian phylogenetic inference. *Systematic Biology 54*, 241–253.

Li, M. and D. B. Dunson (2020). Comparing and weighting imperfect models using d-probabilities. *Journal of the American Statistical Association 115*(531), 1349–1360.

Mallat, S. G. (1999). *A Wavelet Tour of Signal Processing* (2nd ed.). San Diego: Academic Press.

McCullagh, P. and N. G. Polson (2018, 10). Statistical sparsity. *Biometrika 105*(4), 797–814.

Müller, P., G. Parmigiani, and K. Rice (2007). FDR and Bayesian multiple comparisons rules. In J. M. Bernardo, M. J. Bayarri, J. O. Berger, A. Dawid, D. Heckerman, A. F. M. Smith, and M. West (Eds.), *Bayesian Statistics 8*, Oxford, pp. 349–370. Oxford Univ. Press.

Nason, G. (2022). *wavethresh: Wavelets Statistics and Transforms.* R package version 4.7.2.

Parmigiani, G. and L. Inoue (2009). *Decision Theory: Principles and Approaches.* John Wiley & Sons.

Percival, D. B. and A. T. Walden (2006). *Wavelet Methods for Time Series Analysis.* New York: Cambridge University Press.

Pickett, K. M. and C. P. Randle (2005). Strange Bayes indeed: uniform topological priors imply non-uniform clade priors. *Molecular Phylogenetics and Evolution 34*, 203–211.

Rambaut, A., A. J. Drummond, D. Xie, G. Baele, and M. A. Suchard (2018). Posterior summarization in Bayesian phylogenetics using Tracer 1.7. *Systematic Biology 67*(5), 901–904.

Redelings, B. D. and M. A. Suchard (2005). Joint Bayesian estimation of alignment and phylogeny. *Systematic Biology 54*, 401–418.

Ronquist, F. and J. P. Huelsenbeck (2003). MrBayes3: Bayesian phylogenetic inference under mixed models. *Bioinformatics 19*, 1572–1574.

Rouder, J. N., P. L. Speckman, D. Sun, R. D. Morey, and G. Iverson (2009). Bayesian t tests for accepting and rejecting the null hypothesis. *Psychonomic Bulletin & Review 16*, 225–237.

Sanyal, N. and M. A. R. Ferreira (2012). Bayesian hierarchical multi-subject multiscale analysis of functional MRI data. *NeuroImage 63*, 1519–1531.

Sanyal, N. and M. A. R. Ferreira (2017). Bayesian wavelet analysis with nonlocal priors with an application to fMRI data. *Sankhya – Series B 79*, 361–388.

Simmons, M. P., K. M. Pickett, and M. Miya (2004). How meaningful are Bayesian support values? *Molecular Biology and Evolution 21*, 188–199.

Starck, J.-L., F. Murtagh, and J. Fadili (2015). *Sparse Image and Signal Processing: Wavelets and Related Geometric Multiscale Analysis* (2nd ed.). Cambridge university press.

Suchard, M. A. (2005). Stochastic models for horizontal gene transfer: taking a random walk through tree space. *Genetics 170*, 419–431.

Suzuki, Y., G. V. Glazko, and M. Nei (2002). Overcredibility of molecular phylogenies obtained by Bayesian phylogenetics. *Proceedings of the National Academy of Science 99*, 16138–16143.

Vidakovic, B. (1999). *Statistical Modeling by Wavelets.* Wiley.

Wiley, E. O. and B. S. Lieberman (2011). *Phylogenetics: theory and practice of phylogenetic systematics.* John Wiley & Sons.

Williams, J., M. A. R. Ferreira, and T. Ji (2022). BICOSS: Bayesian iterative conditional stochastic search for GWAS. *BMC Bioinformatics 23*(1), 1–14.

Williams, J., S. Xu, and M. A. R. Ferreira (2023). BGWAS: Bayesian variable selection in linear mixed models with nonlocal priors for genome-wide association studies. *BMC Bioinformatics 24*(1), 1–20.

Xu, S., J. Williams, and M. A. R. Ferreira (2023). BG2: Bayesian variable selection in generalized linear mixed models with nonlocal priors for non-Gaussian GWAS data. *BMC Bioinformatics 24*(1), 343.

Yang, Z. and B. Rannala (2005). Branch-length prior influences Bayesian posterior probability of phylogeny. *Systematic Biology 54*, 455–470.

Yao, Y., A. Vehtari, D. Simpson, and A. Gelman (2018). Using stacking to average Bayesian predictive distributions (with discussion). *Bayesian Analysis 13*, 917–1003.

Yu, J., G. Pressoir, W. H. Briggs, I. V. Bi, M. Yamasaki, J. F. Doebley, M. D. McMullen, B. S. Gaut, D. M. Nielsen, J. B. Holland, S. Kresovich, and E. S. Buckler (2006). A unified mixed-model method for association mapping that accounts for multiple levels of relatedness. *Nature Genetics 38*(2), 203–208.

9

Ensembles of Dynamic Multiscale Spatio-Temporal Models

Thais C. O. Fonseca

Universidade Federal do Rio de Janeiro

Marco A. R. Ferreira

Virginia Tech

We propose ensembles of dynamic multiscale spatio-temporal models (DM-STMs) for univariate Gaussian data. Even though DMSTMs allow fast, scalable, and parallelizable computations, the resulting estimates of the latent processes may be spatially blocky. However, in practice many spatio-temporal processes are spatially smooth. To obtain smoother estimates of the latent processes, our ensemble approach considers mixtures of DMSTMs defined on spatially shifted versions of the multiscale partition. However, this is a case that suffers from the M-open multiple optima paradox presented in Chapter 8 (Ferreira et al., 2024). As a consequence, model posterior probabilities tend to incorrectly concentrate on a few DMSTMs of the ensemble and are not reliable weights for Bayesian model averaging. To deal with this issue, we set the model prior probabilities to be the same for all the DMSTMs in the ensemble, and do not update them. In an analysis of data simulated from a spatio-temporal model with a smooth spatial covariance function, when compared to any of the individual DMSTMs in terms of estimating the underlying latent process, the ensemble estimate has a much smaller mean squared error. Finally, an ensemble analysis of temperature data demonstrates the usefulness and flexibility of our ensemble of DMSTMs approach.

DOI: 10.1201/9781032623443-9

9.1 Introduction

We propose ensembles of dynamic multiscale spatio-temporal models (DMSTMs) for univariate Gaussian data. DMSTMs were first developed by Ferreira et al. (2010, 2011). Since then, DMSTMs have been developed for Poisson data (Hoegh et al., 2016; Fonseca and Ferreira, 2017) and for multivariate Gaussian data (Elkhouly and Ferreira, 2021). DMSTMs have been successfully used in several areas of application such as economic analysis (Ferreira et al., 2010), climate science (Fonseca and Ferreira, 2017), sociology (Hoegh et al., 2016), epidemiology (Ferreira et al., 2011), and marketing (Wang et al., 2021). DMSTMs use a generalized Haar multiscale decomposition based on a multiscale partition of the domain of interest. These models allow for temporal dynamics at all scales of resolution. Ferreira (2021) provides a review of spatial and spatio-temporal models based on multiscale factorizations. Even though DMSTMs allow fast, scalable, and parallelizable computations, the spatial portion of their covariance functions are step functions. Thus, their estimates of the latent processes may be spatially blocky. However, many spatio-temporal processes have smooth spatial covariance functions. To ameliorate this issue, here we propose the use of ensembles of DMSTMs. These ensembles are obtained by averaging DMSTMs defined on spatially shifted versions of the multiscale partition. In that context, shifting the origin and averaging over the DMSTMs leads to an implied smooth process with a smooth spatial covariance function.

Because we consider the case when the true data-generating process under study is spatially smooth, none of the individual DMSTMs based on each shifted multiscale partition will be the true model. In the model selection literature, this is known as the M-open case (Yao et al., 2018). Note that, by symmetry, each of the different DMSTMs have the same Kullback-Leibler divergence to the true generating process. Hence, due to the M-open multiple optima (MOMO) paradox presented in Chapter 8 (Ferreira et al., 2024), the posterior probabilities of the different DMSTMs will not converge to a uniform distribution but, instead, will tend to concentrate on a few DMSTMs. Hence, we cannot simply update the prior probabilities of the different DMSTMs with Bayes Theorem and use the resulting posterior probabilities to perform model selection or model averaging.

That is, posterior analysis for ensembles of DMSTMs based on shifting the multiscale structure is not straightforward. Given that we assume that the true data-generating process under study is spatially smooth, and that each DMSTM is equally distant from the true data-generating process, we propose to do not update the prior probabilities of the different DMSTMs. Specifically, we start with uniform prior probabilities for the different DMSTMs equal to the reciprocal of the number of DMSTMs in the ensemble, and we set the posterior probability of each DMSTM to be equal to the prior probability.

That way, in our proposal the ensemble of DMSTMs is a mixture model where the prior weights and the posterior weights are the same. Finally, we use the observed data to separately update the parameters of each DMSTM.

We illustrate the application of our ensemble approach with two examples. First, we present an ensemble analysis of simulated data that shows our approach performs well when estimating the smooth underlying process and appropriately provides uncertainty quantification. Second, we present an ensemble analysis of temperature data in the town of Arraial do Cabo in the State of Rio de Janeiro, Brazil, that demonstrates the usefulness and flexibility of our ensemble of DMSTMs approach.

The remainder of this chapter is organized as follows. Section 2 presents the multiscale decomposition of spatial processes and the resulting disaggregation equations. Section 3 considers a state-space representation of multiscale spatio-temporal models. In Section 4, the ensemble approach is proposed by shifting the original spatial partition. Section 5 presents an application of our ensemble framework to simulated data. Section 6 presents an ensemble DMSTM data analysis of temperature data. Finally, Section 7 provides a brief discussion.

9.2 Multiscale Spatial Model

Spatial covariance functions resulting from multiscale models are step functions given the discrete partition of the spatial domain of interest. However, the original process might be smooth and step functions might not be a good characterization of spatial dependence. Moreover, for multiscale models, the inferred covariance structure will depend on the choice of the spatial origin. To obtain smooth spatial covariance functions that do not depend on the spatial origin, here we propose ensembles that average the multiscale models over shifts of the multiscale partition. The resulting spatial covariance is smooth and does not depend on the origin selected. Other authors have proposed solutions based on the shifting approach. Kolaczyk (1999) proposed a mixture of distributions to estimate the intensity of 1-D Poisson processes based on recursive dyadic partitions of the domain of interest. Tzeng et al. (2005) considered shifted versions of tree structures to obtain predicted surfaces that are smoother that the original one-tree model. Li and Ghosal (2015) proposed a fast translation algorithm for image denoising based on averaging all circulant shifts of the image.

Example 9.1 *Consider a dyadic tree data structure in one dimension with L levels of resolution, that is, each parent node is partitioned into two children. Starting with one node at the coarsest level of resolution and progressing to increasingly finer levels, we obtain 2^{L-1} nodes at the finest level. Let D_{lj} be*

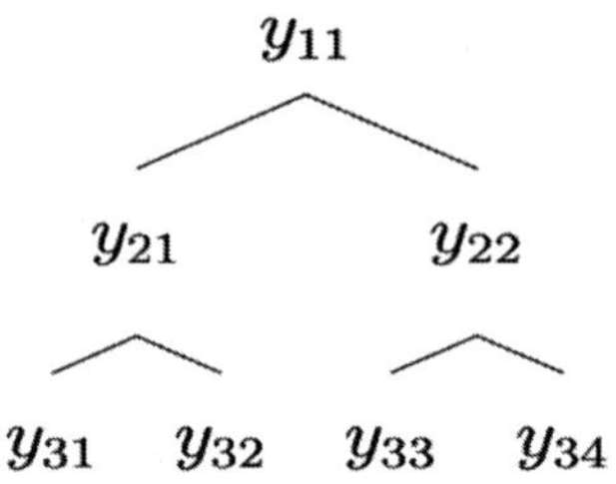

FIGURE 9.1
Multiscale structure for a dyadic tree with 3 levels of resolution.

the set of descendants of subregion j at level l, and $D_{lj} \cap D_{li} = \emptyset, i \neq j$. For example, if we consider $L = 3$ levels of resolutions, then these levels are defined as $l = 1, 2, 3$ and the sets of descendents are $D_{11} = \{(2,1),(2,2)\}$,$D_{21} = \{(3,1),(3,2)\}$, $D_{22} = \{(3,3),(3,4)\}$. Figure 9.1 displays the multiscale structure for this example.

Let y_{Lj}, $j = 1, \ldots, n_L$ be the observed data at the finest level of resolution. Consider a multiscale model that assumes a mass balance condition such that aggregated measurements at the lth level of resolution are recursively defined by $y_{lj} = \sum_{(l+1,j') \in D_{lj}} p_{l+1,j'} y_{l+1,j'}$, where $p_{lj}, l = 1, \ldots, L, j = 1, \ldots, n_l$, are known weights. Analogously, the aggregated mean process is defined by $\mu_{lj} = \sum_{(l+1,j') \in D_{lj}} p_{l+1,j'} \mu_{l+1,j'}$. Let σ^2_{lj} denote the variance of y_{lj}. Additionally, define $\boldsymbol{y}_l = (y_{l1}, \ldots, y_{ln_l})'$, $\boldsymbol{\mu}_l = (\mu_{l1}, \ldots, \mu_{ln_l})'$, $\boldsymbol{\sigma}^2_l = (\sigma^2_{l1}, \ldots, \sigma^2_{ln_l})'$, $\boldsymbol{\rho}_l = (p_{l1}\sigma^2_{l1}, \ldots, p_{ln_l}\sigma^2_{ln_l})'$, and $\boldsymbol{\Sigma}_l = diag(\boldsymbol{\sigma}^2_l)$. Under the assumption of conditional independence of $\boldsymbol{y}_l$ given $\boldsymbol{\mu}_L$ and $\boldsymbol{\sigma}^2_L$, the variance at subregion (l, j) can be recursively computed as $\sigma^2_{lj} = \sum_{(l+1,j') \in D_{lj}} p^2_{l+1,j'} \sigma^2_{l+1,j'}$. Now, conditional on the mean process μ, the joint distribution of the observations at the finest level is multivariate Gaussian. Therefore, using standard results for multivariate normal distributions, the conditional distribution of $\boldsymbol{y}_{D_{lj}}$ given y_{lj} is

$$\boldsymbol{y}_{D_{lj}} \big| y_{lj}, \boldsymbol{\mu}_L, \boldsymbol{\sigma}^2_L \sim N(\boldsymbol{\nu}_{lj} y_{lj} + \boldsymbol{\theta}_{lj}, \boldsymbol{\Omega}_{lj}), \tag{9.1}$$

with $\boldsymbol{\nu}_{lj} = \boldsymbol{\rho}_{D_{lj}} / \sigma^2_{lj}$, $\boldsymbol{\Omega}_{lj} = \boldsymbol{\Sigma}_{D_{lj}} - \sigma^{-2}_{lj} \boldsymbol{\rho}_{D_{lj}} \boldsymbol{\rho}'_{D_{lj}}$, and the disaggregation equation is given by

$$\boldsymbol{\theta}_{lj} = \boldsymbol{\mu}_{D_{lj}} - \boldsymbol{\nu}_{lj} \mu_{lj}. \tag{9.2}$$

Note that the multiscale coefficient vector, $\boldsymbol{\theta}_{lj}$, is the difference between the mean level process at the descendants of subregion B_{lj} and the scaled mean level process at B_{lj}. See Elkhouly and Ferreira (2021) for a connection between multiscale coefficients and spatial directional derivatives.

Let $\boldsymbol{\theta}^e_{lj} = \boldsymbol{y}_{D_{lj}} - \boldsymbol{\nu}_{lj} y_{lj}$ denote the empirical multiscale coefficient vector associated with $\boldsymbol{\theta}_{lj}$. Then, from (9.1) it follows that

$$\boldsymbol{\theta}^e_{lj} | y_{lj}, \boldsymbol{\mu}_L, \boldsymbol{\sigma}^2_L \sim N(\boldsymbol{\theta}_{lj}, \boldsymbol{\Omega}_{lj}), \tag{9.3}$$

which is a singular Gaussian distribution with a singular covariance matrix $\boldsymbol{\Omega}_{lj}$. Thus, conditional on μ_{Lj} and σ^2_{Lj} all the distributions are also Gaussian and the likelihood based on the observations at the finest level of resolution may be factorized as a product with contributions from the several levels of resolution, expressed as

$$\prod_{j=1}^{n_L} p(y_{Lj}|\mu_{Lj}, \sigma^2_{Lj}) \propto p(\boldsymbol{y}_1|\boldsymbol{\mu}_1, \boldsymbol{\Sigma}_1) \prod_{l=1}^{L-1} \prod_{j=1}^{n_l} p(\boldsymbol{y}_{D_{lj}}|y_{lj}, \boldsymbol{\theta}_{lj}, \boldsymbol{\Omega}_{lj}) \tag{9.4}$$

$$= p(\boldsymbol{y}_1|\boldsymbol{\mu}_1, \boldsymbol{\Sigma}_1) \prod_{l=1}^{L-1} \prod_{j=1}^{n_l} p(\boldsymbol{\theta}^e_{lj}|y_{lj}, \boldsymbol{\theta}_{lj}, \boldsymbol{\Omega}_{lj}). \tag{9.5}$$

The constant of proportionality in Equation (9.4) is the Jacobian of the transformation from the finest level $y_{L,1:n_L}$ to the aggregated data — the constant of proportionality depends only on the aggregation weights. In addition, conditional on y_{lj}, the Jacobian of the transformation from $Y_{D_{lj}}$ to $\boldsymbol{\theta}^e_{lj}$ is equal to 1. Thus, the right hand side in Equation (9.4) is exactly equal to the expression in Equation (9.5). Moreover, note that the factorization above is equivalent to a reparametrization of the model, where $\mu_{L1}, \ldots, \mu_{Ln_L}$ are replaced by the mean level process at the coarsest level $\boldsymbol{\mu}_1$ and by the multiscale coefficient vectors $\boldsymbol{\theta}_{lj}$s. In our illustrative example with $L = 3$ levels of resolution, the mean vector at the finest level is $\boldsymbol{\mu}_3 = (\mu_{31}, \mu_{32}, \mu_{33}, \mu_{34})'$ and the reparametrized version in terms of multiscale coefficients is $\Theta = (\mu_{11}, \theta'_{11}, \theta'_{21}, \theta'_{22})'$.

The following theorem shows that the reparametrization from the collection of multiscale coefficients Θ to the mean vector at the finest level $\boldsymbol{\mu}_L$ is a linear operator.

Theorem 9.1 *Consider a multiscale model with a dyadic tree structure and L levels of resolution. Then,* $\boldsymbol{\mu}_L = A\Theta$ *where* $A = (B_1, A_1, \ldots, A_L)$ *and*

$$\begin{aligned} A_L &= diag(1_{m_L}), \\ A_l &= diag(\nu_{l+1,1}, \ldots, \nu_{l+1,n_{l+1}})A_{l+1}, \; l = L-1, \ldots, 1, \\ B_1 &= \nu_1 A_1. \end{aligned}$$

Proof *The proof is by induction. Consider a tree structure model with k levels of resolution and mean vector at the kth level of resolution given by* $\boldsymbol{\mu}_k = (\mu_{k1}, \ldots, \mu_{km_k})'$. *Define the multiscale coefficients at level l by* $\boldsymbol{\theta}_l = (\theta_{l,1}, \ldots, \theta_{l,m_l})$. *Thus, the reparametrized vector in terms of multiscale coefficients is* $\Theta_k = (\mu_{11}, \boldsymbol{\theta}'_1, \boldsymbol{\theta}'_2, \ldots, \boldsymbol{\theta}'_{k-1})'$.

The induction assumption is that $\boldsymbol{\mu}_k = A^{(k)}\Theta_k$, *where* $A^{(k)} = (B_1^{(k)}, A_1^{(k)}, \ldots, A_{k-1}^{(k)})$ *and*

$$\begin{aligned} A_{k-1}^{(k)} &= I_{m_k}, \\ A_l^{(k)} &= diag(\nu_{l+1,1}, \ldots, \nu_{l+1,n_{l+1}})A_{l+1},\ l = k-2, \ldots, 1, \\ B_1^{(k)} &= \nu_1 A_1^{(k)}, \end{aligned}$$

hold for $k \geq 3$ *levels of resolution. For* $k = 2$, $(\mu_{21}, \mu_{22})' = \theta_{11} + \nu_{11}\mu_{11}$.

First, let us check that the induction assumption holds for $k = 3$ *levels of resolution. From the disaggregation equation (9.2) we may write*

$$\begin{cases} (\mu_{21}, \mu_{22})' &= \theta_{11} + \nu_{11}\mu_{11}, \\ (\mu_{31}, \mu_{32})' &= \theta_{21} + \nu_{21}\mu_{21}, \\ (\mu_{33}, \mu_{34})' &= \theta_{22} + \nu_{22}\mu_{22}. \end{cases}$$

That is,

$$\begin{aligned} \boldsymbol{\mu}_3 &= (\theta_{21}, \theta_{22})' + diag(\nu_{21}, \nu_{22})(\mu_{21}, \mu_{22}) & (9.6) \\ &= (\theta_{21}, \theta_{22})' + diag(\nu_{21}, \nu_{22})\theta_{11} + diag(\nu_{21}, \nu_{22})\nu_{11}\mu_{11}. & (9.7) \end{aligned}$$

Thus, $\boldsymbol{\mu}_3 = A^{(3)}\Theta_3$, *where* $A^{(3)} = (B_1^{(3)}, A_1^{(3)}, A_2^{(3)})$ *and*

$$\begin{aligned} B_1^{(3)} &= diag(\nu_{21}, \nu_{22})\nu_{11}, \\ A_1^{(3)} &= diag(\nu_{21}, \nu_{22}), \\ A_2^{(3)} &= I_4. \end{aligned}$$

Thus, the induction assumption holds for $k = 3$.

Next, we prove that the result holds for k+1 levels of resolution. From the disaggregation equation (9.2) we may obtain Θ_{k+1} *recursively as* $\Theta_{k+1} = (\Theta_k, \boldsymbol{\theta}_k)$. *And using the induction assumption and the disaggregation equations, we obtain*

$$\begin{aligned} \boldsymbol{\mu}_{k+1} &= \boldsymbol{\theta}_k + diag\{\boldsymbol{\nu}_{k1}, \ldots, \boldsymbol{\nu}_{k,n_k}\}\boldsymbol{\mu}_k \\ &= \boldsymbol{\theta}_k + diag\{\boldsymbol{\nu}_{k1}, \ldots, \boldsymbol{\nu}_{k,n_k}\}A^{(k)}\Theta_k \\ &= (diag\{\boldsymbol{\nu}_{k1}, \ldots, \boldsymbol{\nu}_{k,n_k}\}A^{(k)} \quad I_{m_k})\Theta_{k+1} \\ &= A^{(k+1)}\Theta_{k+1}. \end{aligned}$$

Therefore, the result follows. □

Thus, inference may be performed for Θ, and the matrix A may be used to recover the mean vector at the finest level $\boldsymbol{\mu}_L$. While it is computationally more efficient to recursively apply the disaggregation equation $\boldsymbol{\mu}_{D_{lj}} = \nu_{lj}\mu_{lj} + \theta_{lj}$ to obtain the mean process at progressively finer resolutions, we can use the fact that $\boldsymbol{\mu}_L = A\Theta$ to obtain the covariance matrix of $\boldsymbol{\mu}_L$ (see Theorem 9.2 in Section 9.3).

Note that the observational model at the finest level of resolution can be written as

$$\mathbf{y}_L \mid \boldsymbol{\mu}_L, V \sim N(\boldsymbol{\mu}_L, V), \tag{9.8}$$

with $V = \text{diag}(\sigma^2_{L1}, \ldots, \sigma^2_{Ln_L})$. Denote by $Z_1 \circ Z_2$ the Hadamard product of Z_1 and Z_2, that is, the element-wise product of two matrices Z_1 and Z_2 of the same dimension. Then, the variances σ^2_{lj} may be defined recursively as $\sigma^2_{lj} = (p_{D_{lj}} \circ p_{D_{lj}})' \sigma^2_{D_{lj}}$, and the weights p_{lj} are such that $\sum_{(l+1,j) \in D_{lj}} p_{l+1,j} = 1$. Finally, note that from Theorem 9.1 the observation equation in the reparametrization is $\mathbf{y}_L \mid \Theta, V \sim N(A\Theta, V)$.

9.3 Dynamic Multiscale Spatio-Temporal Model

In this section, we add a subscript t to indicate time and consider spatio-temporal data. Then, the observation equation at the finest resolution level becomes

$$\boldsymbol{y}_{tL} = \boldsymbol{\mu}_{tL} + \boldsymbol{\epsilon}_{tL}, \; \boldsymbol{\epsilon}_{tL} \sim N(\mathbf{0}, \Sigma_L). \tag{9.9}$$

We apply the multiscale factorization described in Section 9.2 to the data vector $\boldsymbol{y}_{tL}$ at each time point t to obtain the multiscale observation equations

$$\begin{aligned} \boldsymbol{y}_{t1} &= \boldsymbol{\mu}_{t1} + \boldsymbol{\epsilon}_{t1}, \; \boldsymbol{\epsilon}_{t1} \sim N(\mathbf{0}, \boldsymbol{\Sigma}_1), & (9.10) \\ \boldsymbol{\theta}^e_{tlj} &= \boldsymbol{\theta}_{tlj} + \boldsymbol{\epsilon}^\theta_{tlj}, \; \boldsymbol{\epsilon}^\theta_{tlj} \sim N(\mathbf{0}, \Omega_{lj}). & (9.11) \end{aligned}$$

We then assume that the mean level process at the coarsest level $\boldsymbol{\mu}_{t1}$ and the multiscale coefficient $\boldsymbol{\theta}_{tlj}$ evolve through time according to state-space equations. This type of temporal evolution allows fairly flexible behavior such as local linear and quadratic trends, seasonality, and dynamic regression. Here, we consider stationary autoregressive processes of order 1 given by

$$\begin{aligned} \boldsymbol{\mu}_{t1} &= \phi \boldsymbol{\mu}_{t-1,1} + \boldsymbol{\omega}^\mu_{t1}, \; \omega^\mu_{t1} \sim N(\mathbf{0}, \boldsymbol{W}_1), & (9.12) \\ \boldsymbol{\theta}_{tlj} &= \rho_{lj} \boldsymbol{\theta}_{t-1,lj} + \boldsymbol{\omega}^\theta_{tlj}, \; \boldsymbol{\omega}^\theta_{tlj} \sim N(\mathbf{0}, \boldsymbol{\Psi}_{lj}), & (9.13) \end{aligned}$$

where for stationarity $|\phi| < 1$ and $|\rho_{lj}| < 1$, $l = 1, \ldots, L-1$, $j = 1, \ldots, n_l$. Let $\Theta_t = (\boldsymbol{\mu}'_{t1}, \boldsymbol{\theta}'_{t11}, \ldots, \boldsymbol{\theta}'_{t1n_1}, \ldots, \boldsymbol{\theta}'_{t,L-1,1}, \ldots, \boldsymbol{\theta}'_{t,L-1,n_1})'$. In addition, let d_{lj} denote the number of children of subregion (l, j) and $\boldsymbol{I}_d$ denote the $d \times d$ identity matrix. Further, let $\boldsymbol{\Psi}_l = \text{blockdiag}(\boldsymbol{\Psi}_{l1}, \ldots, \boldsymbol{\Psi}_{ln_l})$, $\boldsymbol{W} = \text{blockdiag}(\boldsymbol{W}_1, \boldsymbol{\Psi}_1, \ldots, \boldsymbol{\Psi}_{L-1})$, and $\boldsymbol{\Phi} = \text{blockdiag}(\phi \boldsymbol{I}_{n_1}, \rho_{11} \boldsymbol{I}_{d_{11}}, \ldots, \rho_{1n_1} \boldsymbol{I}_{d_{1n_1}}, \rho_{21} \boldsymbol{I}_{d_{21}}, \ldots, \rho_{2n_2} \boldsymbol{I}_{d_{2n_2}}, \ldots, \rho_{L-1,1} \boldsymbol{I}_{d_{L-1,1}}, \ldots, \rho_{L-1,n_{L-1}} \boldsymbol{I}_{d_{L-1,n_{L-1}}})$. Equations (9.12) and (9.13) may then be rewritten as $\Theta_t \mid \Theta_{t-1}, \Phi, \boldsymbol{W} \sim N(\boldsymbol{\Phi}\Theta_{t-1}, \boldsymbol{W})$. Because of stationarity, standard time series results (Prado et al., 2021) yield the stationary distribution of Θ_t, which is

$$\Theta_t \mid \boldsymbol{\Sigma} \sim N(\mathbf{0}, \boldsymbol{\Sigma}), \tag{9.14}$$

where $\boldsymbol{\Sigma} = \text{blockdiag}(\boldsymbol{W}_1^*, \boldsymbol{\Psi}_1^*, \ldots, \boldsymbol{\Psi}_{L-1}^*)$, $\boldsymbol{W}_1^* = \boldsymbol{W}_1/(1-\phi^2)$, $\boldsymbol{\Psi}_l^* = \text{blockdiag}(\boldsymbol{\Psi}_{l1}^*, \ldots, \boldsymbol{\Psi}_{ln_l}^*)$, and $\boldsymbol{\Psi}_{lj}^* = \boldsymbol{\Psi}_{lj}/(1-\rho_{lj}^2)$. Fundamental to the construction of these multiscale spatio-temporal models is the specification of the evolution covariance matrix $\boldsymbol{\Psi}_{lj}$. This matrix has to be carefully specified to preserve the distinct correlation structure among the elements of the multiscale coefficient vector $\boldsymbol{\theta}_{tlj}$ and to respect the aggregation constraint $P[\boldsymbol{p}'_{D_{lj}}\boldsymbol{\theta}_{tlj} = 0] = 1$.

In terms of the covariance matrix at the coarsest level, we assume $\boldsymbol{W}_1 = \text{diag}(W_{11}, \ldots, W_{1n_1})$, with $W_{1k} = \xi_k \sigma_{1k}^2$. This representation is useful because ξ_k is a signal-to-noise ratio parameter for the temporal evolution at the coarsest level. Thus, it is easier to assign a meaningful prior distribution to ξ_k than to W_{1k}. For the covariance matrix of the evolution innovations of the multiscale coefficient vectors, we assume $\boldsymbol{\Psi}_{lj} = \psi_{lj}\boldsymbol{\Omega}_{lj}$, where ψ_{lj} is a signal-to-noise ratio parameter. Note that this construction both preserves the correlation structure of $\boldsymbol{\theta}_{tlj}$ and implies the aggregation constraint $P[\boldsymbol{p}'_{D_{lj}}\boldsymbol{\theta}_{tlj} = 0] = 1$.

In summary, the model for the temporal evolution of $\boldsymbol{\theta}_{tlj}$ is

$$\boldsymbol{\theta}_{tlj} \mid \boldsymbol{\theta}_{t-1,lj}, \psi_{lj}, \boldsymbol{\Omega}_{lj} \sim N(\rho_{lj}\boldsymbol{\theta}_{t-1,lj}, \psi_{lj}\boldsymbol{\Omega}_{lj}). \tag{9.15}$$

Let D_t denote the information up to time t. Thus, D_0 is the information at time $t = 0$ and $D_t = D_{t-1} \cup \{\boldsymbol{y}_{tL}\}$. The initial conditions are set to $\boldsymbol{\theta}_{0lj} \mid D_0 \sim N(\boldsymbol{0}, C_{0lj}\boldsymbol{\Omega}_{lj})$, where $C_{0lj} = \psi_{lj}/(1-\rho_{lj}^2)$. This results in the following stationary distribution for $\boldsymbol{\theta}_{tlj}$:

$$\boldsymbol{\theta}_{tlj} \mid \psi_{lj}, \boldsymbol{\Omega}_{lj} \sim N(\boldsymbol{0}, \psi_{lj}^*\boldsymbol{\Omega}_{lj}), \tag{9.16}$$

where $\psi_{lj}^* = \psi_{lj}/(1-\rho_{lj}^2)$.

The following theorem provides the stationary covariance matrix of the mean process at the finest level $\boldsymbol{\mu}_{tL}$.

Theorem 9.2 *Consider the multiscale spatio-temporal model with observation equations (9.10) and (9.11), and system equations (9.12) and (9.13). Then, the covariance matrix of the stationary distribution of $\boldsymbol{\mu}_{tL}$ is*

$$Cov(\boldsymbol{\mu}_{tL}) = B_1\boldsymbol{W}_1^*B_1' + \sum_{l=1}^{L-1} A_l\boldsymbol{\Psi}_l^*A_l'. \tag{9.17}$$

Proof *From Theorem 9.1, we obtain* $\text{Cov}(\boldsymbol{\mu}_{tL}) = A\text{Cov}(\Theta_t)A'$ *with* $A = (B_1, A_1, \ldots, A_{L-1})$ *and* $\Theta_t = (\mu_{t1}, \theta'_{t1}, \ldots, \theta'_{t,L-1})'$. *From the system equations (9.12) and (9.13) and the stationarity condition, we obtain* $\boldsymbol{\mu}_{t1} \sim N(\boldsymbol{0}, \boldsymbol{W}_1^*)$ *and* $\boldsymbol{\theta}_{tlj} \sim N(\boldsymbol{0}, \boldsymbol{\Psi}_{lj}^*)$. *Thus,*

$$\begin{aligned}
\text{Cov}(\boldsymbol{\mu}_{tL}) &= (B_1, A_1, \ldots, A_{L-1})\text{blockdiag}(\boldsymbol{W}_1^*, \boldsymbol{\Psi}_1^*, \ldots, \boldsymbol{\Psi}_{L-1}^*) \\
&\quad (B_1, A_1, \ldots, A_{L-1})' \\
&= (B_1\boldsymbol{W}_1^*, A_1\boldsymbol{\Psi}_1^*, \ldots, A_{L-1}\boldsymbol{\Psi}_{L-1}^*)(B_1, A_1, \ldots, A_{L-1})' \\
&= B_1\boldsymbol{W}_1^*B_1' + \sum_{l=1}^{L-1} A_l\boldsymbol{\Psi}_l^*A_l'.
\end{aligned}$$

□

The following theorem provides the cross-covariance matrix of the stationary distribution of $\boldsymbol{\mu}_{tL}$ and $\boldsymbol{\mu}_{t-h,L}$ at the finest level of resolution.

Theorem 9.3 *The cross-covariance matrix of the stationary distribution of* $\boldsymbol{\mu}_{tL}$ *and* $\boldsymbol{\mu}_{t-h,L}$ *is*

$$E[\boldsymbol{\mu}_{tL}\boldsymbol{\mu}'_{t-h,L}] = \phi^{|h|} B_1 \boldsymbol{W}_1^* B_1' + \sum_{l=1}^{L-1} \rho_l^{|h|} A_l \boldsymbol{\Psi}_l^* A_l', \tag{9.18}$$

where $\rho_l^{|h|)} = \mathrm{diag}(\rho_{l1}^{|h|}, \ldots, \rho_{ln_l}^{|h|})$.

Proof *First, let us consider* $h > 0$. *Then,*

$$\begin{aligned}
E[\boldsymbol{\mu}_{tL}\boldsymbol{\mu}'_{t-h,L}] &= E[\Phi\boldsymbol{\mu}_{t-1,L}\boldsymbol{\mu}'_{t-h,L}] \\
&= E[\Phi\Phi\boldsymbol{\mu}_{t-2,L}\boldsymbol{\mu}'_{t-h,L}] \\
&= E[\Phi\Phi\ldots\Phi\boldsymbol{\mu}_{t-h,L}\boldsymbol{\mu}'_{t-h,L}] \\
&= \Phi^h E[\boldsymbol{\mu}_{t-h,L}\boldsymbol{\mu}'_{t-h,L}] \\
&= \Phi^h (B_1, A_1, \ldots, A_{L-1}) diag(\boldsymbol{W}_1^*, \boldsymbol{\Psi}_1^*, \ldots, \boldsymbol{\Psi}_{L-1}^*) \\
&\quad\ (B_1, A_1, \ldots, A_{L-1})' \\
&= \Phi^h (B_1\boldsymbol{W}_1^*, A_1\boldsymbol{\Psi}_1^*, \ldots, A_{L-1}\boldsymbol{\Psi}_{L-1}^*)(B_1, A_1, \ldots, A_{L-1})' \\
&= (\phi^h B_1\boldsymbol{W}_1^*, \rho_1^h A_1\boldsymbol{\Psi}_1^*, \ldots, \rho_{L-1}^h A_{L-1}\boldsymbol{\Psi}_{L-1}^*) \\
&\quad\ (B_1, A_1, \ldots, A_{L-1})' \\
&= \phi^h B_1 \boldsymbol{W}_1^* B_1' + \sum_{l=1}^{L-1} \rho_l^h A_l \boldsymbol{\Psi}_l^* A_l'.
\end{aligned}$$

To prove the result for $h < 0$, *we consider* $h^* = -h = |h|$. *We then obtain* $E[\boldsymbol{\mu}_{t+h^*,L}\boldsymbol{\mu}'_{t,L}]$ *in a development analogous to the development above.* □

Example 9.2 *Consider a spatio-temporal model on a spatial domain that is a 1-D torus, where at each time point observations are obtained at 128 subregions. Assume that the observations depend on a latent mean process according to the observation equation (9.9). In addition, the latent mean process evolves through time with a multivariate random walk where the innovations are spatially correlated according to an exponential covariance function* $\rho(d) = \tau^2 \exp(-\phi d)$. *Now assume that we analyze the observations using a dynamic multiscale spatio-temporal model given by Equations (9.10)–(9.13) with* $L = 8$ *levels of resolution. Focusing only on the spatial part of the correlation structure, Figure 9.2 presents the original smooth spatial covariance function and the covariance function implied by the multiscale model. Note the blockiness features resulting from the multiscale model.*

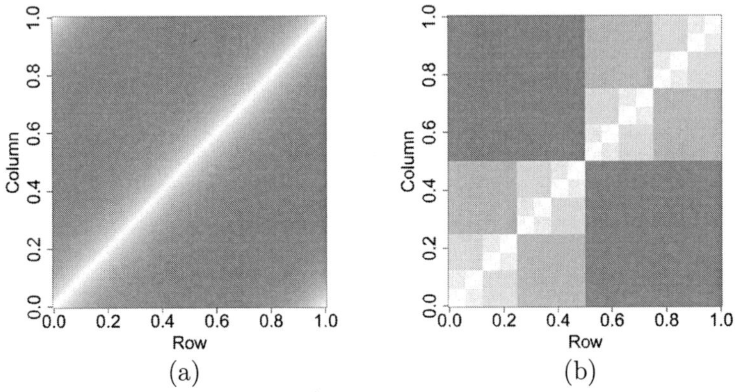

FIGURE 9.2
(a) Original spatial covariance function, and (b) multiscale approximation for 1-D dyadic tree with $L = 8$ levels of resolution.

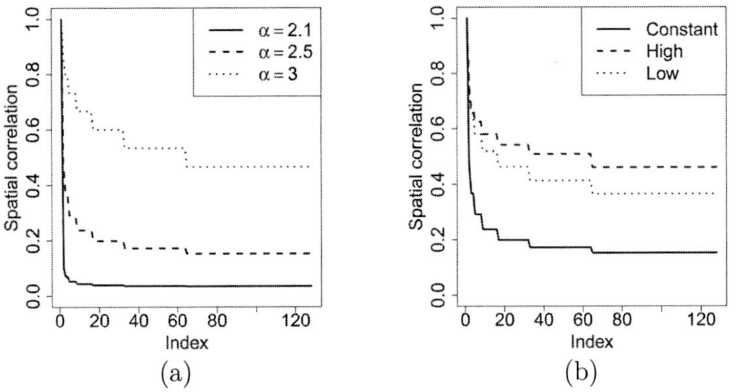

FIGURE 9.3
Spatial covariance at time t for different specifications of $\boldsymbol{W}_1$ and Ψ_l. (a) Varying the signal-to-noise ratio $\psi_{lj} = 2^{-\alpha l}$, $l = 1, 2, \ldots, 8$. (b) Varying the autoregressive coefficients.

The evolution covariance matrices $\boldsymbol{W}_1$ and Ψ_l are fundamental to the spatial and spatio-temporal dependence structure in the multiscale approximation. Figure 9.3 presents different specifications of these parameters in order to illustrate the possible behaviors for the implied covariance functions. Figure 9.3(a) presents the spatial correlation at time t for equal weights for the descendants of each region and varying signal-to-noise ratio, $\xi_i = 1$ and $\psi_{lj} = 2^{-\alpha l}$, $l = 1, 2, \ldots, 8$. The autoregressive coefficients ϕ and ρ_l are all equal to 0.95. Figure 9.3(b) presents the spatio-temporal covariance for signal-to-noise ratio equal $2^{-\alpha l}$ with $\alpha = 2.5$ and three specifications for the autoregressive coefficients labeled constant, low, and high. The constant specification sets all autoregressive coefficients to 0.95. The low specification sets the autoregressive coefficient at the coarsest level to 0.85 and decreases the autoregressive coefficients with increasing levels of resolution. Finally, the high specification sets the autoregressive coefficient at the coarsest level to 0.98 and decreases the autoregressive coefficients with increasing levels of resolution.

Note that the implied spatial covariance functions are step functions. This results from the discrete nature of the discretization of the spatial domain. In two dimensions, this covariance structure will be blocky and usually will not be a good representation of smooth processes. As a possible solution, we propose the use of ensembles of DMSTMs that imply smoother spatial covariance functions.

9.4 Ensembles of DMSTMs

The examples in Section 9.3 illustrate the blocky behavior of the spatial correlations resulting from multiscale models. In what follows, we propose an ensemble approach that aims to obtain smoother spatial structures to represent possibly smooth processes. For simplicity of exposition, we consider the one-dimensional case. Define $\mathcal{T}_0$ to be a partition of the spatial domain of interest with descendant indices in the set $\mathcal{L}_0 = \{(l, j) : l = 0, \ldots, L-1,\ j = 1, \ldots, n_l\}$. Let D_{lj} be the set of descendants of subregion j at level l for the spatial partition with indices in the set $\mathcal{L}_0$, and $D_{lj} \cap D_{li} = \emptyset, i \neq j$. Define $\mathcal{T}_m$ to be the mth shifted version of $\mathcal{T}_0$ such that the descendant indices are in the set

$$\mathcal{L}_m = \{(l, j - m) : (l, j) \in \mathcal{L}_0\}, \tag{9.19}$$

with periodic condition $(l, 0) = (l, n_l)$. Then, $\{\mathcal{T}_m : m \in \mathcal{M}\}$ is a collection of shifted trees in one dimension.

Then, $D_{lj}^{(m)}$ is the set of descendants of subregion j at level l for the shifted tree $\mathcal{T}_m$. Consider the observed data at the finest level of resolution $\mathbf{y}_{tL} = (y_{tL1}, \ldots, y_{tLn_L})'$ and $\mathbf{y}_L = (\mathbf{y}_{1L}', \ldots, \mathbf{y}_{TL}')'$. In what follows, we define the modulation shift operator which takes the element-wise modulation of an integer m.

Definition 9.1 *(Shift operator) The modulation shift operator applied to a vector $\boldsymbol{y}$ of length n is defined by*

$$S_{i'} : \Re^n \to \Re^n,\ (S_{i'}(\boldsymbol{y}))_i = \boldsymbol{y}_{((i-i')\ mod\ n)}. \tag{9.20}$$

Example 9.3 *Consider a tree structure with $L = 3$ levels of resolution. Then, the original tree at the finest level of resolution has the vector of observations $\boldsymbol{y}_{t3} = (y_{t31}, y_{t32}, y_{t33}, y_{t34})'$. Hence, the shifted versions of the tree resulting from the application of (9.20) have vectors of observations*

$$\begin{aligned} S_1(\boldsymbol{y}_{t3}) &= (y_{t34}, y_{t31}, y_{t32}, y_{t33}), \\ S_2(\boldsymbol{y}_{t3}) &= (y_{t23}, y_{t34}, y_{t31}, y_{t32}), \\ S_3(\boldsymbol{y}_{t3}) &= (y_{t32}, y_{t33}, y_{t34}, y_{t31}), \\ S_4(\boldsymbol{y}_{t3}) &= (y_{t31}, y_{t32}, y_{t33}, y_{t34}). \end{aligned}$$

Next, we define the aggregation operator that provides, for each level of resolution, the sums required for likelihood computation in the original dyadic tree.

Definition 9.2 *(Aggregation operator) The aggregation operator for each level of resolution l, with $l = 1, \ldots, L$, on a dyadic tree is defined by*

$$H_l : \Re^{2^l} \to \Re^{2^{l-1}}, \; (H_l(Y))_j = \sum_{j'=2j-1}^{2j} y_{l,j'}, \; j = 1, \ldots, 2^{l-1}. \tag{9.21}$$

Example 9.4 *Consider the case of a dyadic tree structure with $L = 3$ levels of resolution. Then, the data vector at time t at the finest level of resolution is $(y_{t31}, y_{t32}, y_{t33}, y_{t34})$. Thus, the aggregated data vector at resolution level $l = 2$ resulting from the application of (9.21) is*

$$H_2(\boldsymbol{y}_{t3}) = H_2(y_{t31}, y_{t32}, y_{t33}, y_{t34}) = (y_{t31} + y_{t32}, y_{t33} + y_{t34}) = (y_{t21}, y_{t22}),$$

and the aggregated data at resolution level $l = 1$ is

$$H_1(\boldsymbol{y}_{t3}) = y_{t21} + y_{t22} = y_{t11}.$$

To compute the likelihood function for all trees, we apply the aggregation operator in all shifted versions of the original data $\boldsymbol{y}_t$. Thus,

Definition 9.3 *(Aggregation and shift operator) The aggregation and shift operator at level l and shift i applies the aggregation operator for each level of resolution l and each shifted tree i, with $l = 1, \ldots, L$ and $i = 1, \ldots, 2^L$, and it is defined by*

$$C_{l,i} : \Re^{2^l} \to \Re^{2^{l-1}}, \; (C_{l,i}(Y)) = H_l(S_i(Y)). \tag{9.22}$$

At level of resolution l, the resulting aggregated data vector for each shifted version $S_i(\boldsymbol{y}_t)$ has size 2^{l-1}.

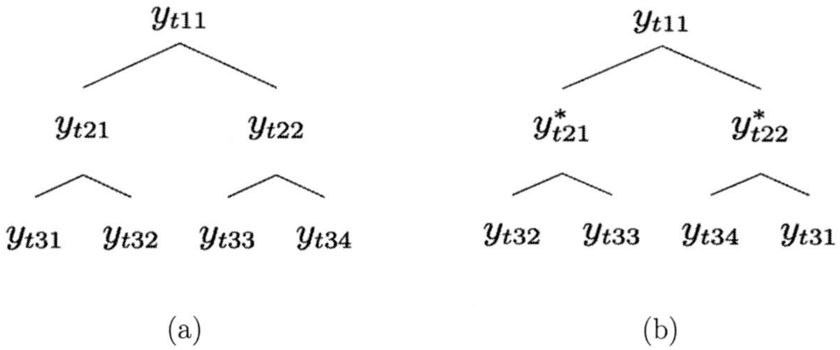

FIGURE 9.4
Data structure for the dyadic illustrative example with 3 levels of resolution for: (a) the original tree; and (b) the tree resulting from the shift operator $S_3(\boldsymbol{y}_{t3})$.

Example 9.5 *Consider the dyadic tree structure with $L = 3$ levels of resolution. Then, the aggregated vectors resulting from the application of (9.22) to the tree obtained with the shift operator $S_3(\boldsymbol{y}_{t3})$ are*

$$\begin{aligned} C_{2,3}(\boldsymbol{y}_{t3}) &= H_2(S_3(\boldsymbol{y}_{t3})) = H_2(y_{t32}, y_{t33}, y_{t34}, y_{t31}) = (y^*_{t21}, y^*_{t22}), \\ C_{1,3}(\boldsymbol{y}_{t3}) &= H_1(S_3(\boldsymbol{y}_{t3})) = H_1(y_{t32}, y_{t33}, y_{t34}, y_{t31}) \\ &= y_{t32} + y_{t33} + y_{t34} + y_{t31} = y_{t11}. \end{aligned}$$

Note that $C_{1,3}(\boldsymbol{y}_{t3}) = y_{t11}$ is the overall total over the entire study region. Figure 9.4 presents the original tree on panel (a) and the tree resulting from the application of the shift operator $S_3(\boldsymbol{y}_{t3})$ on panel (b).

Consider a multiscale model for each tree such that aggregated measurements at the lth level of resolution are recursively defined by $y^{(m)}_{tlj} = \sum_{j' \in D^{(m)}_{lj}} p_{l+1,j'} y^{(m)}_{t,l+1,j'}$, where $p_{lj}, l = 1, \ldots, L, j = 1, \ldots, n_l$, are known weights. Analogously, the aggregated mean process for the m-th tree model is defined by $\mu^{(m)}_{tlj} = \sum_{j' \in D^{(m)}_{lj}} p_{l+1,j'} \mu^{(m)}_{t,l+1,j'}$.

Note that if we shift the original tree $\mathcal{T}_0$ using (9.19) then the parameters of interest will be defined for different sets of parents in each shifted tree $\mathcal{T}_m$ as follows

$$\boldsymbol{\mu}^{(m)}_{tD_{lj}} = \boldsymbol{\theta}^{(m)}_{tlj} + \boldsymbol{\nu}_{tlj} \mu^{(m)}_{tlj}, \; l = 0, 1, \ldots, L-1. \tag{9.23}$$

Now consider the shifted tree $\mathcal{T}_m$, $m \in \mathcal{M}$, and the observational model defined in Equation (9.1). For each shifted tree $\mathcal{T}_m$, the likelihood function may be written as

$$\prod_{t=1}^{T}\prod_{j=1}^{n_L} p(y_{tLj} \mid \mu_{tLj}, \sigma^2_{tLj}, \mathcal{T}_m) \propto \prod_{t=1}^{T} p(\boldsymbol{y}_{t1} \mid \boldsymbol{\mu}_{t1}, \boldsymbol{\Sigma}_{t1}) \prod_{l=1}^{L-1}\prod_{j=1}^{n_l} p(\boldsymbol{y}^{(m)}_{tD_{lj}} \mid y^{(m)}_{tlj}, \boldsymbol{\theta}^{(m)}_{tlj}, \boldsymbol{\Omega}^{(m)}_{tlj}, \mathcal{T}_m). \tag{9.24}$$

For each spatial partition resulting from the shifted tree $\mathcal{T}_m$, assume temporal evolutions such as in Equations (9.12) and (9.13) for the latent states $\Theta^{(m)}_t$. Then, the stationary distribution of $\Theta^{(m)}_t$ in each shifted tree is given by $N(\mathbf{0}, \Sigma^{(m)})$ with $\Sigma^{(m)} = Cov(\Theta^{(m)}_t) = \text{blockdiag}(\boldsymbol{W}^*_1, \boldsymbol{\Psi}^{*(m)}_1, \ldots \boldsymbol{\Psi}^{*(m)}_L)$, $\boldsymbol{\Psi}^{*(m)}_l = \text{blockdiag}(\boldsymbol{\Psi}^{*(m)}_{l1}, \ldots, \boldsymbol{\Psi}^{*(m)}_{ln_l})$.

For each shifted tree $\mathcal{T}_m$, the transformation $A^{(m)}\Theta^{(m)}_t$ recovers the mean process at the finest level of resolution $\boldsymbol{\mu}_{tL}$. We then define the ensemble of DMSTMs to be a weighted average of the DMSTMs implied by each shifted tree. In particular, the stationary distribution of $\boldsymbol{\mu}_{tL}$ based on the ensemble of DMSTMs is

$$p(\boldsymbol{\mu}_{tL}) = \sum_{m\in\mathcal{M}} P_m p(\boldsymbol{\mu}_{tL} \mid \mathcal{T}_m) = \sum_{m\in\mathcal{M}} P_m N(\boldsymbol{\mu}_{tL}; 0, A^{(m)}\Sigma^{(m)}(A^{(m)})'), \tag{9.25}$$

where P_m, $m \in \mathcal{M}$, are weights such that $\sum_{m\in\mathcal{M}} P_m = 1$.

Then, the resulting spatio-temporal covariance for the mean process at the finest level of resolution is given by the following theorem.

Theorem 9.4 *Consider the multiscale spatio-temporal model with disaggregation equation (9.2), observation equation (9.8), and system equation (9.12) for each shifted version of $\mathcal{T}_0$ and prior distribution for μ_{tL} given by a weighted average in (9.25). Then, the stationary covariance matrix of $\boldsymbol{\mu}_{tL}$ at time t is given by*

$$Cov(\boldsymbol{\mu}_{tL}) = B_1\boldsymbol{W}^*_1 B'_1 + \sum_{m\in\mathcal{M}}\sum_{l=1}^{L-1} P_m A^{(m)}_l \boldsymbol{\Psi}^{*(m)}_l (A^{(m)}_l)', \tag{9.26}$$

with $\boldsymbol{\Psi}^{*(m)}_l = \text{blockdiag}(\boldsymbol{\Psi}^{*(m)}_{l,1}, \ldots, \boldsymbol{\Psi}^{*(m)}_{l,n_l})$, $\boldsymbol{W}^*_1 = \boldsymbol{W}_1/(1-\phi^2)$, *and* $\boldsymbol{\Psi}^{*(m)}_{lj} = \boldsymbol{\Psi}^{(m)}_{lj}/(1-\rho^{2(m)}_{lj})$.

Example 9.6 *Let us consider again the spatio-temporal model on a 1-D torus spatial domain from Example 9.2. But here, instead of approximating the process with an 8-resolution-levels dynamic multiscale spatio-temporal model*

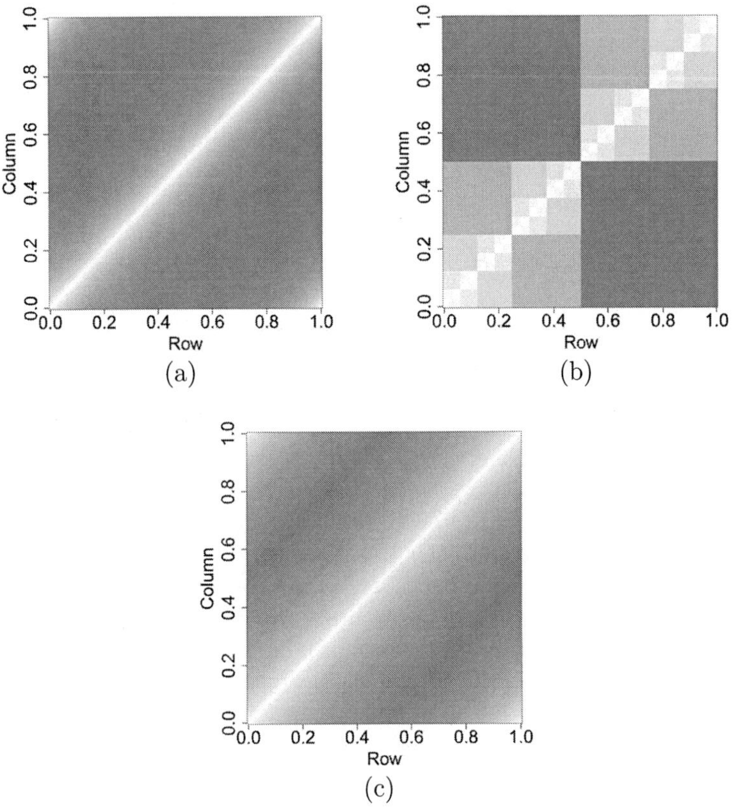

FIGURE 9.5
(a) Original spatial covariance function, (b) multiscale approximation based on one tree, and (c) approximation based on an ensemble of multiscale models. Compared to the multiscale approximation based on one tree, the approximation based on an ensemble of multiscale models is much closer to the original spatial covariance function.

based on one tree structure, we consider an ensemble of models based on distinct $M = 2^{L-2} = 64$ *tree structures. Figure 9.5 presents the heatmap of the spatial covariance matrix for the true smooth process (Panel a), as well as multiscale approximations based on one tree (Panel b) and on an ensemble of multiscale models (Panel c). When compared to the blocky structure resulting from the one-tree-based multiscale approximation, the covariance obtained by the ensemble approximation provides a better representation of the true smooth covariance.*

9.4.1 Posterior Analysis

In this chapter, we assume that the true data generating process under study is spatially smooth, and an ensemble of DMSTMs is used to approximate the true data-generating process. In this case, none of the individual DMSTMs will be the true model – this is an M-open situation (Yao et al., 2018). In addition, by symmetry, the Kullback-Leibler divergences between the true generating process and each of the different DMSTMs obtained by shifting the multiscale structure will be the same. Thus, due to the model selection paradox presented in Chapter 8 (Ferreira et al., 2024), the logarithm of the Bayes factors of each two considered DMSTMs will follow random walks. Thus, as shown in Chapter 8 (Ferreira et al., 2024), the posterior probabilities of the different DMSTMs will not converge but, instead, will tend to concentrate on a few DMSTMs. Hence, we cannot simply update the prior probabilities of the different DMSTMs and use the resulting posterior probabilities to perform model selection or model averaging. That is, posterior analysis for ensembles of DMSTMs based on shifting the multiscale structure is not straightforward.

Given that we assume that the true data-generating process under study is spatially smooth, and that each DMSTM is equally distant from the true data-generating process, we propose to do not update the prior probabilities of the different DMSTM. That is, we set the posterior probability of each DMSTM to be equal to the prior probability which is typically equal to the reciprocal of the number of DMSTMs in the ensemble. That way, the ensemble of DMSTMs is a mixture model where the prior weights and the posterior weights are the same. Finally, we use the observed data to separately update the parameters of each DMSTM.

Conditional on each shifted tree multiscale structure, Theorem 1 of Ferreira et al. (2011) states that the parameters at different levels of resolution are independent a posteriori. Hence, for each DMSTM in the ensemble, inference may be performed simultaneously for different levels of resolution. In addition, when combining the results from the different DMSTMs to obtain an ensemble analysis, we implement fast computations based on the first posterior moments which may be computed using the results obtained for the posterior distribution under each DMSTM.

Conditional on each tree $\mathcal{T}_m$, we may apply Theorem 2 of Ferreira et al. (2011) to obtain that the posterior distribution of θ_{tlj} at time t is

$$\theta_{tlj} \mid D_t, \mathcal{T}_m \sim N(m_{tlj}^{(m)}, C_{tlj}^{(m)}\Omega_{lj}^{(m)}), \tag{9.27}$$

where $C_{tlj}^{(m)} = (1 + (R_{tlj}^{(m)})^{-1})^{-1}$, $m_{tlj}^{(m)} = C_{tlj}^{(m)}(\theta_{tlj}^{e(m)} + (R_{tlj}^{(m)})^{-1} m_{t-1,lj}^{(m)})$, and $R_{tlj}^{(m)} = C_{t-1,lj}^{(m)} + \psi_{lj}$. In addition, the backward sampling distribution of θ_{tlj} at time t is

$$\theta_{tlj} \mid \theta_{t+1,lj}, D_T, \mathcal{T}_m \sim N(h_{tlj}^{(m)}, H_{tlj}^{(m)}\Omega_{tlj}^{(m)}), \tag{9.28}$$

where $H_{tlj}^{(m)} = ((C_{tlj}^{(m)})^{-1} + \psi_{lj}^{-1})^{-1}$ and $h_{tlj}^{(m)} = H_{tlj}^{(m)}((C_{tlj}^{(m)})^{-1} m_{tlj}^{(m)} + \psi_{lj}^{-1}\theta_{t+1,lj}))$, $t = 1, 2, \ldots, T-1$.

To obtain the ensemble analysis, we propose to combine the analysis for each DMSTM in the ensemble using a mixture model where the posterior probabilities of the DMSTMs remain the same as the prior probabilities. Therefore, if we consider the model (9.3), the disaggregation equations (9.2) and the prior distribution (9.25) for θ_{tlj}, then the ensemble distribution of the multiscale coefficient θ_{tlj} given the information up to time t is also a mixture of Gaussian distributions given by

$$p(\theta_{tlj} \mid D_t) = \sum_{m \in \mathcal{M}} P_m N(\theta_{tlj}; m_{tlj}^{(m)}, C_{tlj}^{(m)} \Omega_{lj}^{(m)}), \tag{9.29}$$

where $C_{tlj}^{(m)} = (1 + (R_{tlj}^{(m)})^{-1})^{-1}$, $m_{tlj}^{(m)} = C_{tlj}^{(m)}(\theta_{tlj}^{e(m)} + (R_{tlj}^{(m)})^{-1} m_{t-1,lj}^{(m)})$, and $R_{tlj}^{(m)} = C_{t-1,lj}^{(m)} + \psi_{lj}$. Thus, the ensemble backward sampling equation given the information up to time T is

$$p(\theta_{tlj} \mid \theta_{t+1,lj}, D_T) = \sum_{m \in \mathcal{M}} P_m N(\theta_{tlj}; h_{tlj}^{(m)}, H_{tlj}^{(m)} \Omega_{tlj}^{(m)}), \tag{9.30}$$

where $H_{tlj}^{(m)} = ((C_{tlj}^{(m)})^{-1} + \psi_{lj}^{-1})^{-1}$ and $h_{tlj}^{(m)} = H_{tlj}^{(m)}((C_{tlj}^{(m)})^{-1} m_{tlj}^{(m)} + \psi_{lj}^{-1} \theta_{t+1,lj}))$, $t = 1, 2, \ldots, T-1$.

Note that this formulation yields a closed-form solution to the posterior moments of the ensemble distribution.

9.5 Simulated Example: A 1-D Spatio-Temporal Process

We illustrate our proposed ensemble approach with a 1-D spatio-temporal simulated dataset with $n = 128$ equally spaced locations on the interval $(0, 1]$ for a total of $T = 30$ time points. At time $t = 1$, we simulate one realization of the mean process from a Gaussian spatial process with exponential covariance function $\rho(d) = \tau^2 \exp(-\phi\, d)$ with range parameter $\phi = 10$ and variance $\tau^2 = 1$. We then evolve the mean process over time according to an autoregressive process with an autoregressive coefficient equal to 0.999, where the system innovation at times $t = 2, \ldots, T$, follows a Gaussian spatial process with exponential covariance function with range parameter $\phi = 10$ and variance $\tau^2 = 1$. Then the observed data at each time point are obtained by adding independent measurement error with mean zero and variance $\sigma^2 = 1.5$.

We analyze these spatio-temporal data with an ensemble DMSTM analysis with $L = 8$ levels of resolution. Figure 9.6 presents time series plots of the true mean process (dashed gray line) in subregion 1 at resolution level 2 and ($\mu_{t,2,1}$), in subregion 1 at resolution level 8 ($\mu_{t,8,1}$), and in subregion 10 at resolution level 8 ($\mu_{t,8,10}$), as well as their ensemble estimates (solid black line) and 95% credible intervals (dashed black lines). Note that the latent means at

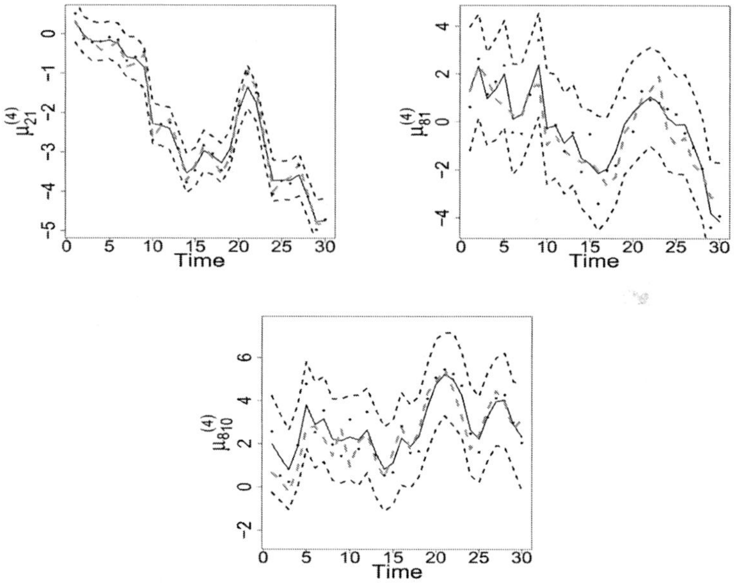

FIGURE 9.6
Simulated spatio-temporal example and an ensemble DMSTM analysis with $L = 8$ levels of resolution. True mean (dashed gray line), ensemble estimate (solid black line), and 95% credible intervals (dashed black lines).

these subregions/resolutions are well estimated by the ensemble approach. In addition, the uncertainty quantification is appropriate with most of the true latent means contained within the 95% credible intervals.

Figure 9.7 presents heatmaps of spatio-temporal observations (Panel (a)), the true mean at the finest resolution level over time and space (Panel (b)), as well as the ensemble estimate (Panel (c)). The ability of our ensemble approach to estimate the mean process at the finest resolution level is remarkable. In particular, the smoothness of the ensemble estimate is visually similar to the smoothness of the true mean process. In addition, the ensemble estimate captures very closely the space-time regions of high and low true mean.

To compare the performance of the ensemble estimate and the estimate based on each tree, we computed the mean squared error for the estimated mean for each tree (m) and for the proposed ensemble point estimator by $MSE(\widehat{\mu}_{Lj}) = (T2^{L-1})^{-1} \sum_{t=1}^{T} \sum_{j=1}^{2^{L-1}} (\widehat{\mu}_{Lj} - \mu_{Lj}^{(true)})^2$. While the MSEs for the individual DMSTM varied from a minimum of 1.11 to a maximum of 1.26 with a median of 1.16, the ensemble estimate resulted in an MSE of 0.98. Therefore, when compared to the estimates based on the individual tree models, the ensemble estimate is substantially better.

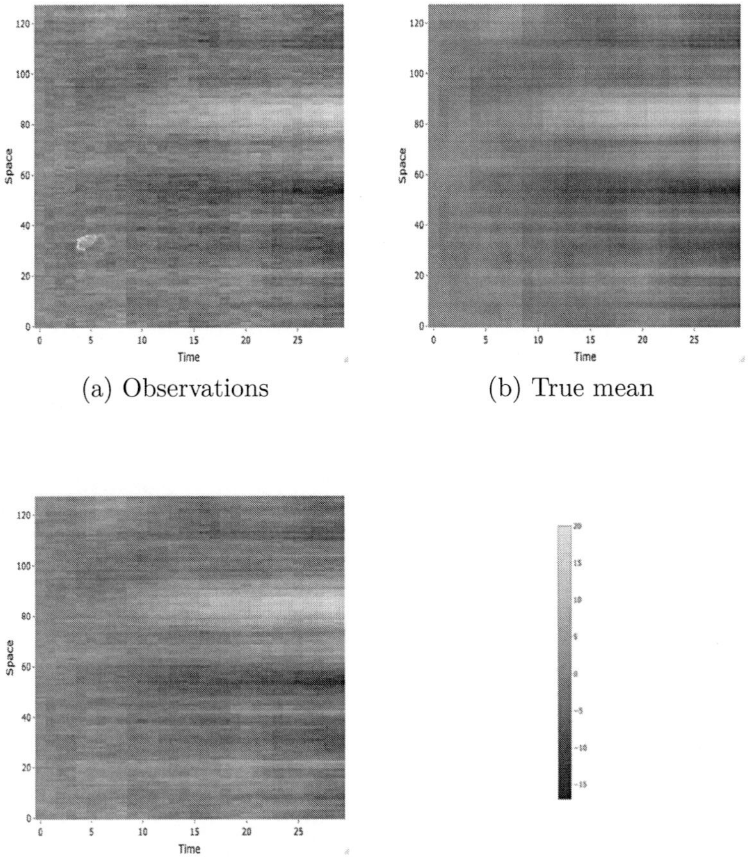

(a) Observations

(b) True mean

(c) Ensemble estimate

FIGURE 9.7
Simulated spatio-temporal example and an ensemble DMSTM analysis with $L = 8$ levels of resolution. Heatmaps of observations, true mean, and ensemble estimate.

9.6 Temperature Data Analysis

This section presents an ensemble DMSTM analysis of weekly average temperature data from a meteorological station in the town of Arraial do Cabo in the State of Rio de Janeiro, Brazil, from the first week of 2007 to the last

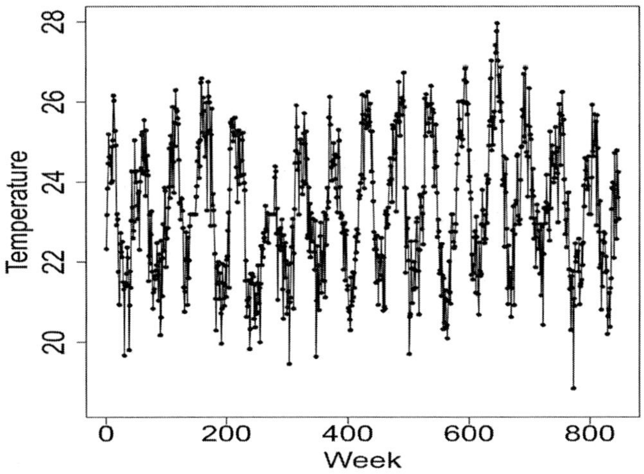

FIGURE 9.8
Observed weekly temperature (^{o}C) from 2007 to 2022.

week of 2022. Specifically, the weekly average temperature is modeled according to the proposed ensemble multiscale framework proposed in Section 9.4. We take a dynamic functional analysis perspective, where we aim to estimate the latent mean temperature as a function of the week at each year. In this perspective, this function of the week varies within each year. In the notation from previous sections, each year is a time point with $t = 2007, \ldots, 2022$, and each week is a subregion at the finest level of resolution with $j = 1, \ldots, 53$. Thus, at the finest level of resolution $L = 7$, for each year t and each week j, the temperature y_{tLj} is observed, $t = 2007, \ldots, 2022$, $j = 1, \ldots, 53$. Figure 9.8 displays the observed temperature in degrees Celsius for each week over the studied years. Note the distinct seasonal effect that we aim to capture with our proposed ensemble approach.

To examine more closely the seasonal pattern, Figure 9.9 presents box-plots for each week. Each of these box-plots includes data for the corresponding week from 2007 to 2022. As expected, there is a marked seasonal pattern, with higher temperatures in the southern hemisphere's summer and lower temperatures in the winter. Recall that, in this application, we are using "weeks" as the subregions. Thus, it is important to note that the temperature in the first weeks is similar to the temperature in the last weeks. This justifies the use in this application of a circulant shift operator to implement our ensemble approach.

To build the multiscale structure, we start with $n_7 = 53$ at the finest level of resolution. If at level l there is an odd number of observations then

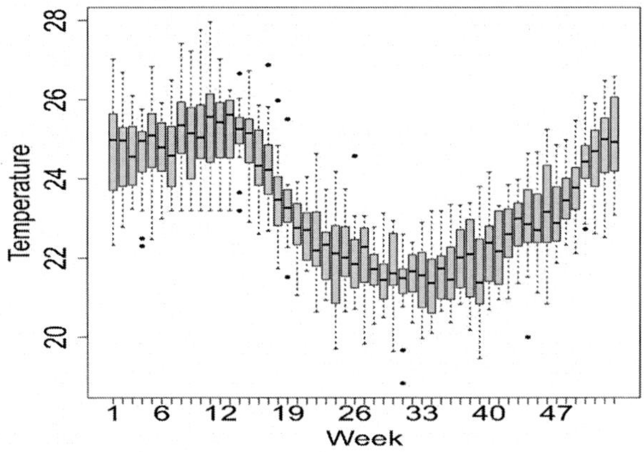

FIGURE 9.9
Observed temperature (^{o}C) data for each week. The box-plot for each week represents data from multiple years.

the mean is computed by applying the aggregation operator for every pair from 1 to $(n_l - 1)/2$ and averaging using equal weights. Observation y_{l,n_l} is repeated at level $l-1$ resulting in $(n_l-1)/2+1$ data points at level $l-1$. If at level l there is an even number of observations then the mean is computed by applying the aggregation operator for every pair from 1 to $n_l/2$ and averaging using equal weights. From levels 7 to 1, we obtain a total of $53, 27, 14, 7, 4, 2$, and 1 observations, respectively.

Figure 9.10 presents the estimated mean process at the coarsest resolution level based on the DMSTM with the original tree structure without any shift. The figure shows the posterior mean (solid line), observed data (points), and 95% credible interval (dashed lines). Importantly, in this application the mean process at the coarsest resolution level represents the annual mean process. An interesting finding for the latent annual mean temperatures at the Arraial do Cabo meteorological station is their increase during the 2010s and their decrease at the beginning of the 2020s.

To shed light on the differences in the results of the ensemble approach when compared to the results of the single DMSTM approach, Figure 9.11 presents time series plots of the estimated mean process temperature in 2015 at the finest level of resolution based on each of the DMSTM models based on shifted trees (solid grey lines), the estimate based on the ensemble approach (solid black line), and the observations (solid black circles). Because in our ensemble approach the posterior probabilities are equal to the prior probabilities which are uniform on the DMSTM model space implied by all possible shifted trees, the ensemble estimate is an average of the estimates from all

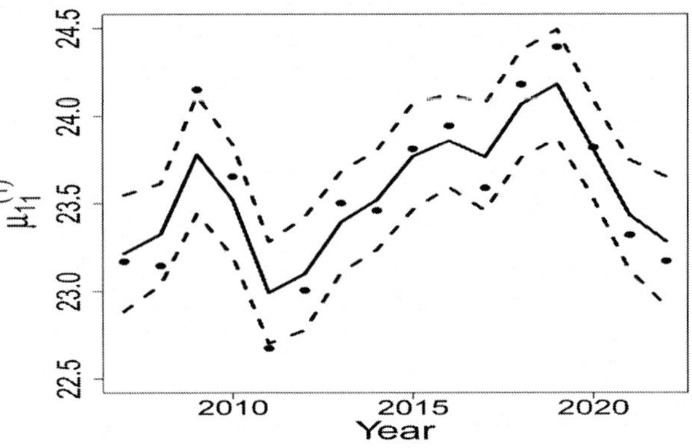

FIGURE 9.10
Weekly temperature application. For DMSTM with original tree structure without any shift, estimated mean process at the coarsest resolution level: Posterior mean (solid line), observed data (points), and 95% credible interval (dashed lines). Note that the mean process at the coarsest resolution level represents the annual mean process.

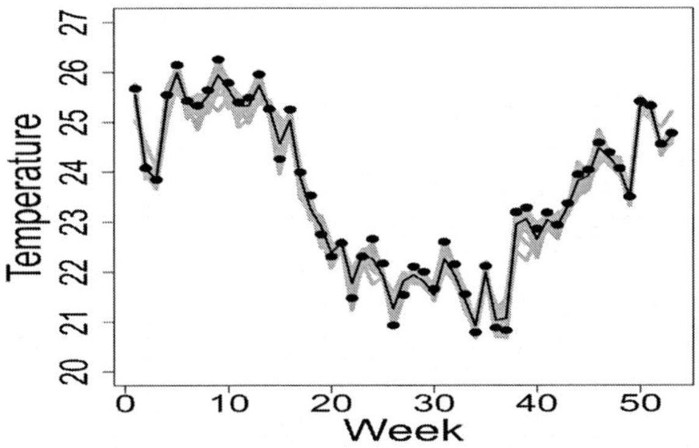

FIGURE 9.11
Weekly temperature application. Estimated mean process temperature in 2015 for all DMSTM models based on shifted trees (solid grey lines), the ensemble estimate (solid black line), and the observations (solid black circles).

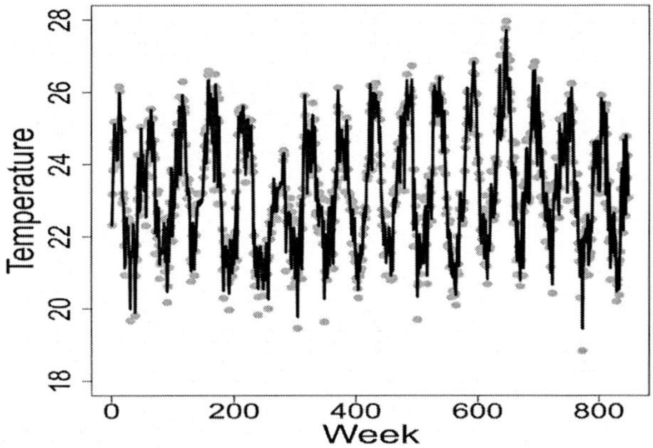

FIGURE 9.12
Weekly temperature application. Ensemble estimate of the mean temperature (solid line) and observed data.

the individual DMSTMs. As a consequence, when compared to the individual DMSTM estimates, the ensemble estimate is much smoother.

Finally, Figure 9.12 presents the ensemble estimate of the temperature mean process (solid line) and the observed data (solid grey circles) across all the study years. The estimate of the mean process is substantially less noisy than the observations. In addition, the year 2012 seems to have had a somehow different pattern than the other years under study, with much lower than average temperatures during the southern hemisphere summer months.

9.7 Conclusions

We have proposed an approach based on ensembles of DMSTMs for spatio-temporal data analysis that is useful in scenarios where the dependence on space is thought to be smooth. In our approach, the ensemble analysis does not depend on the spatial origin of the tree. Our proposal reparameterizes the latent mean process in terms of multiscale coefficients allowing for inference in partitions of the spatial domain. Coupling of the two parameterizations is done by linear matrix operations.

There are several possible directions for future research. A possible extension of this work is to use the stacking framework of Yao et al. (2018) to compute the weights of the multiple DMSTMs. In such a framework, the weights would be optimized such that the resulting ensemble model has optimal predictive ability. Another possible extension is related to the fact that, for simplicity of exposition, we have considered 1-D trees. Extensions to higher dimensions may be developed by considering block aggregation operators. Finally, another possible extension of our ensemble approach is to spatio-temporal Poisson data (Hoegh et al., 2016; Fonseca and Ferreira, 2017).

Bibliography

Elkhouly, M. and M. A. R. Ferreira (2021). Dynamic multiscale spatiotemporal models for multivariate Gaussian data. *Spatial Statistics 41*, 100475.

Ferreira, M. A. R. (2021). Bayesian spatial and spatiotemporal models based on multiscale factorizations. *Wiley Interdisciplinary Reviews: Computational Statistics 13*(2), e1509.

Ferreira, M. A. R., A. I. Bertolde, and S. Holan (2010). Analysis of economic data with multi-scale spatio-temporal models. In A. O'Hagan and M. West (Eds.), *Handbook of Applied Bayesian Analysis*, pp. 295–318. Oxford University Press.

Ferreira, M. A. R., S. H. Holan, and A. I. Bertolde (2011). Dynamic multiscale spatio-temporal models for Gaussian areal data. *Journal of the Royal Statistical Society (Series B: Methodological) 73*, 663–688.

Ferreira, M. A. R., M. A. Jaramillo, and E. B. Hypólito (2024). A model selection paradox with implications to multiscale modeling. In M. A. R. Ferreira (Ed.), *Modeling Spatio-Temporal Data: Markov Random Fields, Objective Bayes, and Multiscale Models*, pp. 195–220. Chapman & Hall/CRC.

Fonseca, T. C. O. and M. A. R. Ferreira (2017). Dynamic multiscale spatiotemporal models for Poisson data. *Journal of the American Statistical Association 112*, 215–234.

Hoegh, A., M. A. R. Ferreira, and S. Leman (2016). Spatiotemporal model fusion: multiscale modelling of civil unrest. *Journal of the Royal Statistical Society: Series C (Applied Statistics) 65*, 529–545.

Kolaczyk, E. D. (1999). Bayesian multiscale models for Poisson processes. *Journal of the American Statistical Association 94*, 920–933.

Li, M. and S. Ghosal (2015). Fast translation invariant multiscale image denoising. *IEEE Transactions on Image Processing 24*(12), 4876–4887.

Prado, R., M. A. R. Ferreira, and M. West (2021). *Time Series: Modeling, Computation, and Inference* (2nd ed.). Boca Raton: CRC Press.

Tzeng, S., H.-C. Huang, and N. Cressie (2005). A fast, optimal spatial-prediction method for massive datasets. *Journal of the American Statistical Association 100*(472), 1343–1357.

Wang, X., J. Pancras, and D. K. Dey (2021). Investigating emergent nested geographic structure in consumer purchases: a Bayesian dynamic multi-scale spatiotemporal modeling approach. *Journal of Applied Statistics 48*, 410–433.

Yao, Y., A. Vehtari, D. Simpson, and A. Gelman (2018). Using stacking to average Bayesian predictive distributions (with discussion). *Bayesian Analysis 13*, 917–1003.

Bibliography

Abramovich, F., T. Sapatinas, and B. W. Silverman (1998). Wavelet thresholding via a Bayesian approach. *Journal of the Royal Statistical Society 60*, 725–749.

Alfaro, M. E., S. Zoller, and F. Lutzoni (2003). Bayes or bootstrap? A simulation study comparing the performance of Bayesian Markov chain Monte Carlo sampling and bootstrapping in assessing phylogenetic confidence. *Molecular Biology and Evolution 20*, 255–266.

Alteri, C., V. Cento, A. Piralla, V. Costabile, M. Tallarita, L. Colagrossi, S. Renica, F. Giardina, F. Novazzi, S. Gaiarsa, et al. (2021). Genomic epidemiology of SARS-CoV-2 reveals multiple lineages and early spread of SARS-CoV-2 infections in Lombardy, Italy. *Nature Communications 12*(1), 434.

Alves, M. B., D. Gamerman, and M. A. R. Ferreira (2010). Transfer functions in dynamic generalized linear models. *Statistical Modelling 10*, 03–40.

Anderson, B. D. and J. B. Moore (1979). *Optimal Filtering.* New Jersey: Prentice Hall.

Anderson, R. M. and R. M. May (1992). *Infectious Diseases of Humans: Dynamics and Control.* New York: Oxford University Press.

Anderson, T. W. (1984). *An Introduction to Multivariate Statistical Analysis* (2nd ed.). New York: John Wiley & Sons.

Anselin, L. (1988). *Spatial Econometrics: Methods and Models.* Springer Science & Business Media.

Apostolopoulos, Y. and S. Sönmez (2007). Demographic and epidemiological perspectives of human movement. In Y. Apostolopoulos and S. Sönmez (Eds.), *Population Mobility and Infectious Disease*, pp. 1–16. New York: Springer.

Assunção, R. M., J. E. Potter, and S. M. Cavenaghi (2002). A Bayesian space varying parameter model applied to estimating fertility schedules. *Statistics in Medicine 21*, 2057–2075.

Assunção, R. M., I. A. Reis, and C. D. L. Oliveira (2001). Diffusion and prediction of Leishmaniasis in a large metropolitan area in Brazil with a Bayesian space-time model. *Statistics in Medicine 20*, 2319–2335.

Bacchetti, P. and A. R. Moss (1989). Incubation period of AIDS in San Francisco. *Nature 338*(6212), 251–253.

Banerjee, S., B. P. Carlin, and A. E. Gelfand (2014). *Hierarchical Modeling and Analysis for Spatial Data* (2nd ed.). Boca Raton, FL: CRC Press.

Berger, J. O. (1985). *Statistical Decision Theory and Bayesian Analysis, Second Ed.* New York: Springer.

Berger, J. O. and J. M. Bernardo (1992a). On the development of reference priors (Disc: p49-60). In J. M. Bernardo, J. O. Berger, A. P. Dawid, and A. F. M. Smith (Eds.), *Bayesian Statistics 4*, pp. 35–49. Oxford University Press.

Berger, J. O. and J. M. Bernardo (1992b). Ordered group reference priors with application to the multinomial problem. *Biometrika 25*, 25–37.

Berger, J. O. and J. M. Bernardo (1992c). Reference priors in a variance components problem. In P. K. Goel and N. S. Iyengar (Eds.), *Bayesian Analysis in Statistics and Econometrics*, pp. 323–340. Berlin: Springer.

Berger, J. O., V. De Oliveira, and B. Sansó (2001). Objective Bayesian analysis of spatially correlated data. *Journal of the American Statistical Association 96*, 1361–1374.

Berk, R. H. (1966). Limiting behavior of posterior distributions when the model is incorrect. *The Annals of Mathematical Statistics 37*, 51–58.

Berliner, L. M., C. K. Wikle, and R. F. Milliff (1999). Multiresolution wavelet analyses in hierarchical Bayesian turbulence models. In P. Müller and B. Vidakovic (Eds.), *Bayesian Inference in Wavelet Based Models*, pp. 341–359. New York: Springer-Verlag.

Bernardinelli, L., D. Clayton, C. Pascutto, C. Montomoli, M. Ghislandi, and M. Songini (1995). Bayesian analysis of space-time variation in disease risk. *Statistics in Medicine 14*, 2433–2443.

Bernardo, J. M. (1979). Reference posterior distributions for Bayesian inference (with discussion). *Journal of the Royal Statistical Society, Series B, Methodological 41*, 113–147.

Bernardo, J. M. and A. F. M. Smith (2000). *Bayesian Theory.* John Wiley & Sons.

Berrocal, V. J., A. E. Gelfand, and D. M. Holland (2010). A spatio-temporal downscaler for output from numerical models. *Journal of Agricultural, Biological, and Environmental Statistics 15*(2), 176–197.

Besag, J. E. (1974). Spatial interaction and the statistical analysis of lattice systems (with discussion). *Journal of the Royal Statistical Society, Series B 36*, 192–236.

Besag, J. E., P. J. Green, D. M. Higdon, and K. L. Mengersen (1995). Bayesian computation and stochastic systems (with discussion). *Statistical Science 10*, 3–66.

Besag, J. E. and C. Kooperberg (1995). On conditional and intrinsic autoregressions. *Biometrika 82*, 733–746.

Besag, J. E., J. York, and A. Mollié (1991). Bayesian image restoration with two applications in spatial statistics (with discussion). *Annals of the Institute of Statistical Mathematics 43*, 1–59.

Best, N., R. Arnold, A. Thomas, L. A. Waller, and E. M. Conlon (1999). Bayesian models for spatially correlated disease and exposure data. *Bayesian Statistics 6*, 65–82.

Best, N., S. Richardson, and A. Thomson (2005). A comparison of Bayesian spatial models for disease mapping. *Statistical Methods in Medical Research 14*(1), 35–59.

Bivand, R. and D. W. S. Wong (2018). Comparing implementations of global and local indicators of spatial association. *TEST 27*(3), 716–748.

Boubel, R. W., D. L. Fox, B. Turner, and A. C. Stern (1994). *Fundamentals of Air Pollution* (3rd ed.). Academic Press.

Bouckaert, R., T. G. Vaughan, J. Barido-Sottani, S. Duchêne, M. Fourment, A. Gavryushkina, J. Heled, G. Jones, D. Kühnert, N. De Maio, et al. (2019). BEAST 2.5: An advanced software platform for Bayesian evolutionary analysis. *PLoS Computational Biology 15*(4), e1006650.

Box, G. E. P. (1979). Robustness in the strategy of scientific model building. In R. L. Launer and G. N. Wilkinson (Eds.), *Robustness in Statistics*, pp. 201–236. Elsevier.

Briggs, W. L., V. E. Henson, and S. F. McCormick (2000). *A Multigrid Tutorial, 2nd edition.* Society for Industrial and Applied Mathematics.

Brower, J. and P. Chalk (2003). The global threat of new and reemerging infectious diseases: Reconciling U.S. national security and public health policy. Technical report, RAND Corporation, Santa Monica, CA.

Brown, P. E., P. J. Diggle, M. E. Lord, and P. C. Young (2001). Space-time calibration of radar-rainfall data. *Applied Statistics 50*, 221–241.

Brown, P. J., N. D. Le, and J. V. Zidek (1994). Multivariate spatial interpolation and exposure to air pollutants. *The Canadian Journal of Statistics 22*, 489–509.

Byun, D. W. and J. K. S. Ching (1999). *Science Algorithms of the EPA Models-3 Community Multiscale Air Quality (CMAQ) Modeling System.* RTP, NC: EPA/600/R-99/030.

Calder, C. A. (2007). Dynamic factor process convolution models for multivariate space-time data with application to air quality assessment. *Environmental and Ecological Statistics 14*(3), 229–247.

Candido, D. S., I. M. Claro, J. G. De Jesus, W. M. Souza, F. R. Moreira, S. Dellicour, T. A. Mellan, L. Du Plessis, R. H. Pereira, F. C. Sales, et al. (2020). Evolution and epidemic spread of SARS-CoV-2 in Brazil. *Science 369*(6508), 1255–1260.

Carlin, B. P. and S. Banerjee (2003). Hierarchical multivariate CAR models for spatio-temporally correlated survival data (with discussion). In J. Bernardo, M. J. Bayarri, J. O. Berger, A. P. Dawid, D. Heckerman, A. F. M. Smith, and M. West (Eds.), *Bayesian Statistics 7*, pp. 45–63. Oxford University Press.

Carter, C. K. and R. Kohn (1994). On Gibbs sampling for state space models. *Biometrika 81*, 541–553.

Carvalho, C. M. and M. West (2007). Dynamic matrix-variate graphical models. *Bayesian Analysis 2*(1), 69–97.

Casella, G. and R. L. Berger (2001). *Statistical Inference* (2nd ed.). Duxbury Press.

Christy, J. R. and W. B. Norris (2004). What may we conclude about global tropospheric temperature trends? *Geophysical Research Letters 31*(L06211), 1–4.

Clyde, M., G. Parmigiani, and B. Vidakovic (1998). Multiple shrinkage and subset selection in wavelets. *Biometrika 85*, 391–402.

Coifman, R. R. and D. L. Donoho (1995). Translation-invariant de-noising. In A. Antoniadis and G. Oppenheim (Eds.), *Wavelets and Statistics*, pp. 125–150. New York, NY: Springer New York.

Cressie, N. and G. Johannesson (2008). Fixed rank Kriging for very large spatial data sets. *Journal of the Royal Statistical Society: Series B (Statistical Methodology) 70*(1), 209–226.

Cressie, N., M. S. Kaiser, M. Daniels, J. Aldworth, J. Lee, S. Lahiri, and L. H. Cox (1999). *Spatial Analysis of Particulate Matter in an Urban Environment.* J. Gomez-Hernandez, A. Soares, and R. Froidevaux, Dordrecht: Kluwer: *GeoEnvII:* Geostatistics for Environmental Applications.

Cressie, N. and T. Read (1985). Do sudden infant deaths come in clusters? *Statistics and Decisions Supplement Issue 2*, 333–349.

Cressie, N. and C. K. Wikle (2011). *Statistics for Spatio-Temporal Data.* New York: Wiley.

Croissant, Y. and G. Millo (2008). Panel data econometrics in R: The plm package. *Journal of Statistical Software 27*(2), 1–43.

Cui, S., S. Guha, M. A. R. Ferreira, and A. N. Tegge (2015). hmmseq: A hidden Markov model for detecting differentially expressed genes from RNA-seq data. *The Annals of Applied Statistics 9*(2), 901–925.

Daubechies, I. (1992). *Ten Lectures on Wavelets.* Philadelphia, PA: SIAM.

De Oliveira, V. and M. A. R. Ferreira (2011). Maximum likelihood and restricted maximum likelihood estimation for a class of Gaussian Markov random fields. *Metrika 74*, 167–183.

Díaz-García, J. A. and R. Gutiérrez-Jáimez (2006). Distribution of the generalised inverse of a random matrix and its applications. *Journal of Statistical Planning and Inference 136*(1), 183–192.

Dobra, A., C. Hans, B. Jones, J. R. Nevins, G. Yao, and M. West (2004). Sparse graphical models for exploring gene expression data. *Journal of Multivariate Analysis 90*(1), 196–212.

Donoho, D. L. and I. M. Johnstone (1995). Adapting to unknown smoothness via wavelet shrinkage. *Journal of the American Statistical Association 90*, 1200–1224.

Donoho, D. L., I. M. Johnstone, G. Kerryacharian, and D. Picard (1995). Wavelet shrinkage: Asymptopia? *Journal of the Royal Statistical Society, Series B 57*, 301–369.

Douglass, D. H., B. D. Pearson, and S. F. Singer (2004). Altitude dependence of atmospheric temperature trends: Climate models versus observation. *Geophysical Research Letters 31*(13), 2–5.

Du Plessis, L., J. T. McCrone, A. E. Zarebski, V. Hill, C. Ruis, B. Gutierrez, J. Raghwani, J. Ashworth, R. Colquhoun, T. R. Connor, et al. (2021). Establishment and lineage dynamics of the SARS-CoV-2 epidemic in the UK. *Science 371*(6530), 708–712.

Eder, B. and S. Yu (2006). A performance evaluation of the 2004 release of Models-3 CMAQ. *Atmospheric Environment 40*, 4811–4824.

Elkhouly, M. (2019). *Bayesian Analysis of Temporal and Spatio-temporal Multivariate Environmental Data.* Ph.D. thesis, Virginia Polytechnic Institute and State University.

Elkhouly, M. and M. A. R. Ferreira (2021). Dynamic multiscale spatiotemporal models for multivariate Gaussian data. *Spatial Statistics 41*, 100475.

Elkhouly, M., S. E. Zick, and M. A. R. Ferreira (2023). Long term temporal trends in synoptic-scale weather conditions favoring significant tornado occurrence over the central United States. *PLoS One 18*(2), e0281312.

Epstein, P. (2010). The ecology of climate change and infectious diseases: comment. *Ecology 91*(3), 925–928.

Fahrmeir, L. and G. Tutz (2001). *Multivariate Statistical Modelling Based on Generalized Linear Models* (2nd ed.). New York: Springer.

Fedorenko, R. P. (1961). A relaxation method for solving elliptic difference equations. *Zhurnal Vychislitelnoj Matematyki i Matematicheskoj Fiziki 1*, 922–927. Translation appeared in: USSR Computational Mathematics and Mathematical Physics (1962), vol 1, pp 1092–1096.

Feller, W. (1968). *An Introduction to Probability Theory and its Applications*, Volume 1, 3rd ed. Wiley.

Ferguson, T. S. (1973). A Bayesian analysis of some nonparametric problems. *The Annals of Statistics 1*(2), 209–230.

Ferreira, M. A. R. (2002). *Bayesian Multi-scale Modelling.* Ph. D. thesis, Duke University.

Ferreira, M. A. R. (2011). Invited discussion on "Characterizing uncertainty of future climate change projections using hierarchical Bayesian models by Tebaldi, Smith and Sansó". In J. M. Bernardo, M. J. Bayarri, J. O. Berger, A. P. Dawid, D. Heckerman, A. F. M. Smith, and M. West (Eds.), *Bayesian Statistics 9*, pp. 639–658. Oxford: Oxford University Press.

Ferreira, M. A. R. (2015). Inhomogeneous evolutionary MCMC for Bayesian optimal sequential environmental monitoring. *Environmental and Ecological Statistics 22*(4), 705–724.

Ferreira, M. A. R. (2019). The limiting distribution of the Gibbs sampler for the intrinsic conditional autoregressive model. *Brazilian Journal of Probability and Statistics 33*(4), 734–744.

Ferreira, M. A. R. (2021). Bayesian spatial and spatiotemporal models based on multiscale factorizations. *Wiley Interdisciplinary Reviews: Computational Statistics 13*(2), e1509.

Ferreira, M. A. R. (2024). Proper Gaussian Markov random fields. In M. A. R. Ferreira (Ed.), *Modeling Spatio-Temporal Data: Markov Random Fields, Objective Bayes, and Multiscale Models*, pp. 1–22. Chapman & Hall/CRC.

Ferreira, M. A. R., A. I. Bertolde, and S. Holan (2010). Analysis of economic data with multi-scale spatio-temporal models. In A. O'Hagan and M. West (Eds.), *Handbook of Applied Bayesian Analysis*, pp. 295–318. Oxford University Press.

Ferreira, M. A. R., Z. Bi, M. West, H. Lee, and D. Higdon (2003). Multi-scale modelling of 1-D permeability fields. In J. Bernardo, M. J. Bayarri, J. O. Berger, A. P. Dawid, D. Heckerman, A. F. M. Smith, and M. West (Eds.), *Bayesian Statistics 7*, pp. 519–527. Oxford University Press.

Ferreira, M. A. R. and V. De Oliveira (2007). Bayesian reference analysis for Gaussian Markov Random Fields. *Journal of Multivariate Analysis 98*, 789–812.

Ferreira, M. A. R. and D. Gamerman (2000). Dynamic generalized linear models. In G. Dey and Mallick (Eds.), *Generalized Linear Models: a Bayesian Perspective*, New York, pp. 57–72. Marcel Dekker.

Ferreira, M. A. R., D. Gamerman, and H. S. Migon (1997). Bayesian dynamic hierarchical models: Covariance matrices estimation and non-normality. *Brazilian Journal of Probability and Statistics 11*, 67–79.

Ferreira, M. A. R., D. Higdon, H. K. H. Lee, and M. West (2005). Multiscale random field models. Technical report, UFRJ - DME.

Ferreira, M. A. R., S. H. Holan, and A. I. Bertolde (2011). Dynamic multi-scale spatio-temporal models for Gaussian areal data. *Journal of the Royal Statistical Society (Series B: Methodological) 73*, 663–688.

Ferreira, M. A. R. and M. A. Jaramillo (2014). Bayesian multiscale phylogenetics. *Journal of the Indian Society of Agricultural Statistics 68*, 285–292.

Ferreira, M. A. R., M. A. Jaramillo, and E. B. Hypólito (2024). A model selection paradox with implications to multiscale modeling. In M. A. R. Ferreira (Ed.), *Modeling Spatio-Temporal Data: Markov Random Fields, Objective Bayes, and Multiscale Models*, pp. 195–220. Chapman & Hall/CRC.

Ferreira, M. A. R. and H. K. H. Lee (2007). *Multiscale Modeling: A Bayesian Perspective.* Springer Series in Statistics. New York: Springer.

Ferreira, M. A. R., E. M. Porter, and C. T. Franck (2021). Fast and scalable computations for Gaussian hierarchical models with intrinsic conditional autoregressive spatial random effects. *Computational Statistics and Data Analysis 162*, 107264.

Ferreira, M. A. R. and E. Salazar (2014). Bayesian reference analysis for exponential power regression models. *Journal of Statistical Distributions and Applications 1*(1), 1–12.

Ferreira, M. A. R. and N. Sanyal (2014). Bayesian optimal sequential design for nonparametric regression via inhomogeneous evolutionary MCMC. *Statistical Methodology 18*, 131–141.

Ferreira, M. A. R. and M. A. Suchard (2008). Bayesian analysis of elapsed times in continuous-time Markov chains. *Canadian Journal of Statistics 36*, 355–368.

Ferreira, M. A. R. and J. C. Vivar (2024). Spatio-temporal models for Poisson areal data. In M. A. R. Ferreira (Ed.), *Modeling Spatio-Temporal Data: Markov Random Fields, Objective Bayes, and Multiscale Models*, pp. 69–89. Chapman & Hall/CRC.

Ferreira, M. A. R., M. West, H. K. H. Lee, and D. Higdon (2006). Multi-scale and hidden resolution time series models. *Bayesian Analysis 1*, 947–968.

Fonseca, T. C. O. and M. A. R. Ferreira (2017). Dynamic multiscale spatiotemporal models for Poisson data. *Journal of the American Statistical Association 112*, 215–234.

Fonseca, T. C. O. and M. A. R. Ferreira (2024). Ensembles of dynamic multiscale spatio-temporal models. In M. A. R. Ferreira (Ed.), *Modeling Spatio-Temporal Data: Markov Random Fields, Objective Bayes, and Multiscale Models*, pp. 221–245. Chapman & Hall/CRC.

Fonseca, T. C. O., M. A. R. Ferreira, and H. S. Migon (2008). Objective Bayesian analysis for the Student-t regression model. *Biometrika 95*, 325–333.

Fonseca, T. C. O., H. S. Migon, and M. A. R. Ferreira (2012). Bayesian analysis based on the Jeffreys prior for the hyperbolic distribution. *Brazilian Journal of Probability and Statistics 26*, 327–343.

Frühwirth-Schnatter, S. (1994). Data augmentation and dynamic linear models. *Journal of Time Series Analysis 15*(2), 183–202.

Frühwirth-Schnatter, S. (1995). Bayesian model discrimination and Bayes factors for linear Gaussian state-space models. *Journal of the Royal Statistical Society, Series B 57*, 237–246.

Frühwirth-Schnatter, S. (2006). *Finite mixture and Markov switching models.* Springer.

Fuentes, M. and A. Raftery (2005). Model evaluation and spatial interpolation by Bayesian combination of observations with outputs from numerical models. *Biometrics 61*, 36–45.

Gamerman, D. and H. F. Lopes (2006). *Markov Chain Monte Carlo: Stochastic Simulation for Bayesian Inference* (2nd ed.). Boca Raton, FL: Chapman and Hall/CRC.

Gamerman, D., A. R. B. Moreira, and H. Rue (2003). Space-varying regression models: specifications and simulation. *Computational Statistics and Data Analysis 42*, 513–533.

Gao, L., S. Banerjee, and B. Ritz (2023). Spatial difference boundary detection for multiple outcomes using Bayesian disease mapping. *Biostatistics 24*(4), 922–944.

Garside, L. M. and D. J. Wilkinson (2003). Dynamic lattice-Markov spatio-temporal models for environmental data. In J. Bernardo, J. Berger, A. Dawid, and A. Smith (Eds.), *Bayesian Statistics 7*, pp. 535–542. Oxford: Oxford University Press.

Gelfand, A. E., A. M. Schmidt, S. Banerjee, and C. Sirmans (2004). Nonstationary multivariate process modeling through spatially varying coregionalization. *Test 13*(2), 263–312.

Gelfand, A. E. and A. F. M. Smith (1990). Sampling-based approaches to calculating marginal densities. *Journal of the American Statistical Association 85*(410), 398–409.

Gelfand, A. E. and P. Vounatsou (2003). Proper multivariate conditional autoregressive models for spatial data analysis. *Biostatistics 4*(1), 11–15.

Gelhar, L. W. (1993). *Stochastic Subsurface Hydrology.* Englewood Cliffs, NJ: Prentice-Hall.

Gelman, A. and D. B. Rubin (1992). Inference from iterative simulation using multiple sequences. *Statistical Science 7*, 457–511.

Ghosh, J. K., M. Delampady, and T. Samanta (2006). *An Introduction to Bayesian Analysis, Theory and Methods.* New York: Springer.

Gidas, B. (1989). A renormalization group approach to image processing problems. *IEEE Transactions on Pattern Analysis and Machine Intelligence 11*, 164–180.

Goicoa, T., M. D. Ugarte, J. Etxeberria, and A. F. Militino (2016). Age–space–time CAR models in Bayesian disease mapping. *Statistics in Medicine 35*(14), 2391–2405.

Goodman, J. and A. D. Sokal (1989). Multigrid Monte Carlo method. *Physical Review Letters D 40*, 2035–2072.

Gotway, C. and L. Young (2002). Combining incompatible spatial data. *Journal of the American Statistical Association 97*, 632–648.

Haar, A. (1910). Zur theorie der orthogonalen funktionensysteme. *Mathematische Annalen 69*, 331–371.

Hanson, T. E. (2006). Inference for mixtures of finite Pólya tree models. *Journal of the American Statistical Association 101*(476), 1548–1565.

Hanson, T. E. and W. O. Johnson (2002). Modeling regression error with a mixture of Pólya trees. *Journal of the American Statistical Association 97*(460), 1020–1033.

Hawking, S. (2002). *On the Shoulders of Giants: The Great Works of Physics and Astronomy.* Philadelphia: Running Press.

He, D., D. Sun, and L. He (2021). Objective Bayesian analysis for the Student-t linear regression. *Bayesian Analysis 16*, 129–145.

Higdon, D. (1998). A process-convolution approach to modeling temperatures in the North Atlantic Ocean. *Journal of Environmental and Ecological Statistics 5*, 173–190.

Higdon, D., J. Swall, and J. Kern (1999). Non-stationary spatial modeling. In J. Bernardo, J. Berger, A. Dawid, and A. Smith (Eds.), *Bayesian Statistics 6.* London: Oxford University Press.

Higdon, D. M. (2002). Space and space-time modeling using process convolutions. In C. W. Anderson, V. Barnett, P. C. Chatwin, and A. H. El-Shaarawi (Eds.), *Quantitative Methods for Current Environmental Issues*, pp. 37–54. New York: Springer-Verlag.

Hoegh, A., M. A. R. Ferreira, and S. Leman (2016). Spatiotemporal model fusion: multiscale modelling of civil unrest. *Journal of the Royal Statistical Society: Series C (Applied Statistics) 65*, 529–545.

Holmes, E. E., E. J. Ward, and K. Wills (2012). MARSS: Multivariate Autoregressive State-space Models for Analyzing Time-Series Data. *R Journal 4*(1), 11–19.

Huelsenbeck, J. P. and B. Rannala (2004). Frequentist properties of Bayesian posterior probabilities of phylogenetic trees under simple and complex substitution models. *Systematic Biology 53*, 904–913.

Huelsenbeck, J. P. and F. Ronquist (2001). MrBayes: Bayesian inference of phylogenetic trees. *Bioinformatics 17*, 754–755.

Huerta, G., B. Sansó, and J. R. Stroud (2004). A spatiotemporal model for Mexico City ozone levels. *Journal of the Royal Statistical Society, Series C 53*, 231–248.

Husmeier, D. and G. McGuire (2003). Detecting recombination in 4-taxa DNA sequence alignments with Bayesian hidden Markov models and Markov chain Monte Carlo. *Molecular Biology and Evolution 20*, 315–337.

Hypólito, E. B. (2005). Uma resposta Bayesiana ao paradoxo de Suzuki (in Portuguese). Master's thesis, Departamento de Métodos Estatísticos, Universidade Federal do Rio de Janeiro.

Jara, A. and T. E. Hanson (2011). A class of mixtures of dependent tail-free processes. *Biometrika 98*(3), 553–566.

Jeffrey, R. C. (1988). Conditioning, kinematics, and exchangeability. In B. Skyrms and W. L. Harper (Eds.), *Causation, Chance, and Credence*, Volume 1, pp. 221–255. Dordrecht: Kluwer.

Jeffreys, H. (1961). *Theory of Probability*. London: Oxford University Press.

Johannesson, G. and N. Cressie (2004). Variance-covariance modeling and estimation for multi-resolution spatial models. In *geoENV IV – Geostatistics for Environmental Applications*, pp. 319–330. Springer.

Johannesson, G., N. Cressie, and H. Huang (2007). Dynamic multi-resolution spatial models. *Environmental and Ecological Statistics 14*, 5–25.

Jun, M. and M. L. Stein (2004). Statistical comparison of observed and CMAQ modeled daily sulfate levels. *Atmospheric Environment 38*, 4427–4436.

Kaiser, M. S., M. J. Daniels, K. Furakawa, and P. Dixon (2002). Analysis of particulate matter air pollution using Markov random field models of spatial dependence. *Environmetrics 13*, 615–628.

Kalnay, E., M. Kanamitsu, R. Kistler, W. Collins, D. Deaven, L. Gandin, M. Iredell, S. Saha, G. White, J. Woollen, et al. (1996). The NCEP/NCAR 40-year reanalysis project. *Bulletin of the American Meteorological Society 77*(3), 437–472.

Kang, H. M., J. H. Sul, S. K. Service, N. A. Zaitlen, S. Y. Kong, N. B. Freimer, C. Sabatti, and E. Eskin (2010). Variance component model to account for sample structure in genome-wide association studies. *Nature Genetics 42*(4), 348–354.

Kang, H. M., N. A. Zaitlen, C. M. Wade, A. Kirby, D. Heckerman, M. J. Daly, and E. Eskin (2008). Efficient control of population structure in model organism association mapping. *Genetics 178*(3), 1709–1723.

Karpman, D., M. A. R. Ferreira, and C. K. Wikle (2013). A point process model for tornado report climatology. *Stat 2*(1), 1–8.

Kass, R. E. and A. E. Raftery (1995). Bayes factors. *Journal of the American Statistical Association 90*, 773–795.

Katzfuss, M. (2013). Bayesian nonstationary spatial modeling for very large datasets. *Environmetrics 24*(3), 189–200.

Katzfuss, M. (2017). A multi-resolution approximation for massive spatial datasets. *Journal of the American Statistical Association 112*(517), 201–214.

Katzfuss, M. and W. Gong (2020). A class of multi-resolution approximations for large spatial datasets. *Statistica Sinica 30*(4), 2203–2226.

Keefe, M. J., M. A. R. Ferreira, and C. T. Franck (2018). On the formal specification of sum-zero constrained intrinsic conditional autoregressive models. *Spatial Statistics 24*, 54 – 65.

Keefe, M. J., M. A. R. Ferreira, and C. T. Franck (2019). Objective Bayesian analysis for Gaussian hierarchical models with intrinsic conditional autoregressive priors. *Bayesian Analysis 14*(1), 181–209.

Kibria, B., M. Golam, L. Sun, J. Zidek, and N. Le (1997). Bayesian spatial prediction of random space-time fields with application to mapping $PM_{2.5}$ exposure. *Journal of American Statistica Association 97*, 112–124.

Kim, H., D. Sun, and R. K. Tsutakawa (2001). A bivariate Bayes method for improving the estimates of mortality rates with a twofold conditional autoregressive model. *Journal of the American Statistical Association 96*, pp. 1506–1521.

King, M. J. and A. Datta-Gupta (1998). Streamline simulation: A current perspective. *In Situ 22*, 91–140.

Knorr-Held, L. (2000). Bayesian modelling of inseparable space-time variation in disease risk. *Statistics in Medicine 19*, 2555–2567.

Knorr-Held, L. and S. Richardson (2003). A hierarchical model for space-time surveillance data on meningococcal disease incidence. *Applied Statistics 52*, 169–183.

Kolaczyk, E. D. (1999). Bayesian multiscale models for Poisson processes. *Journal of the American Statistical Association 94*, 920–933.

Kolaczyk, E. D. and H. Huang (2001). Multiscale statistical models for hierarchical spatial aggregation. *Geographical Analysis 33*, 95–118.

Lafferty, K. D. (2009). The ecology of climate change and infectious diseases. *Ecology 90*(4), 888–900.

Landim, F. and D. Gamerman (2000). Dynamic hierarchical models: an extension to matrix-variate observations. *Computational Statistics & Data Analysis 35*(1), 11–42.

Lavine, M. (1992). Some aspects of Pólya tree distributions for statistical modelling. *The Annals of Statistics 20*(3), 1222–1235.

Lawson, A. B. (2018). *Bayesian Disease Mapping: Hierarchical Modeling in Spatial Epidemiology* (3rd ed.). Boca Raton: Chapman and Hall/CRC.

Lee, D. (2011). A comparison of conditional autoregressive models used in Bayesian disease mapping. *Spatial and Spatio-temporal Epidemiology 2*(2), 79 – 89.

Lee, D. (2013). CARBayes: An R package for Bayesian spatial modeling with conditional autoregressive priors. *Journal of Statistical Software 55*(13), 1–24.

Lee, H. K. H., D. Higdon, Z. Bi, M. A. R. Ferreira, and M. West (2002). Markov random field models for high-dimensional parameters in simulations of fluid flow in porous media. *Technometrics 44*(3), 230–241.

Leroux, B. G., X. Lei, and N. Breslow (2000). Estimation of disease rates in small areas: a new mixed model for spatial dependence. In M. E. Halloran and D. Berry (Eds.), *Statistical Models in Epidemiology, the Environment, and Clinical Trials*, pp. 179–191. Springer.

Lewis, P. O., M. T. Holder, and K. E. Holsinger (2005). Polytomies and Bayesian phylogenetic inference. *Systematic Biology 54*, 241–253.

Li, M. and D. B. Dunson (2020). Comparing and weighting imperfect models using d-probabilities. *Journal of the American Statistical Association 115*(531), 1349–1360.

Li, M. and S. Ghosal (2015). Fast translation invariant multiscale image denoising. *IEEE Transactions on Image Processing 24*(12), 4876–4887.

Lindley, D. V. (1957). A statistical paradox. *Biometrika 44*, 187–192.

Liu, J. S. and C. Sabatti (2000). Generalised Gibbs sampler and multigrid Monte Carlo for Bayesian computation. *Biometrika 87*, 353–369.

Logan, J. R., C. Bauer, J. Ke, H. Xu, and F. Li (2020). Models for small area estimation for census tracts. *Geographical Analysis 52*(3), 325–350.

Lopes, H., D. Gamerman, and E. Salazar (2011). Generalized spatial dynamic factor models. *Computational Statistics & Data Analysis 55*, 1319–1330.

Lopes, H. F., E. Salazar, and D. Gamerman (2008). Spatial dynamic factor analysis. *Bayesian Analysis 3*(4), 759–792.

Ma, L. (2017a). Adaptive shrinkage in Pólya tree type models. *Bayesian Analysis 12*(3), 779–805.

Ma, L. (2017b). Recursive partitioning and multi-scale modeling on conditional densities. *Electronic Journal of Statistics 11*(1), 1297–1325.

Ma, L. and J. Soriano (2018). Analysis of distributional variation through multi-scale beta-binomial modeling. *Journal of Computational and Graphical Statistics 27*, 529–541.

MacNab, Y. C. and C. B. Dean (2001). Autoregressive spatial smoothing and temporal spline smoothing for mapping rates. *Biometrics 57*, 949–956.

Mallat, S. G. (1989a). Multiresolution approximations and the wavelet orthonormal bases of l^2(r). *Transactions of the American Mathematical Society 315*, 69–87.

Mallat, S. G. (1989b). A theory for multiresolution signal decomposition: the wavelet representation. *IEEE Transactions on Pattern Analysis and Machine Intelligence 11*, 674–693.

Mallat, S. G. (1999). *A Wavelet Tour of Signal Processing* (2nd ed.). San Diego: Academic Press.

Mardia, K. V. (1988). Multi-dimensional multivariate Gaussian Markov random fields with application to image processing. *Journal of Multivariate Analysis 24*(2), 265–284.

McCullagh, P. and J. A. Nelder (1989). *Generalized Linear Models* (2nd. ed.), Volume 37 of *Monographs on Statistics and Applied Probability.* London: Chapman and Hall.

McCullagh, P. and N. G. Polson (2018, 10). Statistical sparsity. *Biometrika 105*(4), 797–814.

Mercer, L. D., J. Wakefield, A. Pantazis, A. M. Lutambi, H. Masanja, and S. Clark (2015). Space-time smoothing of complex survey data: small area estimation for child mortality. *The Annals of Applied Statistics 9*(4), 1889.

Migon, H. S., D. Gamerman, H. F. Lopes, and M. A. R. Ferreira (2005). Dynamic models. In D. Dey and C. R. Rao (Eds.), *Bayesian Thinking, Modeling and Computation*, Handbook of Statistics, pp. 553–588. North Holland.

Mills, T. C. (2007). Time series modelling of two millennia of northern hemisphere temperatures: Long memory or shifting trends? *Journal of the Royal Statistical Society - Series A 170*(1), 83–94.

Min, X., D. Sun, C. Z. He, and M. Schootman (2010). A Bayesian hierarchical model of nontraumatic lower-extremity amputation rates. *Spatial and Spatio-temporal Epidemiology 1*, 169–176.

Mugglin, A. S., N. Cressie, and I. Gemmell (2002). Hierarchical statistical modelling of influenza epidemic dynamics in space and time. *Statistics in Medicine 21*, 2703–2721.

Muirhead, R. J. (1982). *Aspects of Multivariate Statistical Theory.* New York: Wiley.

Müller, P., A. Erkanli, and M. West (1996). Bayesian curve fitting using multivariate normal mixtures. *Biometrika 83*(1), 67–79.

Müller, P., G. Parmigiani, and K. Rice (2007). FDR and Bayesian multiple comparisons rules. In J. M. Bernardo, M. J. Bayarri, J. O. Berger, A. Dawid, D. Heckerman, A. F. M. Smith, and M. West (Eds.), *Bayesian Statistics 8*, Oxford, pp. 349–370. Oxford Univ. Press.

Nason, G. (2022). *wavethresh: Wavelets Statistics and Transforms.* R package version 4.7.2.

Nychka, D., S. Bandyopadhyay, D. Hammerling, F. Lindgren, and S. Sain (2015). A multiresolution Gaussian process model for the analysis of large spatial datasets. *Journal of Computational and Graphical Statistics 24*(2), 579–599.

O'Hagan, A. (1995). Fractional Bayes factors for model comparison. *Journal of the Royal Statistical Society. Series B (Methodological) 57*(1), 99–138.

Parmigiani, G. and L. Inoue (2009). *Decision Theory: Principles and Approaches.* John Wiley & Sons.

Penrose, R. (1955). A generalized inverse for matrices. *Mathematical Proceedings of the Cambridge Philosophical Society 51*(3), 406–413.

Percival, D. B. and A. T. Walden (2006). *Wavelet Methods for Time Series Analysis.* New York: Cambridge University Press.

Perlwitz, J., M. Hoerling, and R. Dole (2015). Arctic tropospheric warming: Causes and linkages to lower latitudes. *Journal of Climate 28*(6), 2154–2167.

Pettitt, A. N., I. S. Weir, and A. G. Hart (2002). A conditional autoregressive Gaussian process for irregularly spaced multivariate data with application to modelling large sets of binary data. *Statistics and Computing 12*, 353–367.

Pickett, K. M. and C. P. Randle (2005). Strange Bayes indeed: uniform topological priors imply non-uniform clade priors. *Molecular Phylogenetics and Evolution 34*, 203–211.

Plummer, M., N. Best, K. Cowles, and K. Vines (2006). CODA: convergence diagnosis and output analysis for MCMC. *R News 6*(March), 7–11.

Pole, A. and M. West (1989). Reference analysis of the dynamic linear model. *Journal of Time Series Analysis 10*, 131–147.

Pope III, A. C., R. T. Burnett, M. J. Thun, E. E. Calle, D. Krewski, K. Ito, and G. D. Thurston (2002). Lung cancer, cardiopulmonary mortality and long-term exposure to fine particulate air pollution. *Journal of American Medical Association 292*, 2372–2378.

Porter, E. M., C. T. Franck, and M. A. R. Ferreira (2023). Objective Bayesian model selection for spatial hierarchical models with intrinsic conditional autoregressive priors. *Bayesian Analysis 19*, accepted for publication.

Porter, E. M., C. T. Franck, and M. A. R. Ferreira (2024). Gaussian spatial hierarchical models with ICAR priors. In M. A. R. Ferreira (Ed.), *Modeling Spatio-Temporal Data: Markov Random Fields, Objective Bayes, and Multiscale Models*, pp. 23–46. Chapman & Hall/CRC.

Porter, E. M., M. J. Keefe, C. T. Franck, and M. A. R. Ferreira (2018). *ref.ICAR: Objective Bayes Intrinsic Conditional Autoregressive Model for Areal Data.* CRAN. R package version 1.0.

Prado, R., M. A. R. Ferreira, and M. West (2021). *Time Series: Modeling, Computation, and Inference* (2nd ed.). Boca Raton: CRC Press.

Press, W. H., S. A. Teukolsky, W. T. Vetterling, and B. P. Flannery (1992). *Numerical Recipes in C* (2nd ed.). Cambridge: Cambridge University Press.

Quintana, J. M. and M. West (1987). An analysis of international exchange rates using multivariate DLM's. *Journal of the Royal Statistical Society: Series D (The Statistician) 36*(2-3), 275–281.

R Core Team (2023). *R: A Language and Environment for Statistical Computing.* Vienna, Austria: R Foundation for Statistical Computing.

Rambaut, A., A. J. Drummond, D. Xie, G. Baele, and M. A. Suchard (2018). Posterior summarization in Bayesian phylogenetics using Tracer 1.7. *Systematic Biology 67*(5), 901–904.

Redelings, B. D. and M. A. Suchard (2005). Joint Bayesian estimation of alignment and phylogeny. *Systematic Biology 54*, 401–418.

Reich, B. J., J. S. Hodges, and V. Zadnik (2006). Effects of residual smoothing on the posterior of the fixed effects in disease-mapping models. *Biometrics 62*(4), 1197–1206.

Robert, C. P. (2001). *The Bayesian Choice* (2nd ed.). New York: Springer.

Robert, C. P. and G. Casella (2004). *Monte Carlo Statistical Methods* (2nd ed.). Springer-Verlag.

Ronquist, F. and J. P. Huelsenbeck (2003). MrBayes3: Bayesian phylogenetic inference under mixed models. *Bioinformatics 19*, 1572–1574.

Rossini, A. J., L. Tierney, and N. Li (2007). Simple parallel statistical computing in R. *Journal of Computational and Graphical Statistics 16*, 399–420.

Rouder, J. N., P. L. Speckman, D. Sun, R. D. Morey, and G. Iverson (2009). Bayesian t tests for accepting and rejecting the null hypothesis. *Psychonomic Bulletin & Review 16*, 225–237.

Royle, A. and M. Berliner (1999). A hierarchical approach to multivariate spatial modeling and prediction. *Journal of Agricultural, Biological and Environmental Statistics 4*, 29–56.

Rue, H. and L. Held (2005). *Gaussian Markov Random Fields: Theory and Applications.* CRC Press.

Rue, H., S. Martino, and N. Chopin (2009). Approximate Bayesian inference for latent Gaussian models by using integrated nested Laplace approximations. *Journal of the Royal Statistical Society: Series B (Statistical Methodology) 71*(2), 319–392.

Ruiz-Cárdenas, R., M. A. R. Ferreira, and A. M. Schmidt (2010). Stochastic search algorithms for optimal monitoring network designs. *Environmetrics 21*, 102–112.

Ruiz-Cárdenas, R., M. A. R. Ferreira, and A. M. Schmidt (2012). Evolutionary Markov chain Monte Carlo algorithms for optimal monitoring network designs. *Statistical Methodology (Special Issue on Astrostatistics and Spatial Statistics) 9*, 185–194.

Sahu, S. K. and K. V. Mardia (2005). A Bayesian Kriged-Kalman model for short-term forecasting of air pollution levels. *Journal of the Royal Statistical Society, Series C 54*, 223–244.

Salazar, E. and M. A. R. Ferreira (2011). Temporal aggregation of lognormal autoregressive processes. *Journal of Time Series Analysis 32*, 661–671.

Salazar, E., M. A. R. Ferreira, and H. S. Migon (2012). Objective Bayesian analysis for exponential power regression models. *Sankhya – Series B 74*, 107–125.

Sanyal, N. and M. A. R. Ferreira (2012). Bayesian hierarchical multi-subject multiscale analysis of functional MRI data. *NeuroImage 63*, 1519–1531.

Sanyal, N. and M. A. R. Ferreira (2017). Bayesian wavelet analysis with nonlocal priors with an application to fMRI data. *Sankhya – Series B 79*, 361–388.

Schmid, V. and L. Held (2004). Bayesian extrapolation of space-time trends in cancer registry data. *Biometrics 60*, 1034–1042.

Schmidt, A. M. and A. E. Gelfand (2003). A Bayesian coregionalization approach for multivariate pollutant data. *Journal of Geophysical Research: Atmospheres 108*, D24.

Shaddick, G. and J. Wakefield (2002). Modelling daily multivariate pollutant data at multiple sites. *Journal of the Royal Statistica Society, Series C 51*, 351–372.

Shin, H. and M. A. R. Ferreira (2023). Dynamic ICAR spatiotemporal factor models. *Spatial Statistics 56*, 100763.

Simmons, M. P., K. M. Pickett, and M. Miya (2004). How meaningful are Bayesian support values? *Molecular Biology and Evolution 21*, 188–199.

Siotani, M., T. Hayakawa, and Y. Fujikoshi (1985). *Modern Multivariate Statistical Analysis*. Columbus: American Sciences Press.

Smith, R. L., S. Kolenikov, and L. H. Cox (2003). Spatio-temporal modeling of $PM_{2.5}$ data with missing values. *Journal of Geophysical Research-Atmospheres 108 D24 9004*, doi:https://doi.org/10.1029/2002JD002914.

Spiegelhalter, D. J., N. G. Best, B. P. Carlin, and A. Van Der Linde (2002). Bayesian measures of model complexity and fit. *Journal of the Royal Statistical Society: Series B (Statistical Methodology) 64*(4), 583–639.

Starck, J.-L., F. Murtagh, and J. Fadili (2015). *Sparse Image and Signal Processing: Wavelets and Related Geometric Multiscale Analysis* (2nd ed.). Cambridge university press.

Stein, M. L. (2008). A modeling approach for large spatial datasets. *Journal of the Korean Statistical Society 37*(1), 3–10.

Stein, M. L. (2014). Limitations on low rank approximations for covariance matrices of spatial data. *Spatial Statistics 8*(C), 1–19.

Stroud, J. R., P. Müller, and B. Sansó (2001). Dynamic models for spatiotemporal data. *Journal of the Royal Statistical Society, Series B 63*, 673–689.

Suchard, M. A. (2005). Stochastic models for horizontal gene transfer: taking a random walk through tree space. *Genetics 170*, 419–431.

Sun, D. and J. O. Berger (2007). Objective Bayesian analysis for the multivariate normal model. In J. M. Bernardo, M. Bayarri, J. Berger, A. Dawid, D. Heckerman, A. Smith, and M. West (Eds.), *Bayesian Statistics 8*, pp. 525–547. Oxford: Oxford University Press.

Sun, D., R. K. Tsutakawa, H. Kim, and Z. He (2000). Spatio-temporal interaction with disease mapping. *Statistics in Medicine 19*, 2015–2035.

Sun, D., R. K. Tsutakawa, and P. L. Speckman (1999). Posterior distribution of hierarchical models using CAR(1) distributions. *Biometrika 86*, 341–350.

Sun, L., J. V. Zidek, N. D. Le, and H. Ozkaynak (2000). Interpolating Vancouver's daily ambient PM_{10} field. *Environmetrics 11*, 651–663.

Sun, Y., B. Li, and M. G. Genton (2012). Geostatistics for large datasets. In E. Porcu, J. M. Montero, and M. Schlather (Eds.), *Advances and Challenges in Space-Time Modelling of Natural Events*, pp. 55–77. Springer.

Suzuki, Y., G. V. Glazko, and M. Nei (2002). Overcredibility of molecular phylogenies obtained by Bayesian phylogenetics. *Proceedings of the National Academy of Science 99*, 16138–16143.

Swall, J. L. and K. M. Foley (2009). The impact of spatial correlation and incommensurability on model evaluation. *Atmospheric Environment 43*, 1204–1217.

Tebaldi, C., R. L. Smith, and B. Sansó (2011). Characterizing uncertainty of future climate change projections using hierarchical Bayesian models. In J. M. Bernardo, M. J. Bayarri, J. O. Berger, A. P. Dawid, D. Heckerman, A. F. M. Smith, and M. West (Eds.), *Bayesian Statistics 9*, pp. 639–658. Oxford: Oxford University Press.

Terzopoulos, D. (1986). Image analysis using multigrid relaxation methods. *IEEE Transactions on Pattern Analysis and Machine Intelligence PAMI-8*, 129–139.

Tian, B., E. J. Fetzer, B. H. Kahn, J. Teixeira, E. Manning, and T. Hearty (2013). Evaluating CMIP5 models using AIRS tropospheric air temperature and specific humidity climatology. *Journal of Geophysical Research: Atmospheres 118*(1), 114–134.

Tzala, E. and N. Best (2008). Bayesian latent variable modelling of multivariate spatio-temporal variation in cancer mortality. *Statistical Methods in Medical Research 17*, 97–118.

Tzeng, S., H.-C. Huang, and N. Cressie (2005). A fast, optimal spatial-prediction method for massive datasets. *Journal of the American Statistical Association 100*(472), 1343–1357.

Vasco, D. W. and A. Datta-Gupta (1999). Asymptotic solutions for solute transport: A formalism for tracer tomography. *Water Resources Research 35*, 1–16.

Ver Hoef, J. M., E. E. Peterson, M. B. Hooten, E. M. Hanks, and M.-J. Fortin (2018). Spatial autoregressive models for statistical inference from ecological data. *Ecological Monographs 88*(1), 36–59.

Vidakovic, B. (1999). *Statistical Modeling by Wavelets*. Wiley.

Vivar, J. C. and M. A. R. Ferreira (2009). Spatio-temporal models for Gaussian areal data. *Journal of Computational and Graphical Statistics 18*, 658–674.

Waller, L. A., B. P. Carlin, H. Xia, and A. E. Gelfand (1997). Hierarchical spatio-temporal mapping of disease rates. *Journal of the American Statistical Association 92*, 607–617.

Walsh, S., M. A. R. Ferreira, D. Higdon, and S. E. Zick (2023). A Bayesian hierarchical model framework to quantify uncertainty of tropical cyclone precipitation forecasts. *The Annals of Applied Statistics 17*(3), 1984–2001.

Wang, X., J. Pancras, and D. K. Dey (2021). Investigating emergent nested geographic structure in consumer purchases: a Bayesian dynamic multi-scale spatiotemporal modeling approach. *Journal of Applied Statistics 48*, 410–433.

Watanabe, S. (2010). Asymptotic equivalence of Bayes cross validation and widely applicable information criterion in singular learning theory. *Journal of Machine Learning Research 11*(12), 3571–3594.

Wentz, F. J. and M. Schabel (1998). Effects of orbital decay on satellite-derived lower-tropospheric temperature trends. *Nature 394*(6694), 661–664.

West, M. and P. J. Harrison (1997). *Bayesian Forecasting and Dynamic Models* (2nd. ed.). New York: Springer.

West, M., P. J. Harrison, and H. S. Migon (1985). Dynamic generalized linear models and Bayesian forecasting (with discussion). *Journal of the American Statistical Association 80*, 73–96.

Wickham, H. (2016). *ggplot2: Elegant Graphics for Data Analysis.* Springer-Verlag New York.

Wikle, C. K. (2003). Hierarchical models in environmental science. *International Statistical Review 71*, 181–199.

Wikle, C. K., L. M. Berliner, and N. Cressie (1998). Hierarchical Bayesian space-time models. *Environmental and Ecological Statistics 5*, 117–154.

Wikle, C. K. and M. B. Hooten (2010). A general science-based framework for dynamical spatio-temporal models. *Test 19*, 417–451.

Wikle, C. K., R. F. Milliff, D. Nychka, and L. M. Berliner (2001). Spatio-temporal hierarchical Bayesian modeling: Tropical ocean surface winds. *Journal of the American Statistical Association 96*, 382–397.

Wiley, E. O. and B. S. Lieberman (2011). *Phylogenetics: theory and practice of phylogenetic systematics.* John Wiley & Sons.

Williams, J., M. A. R. Ferreira, and T. Ji (2022). BICOSS: Bayesian iterative conditional stochastic search for GWAS. *BMC Bioinformatics 23*(1), 1–14.

Williams, J., S. Xu, and M. A. R. Ferreira (2023). BGWAS: Bayesian variable selection in linear mixed models with nonlocal priors for genome-wide association studies. *BMC Bioinformatics 24*(1), 1–20.

Wong, W. H. and L. Ma (2010). Optional Pólya tree and Bayesian inference. *The Annals of Statistics 38*(3), 1433–1459.

Wu, H.-H., M. A. R. Ferreira, M. Elkhouly, and T. Ji (2020). Nonlocal priors for variable selection in generalized linear models. *Sankhya – Series A 82-A*, 147–185.

Wu, H.-H., M. A. R. Ferreira, and M. E. Gompper (2016). Consistency of hyper-g-prior-based Bayesian variable selection for generalized linear models. *Brazilian Journal of Probability and Statistics 30*, 691–709.

Xu, S., M. A. R. Ferreira, E. M. Porter, and C. T. Franck (2023). Bayesian model selection for generalized linear mixed models. *Biometrics 79*(4), 3266–3278.

Xu, S., J. Williams, and M. A. R. Ferreira (2023). BG2: Bayesian variable selection in generalized linear mixed models with nonlocal priors for non-Gaussian GWAS data. *BMC Bioinformatics 24*(1), 343.

Yang, Z. and B. Rannala (2005). Branch-length prior influences Bayesian posterior probability of phylogeny. *Systematic Biology 54*, 455–470.

Yao, Y., A. Vehtari, D. Simpson, and A. Gelman (2018). Using stacking to average Bayesian predictive distributions (with discussion). *Bayesian Analysis 13*, 917–1003.

Yu, J., G. Pressoir, W. H. Briggs, I. V. Bi, M. Yamasaki, J. F. Doebley, M. D. McMullen, B. S. Gaut, D. M. Nielsen, J. B. Holland, S. Kresovich, and E. S. Buckler (2006). A unified mixed-model method for association mapping that accounts for multiple levels of relatedness. *Nature Genetics 38*(2), 203–208.

Zammit-Mangion, A. and J. Rougier (2020). Multi-scale process modelling and distributed computation for spatial data. *Statistics and Computing 30*(6), 1609–1627.

Zhang, S., D. Sun, C. Z. He, and M. Schootman (2006). A Bayesian semiparametric model for colorectal cancer incidences. *Statistics in Medicine 25*, 285–309.

Zidek, J. V., L. Sun, N. D. Le, and H. Ozkaynak (2002). Contending with space-time interaction in the spatial prediction of pollution: Vancouver's hourly ambient $pm_{2.5}$ field. *Environmetrics 13*, 595–613.

Author Index

Subject Index